NONASSOCIATIVE ALGEBRAIC MODELS

PROCEEDINGS OF THE WORKSHOP

ON

NONASSOCIATIVE ALGEBRAIC MODELS

held at

Universidad de Zaragoza, Zaragoza, Spain
April, 1989

Edited by

Santos González and Hyo Chul Myung

NOVA SCIENCE PUBLISHERS, INC.
New York

Nova Science Publishers, Inc.
6080 Jericho Turnpike, Suite 207
Commack, New York 11725

Santos González
Departamento de Matematicas
Universidad de Zaragoza
50009 Zaragoza, Spain

Hyo Chul Myung
Department of Mathematics
University of Northern Iowa
Cedar Falls, Iowa 50614 U.S.A.

Book Production Manager: Janet Glanzman White
Graphics: Elenor Kallberg, Christopher Concannon, and Michael A. Masotti

Library of Congress Cataloging-in-Publication Data available upon request

ISBN 1-56072-050-6

Printed in the United States of America

Table of Contents

PREFACE

In recent years there has been a renewed interest in the study of nonassociative algebras, not only for its own sake, but also its applications to other areas of mathematics as well as to the physical and biological sciences. In particular, during the last decade, considerable research activities in nonassociative algebras have taken place at several Spanish universities, notably at Universities of Granada, Málaga, Madrid and Zaragoza. As an outgrowth of these research activities, in April 1989, the University of Zaragoza organized the Workshop on Nonassociative Algebraic Models in cooperation with Professor Efim Zel'manov's visit to the University from the Institute of Mathematics, Novosibirsk, Russia.

Zel'manov's visit to the University was concurrent with his positive solution to the long standing "Restricted Burnside Problem" and he presented his solution for the first time at this workshop. Most appropriately for the workshop, and perhaps surprisingly, his solution employed the theory of Jordan algebras.

The workshop attracted participants from the aforementioned universities and from the neighboring countries of France, Portugal and Morocco. In addition to the formal presentations, there were informal seminars and opportunities to discuss new ideas and open problems of mutual interest in nonassociative algebras.

We were very pleased with the positive reception that the workshop received from both the local community and the scientific community at large. The workshop was quite successful in stimulating the interest of young researchers as well as those who are established in the field.

The present volume contains selected papers which were presented at the workshop. We are grateful to professors K. McCrimmon and J.M. Osborn for their assistance and comments. Their work contributed significantly to the final preparation of the manuscripts.

We are most pleased to acknowledge the generous financial support of the following institutions: Presidency of the University of Zaragoza; Faculty of Sciences; Department of Mathematics; Dirección General de Investigación Cientifica y Técnica (under the project 87-0054); and Diputación General de Aragón.

Zaragoza and Cedar Falls
April, 1991

S. González and H.C. Myung

LIST OF PARTICIPANTS

1. Albuquerque, Helena, Universidade de Coimbra, Portugal
2. Anquela, José Angel, Universidad de Zaragoza, Spain
3. Benito Clavijo, Pilar, Colegio Universitario de la Rioja, Spain
4. Bouzraa, Habib, Université Hassan II, Casablanca, Morocco
5. Cabrera, Miguel, Universidad de Granada, Spain
6. Camarinha, Margarida, Universidade de Coimbra, Portugal
7. Castellón, Alberto, Universidad de Málaga, Spain
8. Cortés, Teresa, Universidad de Zaragoza, Spain
9. Cuenca Mira, José Antonio, Universidad de Málaga, Spain
10. Elduque, Alberto, Universidad de Zaragoza, Spain
11. Essannouni, Hassane, Université Mohamed V, Rabat, Morocco
12. Fernández López, Antonio, Universidad de Málaga, Spain
13. Galé, José Esteban, Universidad de Zaragoza, Spain
14. García Martín, Amable, Universidad de Málaga, Spain
15. García Rus, Eulalia, Universidad de Málaga, Spain
16. González, Santos, Universidad de Zaragoza, Spain
17. Jiménez Garijo, Pedro, Universidad de Granada, Spain
18. Kaidi, Amin, Université Mohamed V, Rabat, Morocco
19. Laliena, Jesús, Colegio Universitario de la Rioja, Spain
20. Martín, Cándido, Universidad de Málaga, Spain
21. Martínez Moreno, Juan, Universidad de Granada, Spain
22. Martínez, Consuelo, Universidad de Zaragoza, Spain
23. Micali, Artibano, Université de Sciences et Techniques du Languedoc, Montpellier, France
24. Montaner, Fernando, Universidad de Zaragoza, Spain
25. Pérez de Vargas, Alberto, Universidad Complutense de Madrid, Spain
26. Rachidi, Mustapha, Université Mohamed V, Rabat, Morocco
27. Rochdi, Abdellatif, Université Hassan II, Casablanca, Morocco
28. Rodríguez Palacios, Angel, Universidad de Granada, Spain
29. Sánchez Sánchez, Antonio, Universidad de Málaga, Spain
30. López Sánchez, Jesús, Universidad Complutense de Madrid, Spain
31. Siles Molina, Mercedes, Universidad de Málaga, Spain
32. Varea Agudo, Jesús, I.B. Marcilla, Navarra, Spain
33. Varea Agudo, Vicente, Universidad de Zaragoza, Spain
34. Vicente Matilla, Pilar, I.B. Juan de Encina, León, Spain
35. Villena, Armando, Universidad de Granada, Spain
36. Zel'manov, Efim, Institute of Mathematics, Novosibirsk, Russia

Nonassociative
Algebraic
Models

ISBN 1-56072-050-6

Some Counter-Examples in the Theory of Jordan Algebras

Y.A. Medvedev
Institute of Mathematics, Siberian Division, Soviet Academy of Sciences, Novosibirsk 630090, Russia

E.I. Zel'manov
Department of Mathematics, Yale University, Box 2155 Yale Station, New Haven, CT 06520, USA

ABSTRACT.

For a study of *infinite-dimensional algebras* in a variety it seems to be crucially important to know the structure of *finite-dimensional superalgebras* corresponding to that variety. Often complicated phenomena in the variety become transparent if we reveal the superalgebras which are hidden behind them. Some merits of this approach were demonstrated in [8], [**14-16**]. In this paper we consider two old problems in the theory of Jordan algebras and exhibit superalgebras which cast light on them.

Throughout the paper, by a *Jordan algebra* we mean a linear Jordan algebra over a field F of characteristic $\not\approx 2$. Along with the bilinear product xy an important role in Jordan algebras is played by the trilinear product

$$\{x, y, z\} = (xy)z + x(yz) - (xz)y.$$

An element a of a Jordan algebra J is called an ***absolute zero divisor*** (cf. [1]) if $a^2 = \{a, J, a\} = 0$.

ABSOLUTE ZERO DIVISOR PROBLEM. *Is the ideal $I_J(a)$ in J generated by an absolute zero divisor a nilpotent?*

S.V. Pchelintsev [7] answered this question in the negative: he constructed a prime Jordan algebra with nonzero absolute zero divisors (a ***Pchelintsev Monster***). However, in [10] it was shown that $I_J(a)$ is ***locally nilpotent***, while in [6] it was shown that $I_J(a)$ is ***nilpotent*** for ***finitely generated*** J.

Our second problem concerns the free Jordan algebra FJ of countable rank. The standard Amitsur argument shows that the Jacobson radical $Rad(FJ)$ of FJ is nil. In [11] it was proved that $Rad(FJ)$ consists of all those elements which vanish on all special Jordan algebras and on the 27-dimensional exceptional simple Albert algebra. However, it remained unclear whether $Rad(FJ)$ was nonzero ([1],[9]). In [5] one of the authors constructed a Jordan algebra on which an element of $Rad(FJ)$ was nonzero, thus establishing that $Rad(FJ) \not\approx 0$.

RADICAL SOLVABILITY PROBLEM. *Is the radical $Rad(FJ)$ solvable?*

We remark that in [8] it was shown that the radical of the free ***alternative*** algebra over a field of characteristic zero is nilpotent.

In §1 below we recall some basic definitions concerning superalgebras, Jordan superalgebras. Then we show in §§2, 3 how Jordan superalgebras of Poisson brackets (or of vector fields) are responsible for the existence of Pchelintsev Monsters. In §4 we show that the radical $Rad(FJ)$ is unsolvable because of the existence of the 10-dimensional simple Kac superalgebra.

§1. Superalgebras.

By a ***superalgebra*** we mean an arbitrary Z_2-graded algebra $A = A_0 \oplus A_1$, $A_iA_j \subseteq A_{i+j}$ $(i, j \in Z_2)$. For example, the Grassman algebra $G = \langle 1, e_i \mid e_i^2 = 0, e_ie_j + e_je_i = 0 \text{ for } i, j = 1, 2, \dots \rangle$ carries the natural Z_2-grading $G = G_0 \oplus G_1$, where $e_{i_1} \dots e_{i_r} \in G_0$ if r is even and $e_{i_1} \dots e_{i_r} \in G_1$ if r is odd. The subalgebra $G(A) = G_0 \otimes A_0 + G_1 \otimes A_1$ of the tensor product $G \otimes A$ is called the ***Grassman envelope*** of the superalgebra A.

Let Γ be a homogeneous variety of linear algebras. Say that a superalgebra $A = A_0 \oplus A_1$ is a ***Γ-superalgebra*** if $G(A) \in \Gamma$. It is easy to see that $G(A) \in \Gamma$ if and only if A satisfies a certain system of graded identities (depending on the variety). For example, an *associative superalgebra* is just an associative Z_2-graded algebra. A superalgebra is *super-commutative* if it satisfies the graded identity

$$x_\alpha y_\beta = (-1)^{\alpha\beta} y_\beta x_\alpha$$

for all $x_\alpha \in A_\alpha, y_\beta \in A_\beta$; a superalgebra is a *Lie superalgebra* if it satisfies the super-anticommutative and super-Jacobi identities

$$\begin{aligned}\langle x_\alpha y_\beta\rangle &= -(-1)^{\alpha\beta}\langle y_\beta x_\alpha\rangle,\\ (-1)^{\alpha\gamma}\langle x_\alpha, \langle y_\beta, z_\gamma\rangle\rangle &+ (-1)^{\alpha\beta}\langle y_\beta, \langle z_\gamma, x_\alpha\rangle\rangle\\ &+ (-1)^{\beta\gamma}\langle z_\gamma, \langle x_\alpha, y_\beta\rangle\rangle = 0,\end{aligned}$$

for all $x_\alpha \in A_\alpha, y_\beta \in A_\beta, z_\gamma \in A_\gamma$. The graded identities defining *Jordan superalgebras* (without 6-torsion, cf. [4]) are super-commutativity and the super-Jordan identity

$$\begin{aligned}x_\alpha y_\beta = {} & (-1)^{\alpha\beta} y_\beta x_\alpha + (-1)^{\alpha\gamma}(x_\alpha y_\beta)(z_\gamma t_\delta) + (-1)^{\alpha\beta}(y_\beta z_\gamma)(x_\alpha t_\delta)\\ & + (-1)^{\beta\gamma}(z_\gamma x_\alpha)(y_\beta t_\delta) = (-1)^{\alpha\gamma} x_\alpha((y_\beta z_\gamma)t_\delta)\\ & + (-1)^{\alpha\beta} y_\beta((z_\gamma x_\alpha)t_\delta) + (-1)^{\beta\gamma} z_\gamma((x_\alpha y_\beta)t_\delta).\end{aligned}$$

Let us consider now some examples of Jordan superalgebras.

Example 1 (Special Superalgebra). Let $A = A_0 \oplus A_1$ be an associative superalgebra and define the new multiplication

$$a_\alpha \circ b_\beta = a_\alpha b_\beta + (-1)^{\alpha\beta} b_\beta a_\alpha.$$

Then $(A, \circ)$ is a Jordan superalgebra which we will denote by $A^{(+)}$. A superalgebra is special if it is embeddable in some $A^{(+)}$, otherwise it is exceptional.

Example 2 (Hermitian Superalgebra). A graded linear map $* : A \to A$ of an associative superalgebra A is called a *superinvolution* if $(a_\alpha^*)^* = a_\alpha, (a_\alpha b_\beta)^* = (-1)^{\alpha\beta} b_\beta^* a_\alpha^*$ for any $a_\alpha \in A_\alpha, b_\beta \in A_\beta$. Then the subspace of $*$-invariant elements $H(A,*) = H(A_0,*) \oplus H(A_1,*)$ is a sub-superalgebra of $A^{(+)}$.

Example 3 (10 - Dimensional Kac Superalgebra). The following 10-dimensional Jordan superalgebra $J = J_0 \oplus J_1$ of V.G. Kac (cf. [4], [3]) is simple and exceptional (cf. §4 below). Let $a_1, a_2, a_3, a_4, a_5, a_6$ be a basis of J_0 and let b_1, b_2, b_3, b_4 be a basis of J_1. The multiplication in J is defined by

$$\begin{aligned}& a_1 a_i = a_i \ (i = 1, \ldots, 5), a_1 b_i = (1/2) b_i \ (i = 1, \ldots, 4),\\ & a_2 a_3 = a_1,\ a_2 b_3 = b_1,\ a_2 b_4 = b_2,\ a_3 b_1 = (1/2) b_3,\\ & a_3 b_2 = (1/2) b_4,\ a_4 a_5 = a_1,\ a_4 b_2 = b_1,\ a_4 b_4 = -b_3,\\ & a_5 b_1 = (1/2) b_2,\ a_5 b_3 = -(1/2) b_4,\ a_6 a^2 = a_6,\\ & a_6 b_i = (1/2) b_i \ (i = 1, \ldots, 4),\ b_1 b_2 = (1/2) a_2,\\ & b_1 b_3 = -(1/2) a_4,\ b_1 b_4 = (1/2) a_1 - (3/2) a_6,\\ & b_2 b_3 = -(1/2) a_1 + (3/2) a_6,\ b_2 b_4 = a_5,\ b_3 b_4 = a_3\end{aligned}$$

where all products which are not obtained from these by transpositions of factors are equal to zero.

Example 4 (Superalgebra of Vector Fields on the Line). Let us consider the polynomial algebra $K[x]$ with the usual derivative $\delta(x^i) = ix^{i-1}$. Let both J_0 and J_1 be the copies of $K[x]$,

$$J = K[x] \oplus \overline{K[x]}$$

the multiplication in $J = J_0 \oplus J_1$ being defined by

$$\bar{f}g = f\bar{g} = \overline{fg}\ ,\ \bar{f}\bar{g} = \delta(f)g - f\delta(g).$$

Then J is a Jordan superalgebra.

Example 5 (Superalgebra of Poisson Brackets). Now let $K[X,Y]$ be the polynomial algebra in countable many variables x_i, y_i, $i = 1, 2, \ldots$ and $J = K[X,Y] \oplus \overline{K[X,Y]}$ where

$$f\bar{g} = \bar{f}g = \overline{fg}\ ,\ \bar{f}\bar{g} = \sum_i \{\partial_{x_i}(f)\partial_{y_i}(g) - \partial_{y_i}(f)\partial_{x_i}(g)\}$$

is the Poisson bracket (∂_{x_i} the partial derivative with respect to x_i).

If char $= 0$ then both Jordan superalgebras of Examples 4,5 are infinite-dimensional and simple. If char $K = p > 0$, then instead of ordinary polynomials we should consider the truncated polynomial algebras $K_p[x] = K[x]/\langle x^p\rangle$, $K_p[X,Y] = K[X,Y]/\langle x_i^p, y_i^p \mid i = 1, 2, \ldots\rangle$ to obtain a simple Jordan superalgebra.

§2. Pchelintsev Monsters and the Jordan Superalgebra of Vector Fields on the Line.

Let K be a field of characteristic 0 and let $K[x]$ be the polynomial algebra in one variable. Consider the Jordan superalgebra $J = J_0 \oplus J_1$ of vector fields on the line (Example 4):

$$J_0 = K[x],\ J_1 = \overline{K[x]},\ f\bar{g} = \overline{fg},\ \bar{f}\bar{g} = \delta(f)g - f\delta(g).$$

THEOREM 1. *If J is the Jordan superalgebra of vector fields on the line, then the free algebra in the variety generaed by the Grassman envelope $G(J)$ is a prime degenerate Jordan algebra over K of characteristic* 0.

In order to establish this theorem, we shall need a more explicit construction of the above-mentioned free algebra. Consider the associative algebra F presented by generators $e_{ij}^{(\alpha)}$ $(0 \le i, j \le \infty, \alpha = 0, 1)$ and relations $e_{ij}^{(\alpha)} e_{pq}^{(\beta)} = (-1)^{\alpha\beta} e_{pq}^{(\beta)} e_{ij}^{(\alpha)}$ (in particular, $e_{ij}^{(1)2} = 0$). The algebra F has a natural Z_2-grading such that $e_{ij}^{(\alpha)} \in F_\alpha$, and moreover $F = F_0 \oplus F_1$ is the free associative supercommutative superalgebra on a countable number of even and odd generators.

Let $F(J) = F_0 \otimes J_0 + F_1 \otimes J_1$ be the F-envelope of J. Every element of $F(J)$ can be presented in the form of a finite sum (polynomial) $\Sigma_i a_i x^i + \Sigma_i b_i \overline{x^i}$, where $a_i \in F_0$, $b_i \in F_1$. Along with $F(J)$ let us consider the Jordan algebra $F((J))$ of infinite sums (power series)

$$\sum_{i=0}^{i=\infty} a_i x^i + \sum_{i=0}^{i=\infty} b_i \overline{x^i}$$

for $a_i \in F_0$, $b_i \in F_1$. Clearly $F((J))$ is a completion of $F(J)$ in the natural topology.

We claim that the subalgebra A of $F((J))$ generated by the elements

$$X_k = \sum_{i=0}^{i=\infty} e_{ki}^{(0)} x^i + \sum_{i=0}^{i=\infty} e_{ki}^{(1)} \overline{x^i}, \quad (k = 1, 2, \ldots)$$

together with the map $x_k \to X_k$ is free on the set $X = \{x_k \mid k = 1, 2 \ldots\}$ in the variety generated by $G(J)$. Indeed, every map $X \to F((J))$ can be extended to a homomorphism $A \to F((J))$. This shows that A is free in the variety it generates, and that this variety is the same as that generated by $F((J))$, hence by $F(J)$ [since its completion is $F((J))$], hence by $G(J)$ [since F is a scalar extension of G].

For an element $a = \sum_{i=0}^{i=\infty} a_i^{(0)} x^i + \sum_{i=0}^{i=\infty} a_i^{(1)} \overline{x^i}$, $a_i^{(\alpha)} \in F_\alpha$, denote by $d(a)$ the lowest "power" of x or $\bar{x}$ that appears in a, i.e. the minimal i such that either $a_i^{(1)} \neq 0$ or $a_i^{(0)} \neq 0$.

LEMMA 1. *If a Jordan homogeneous polynomial $f(x_1, \ldots, x_n)$ is not identically zero on $F((J))$ then for any $m \geq 1$ there exist elements $u_1, \ldots, u_n \in F((J))$, such that $f(u_1, \ldots, u_n) \neq 0$ and $d(u_i) \geq m$ $(i = 1, 2, \ldots, n)$.*

PROOF: Fix $m \geq 1$. If some partial linearization f' of f is not identically zero on $F((J))$ then [since the values $f'(v_1, \ldots, v_r)$ are linear combinations of values $f(u_1, \ldots, u_n)$ at arguments u_i which are linear combinations of arguments v_j of f'], it is sufficient to prove the lemma for f'. Hence we may assume that f is additive in each argument, $f(x_1, \ldots, x_i + x_i', \ldots, x_n) = f(x_1, \ldots, x_i, \ldots, x_n) + f(x_1, \ldots, x_i', \ldots, x_n)$ for all x_i, x_i' in $F((J))$, but doesn't vanish identically [if $f' = f(x_1, \ldots, x_i + x_i', \ldots, x_n) - f(x_1, \ldots, x_i, \ldots, x_n) - f(x_1, \ldots, x_i', \ldots, x_n)$ vanishes on $F((J))$, then f is already additive in x_i, otherwise replace f by f' of lower degree in x_i and x_i', and repeat]. Note that we are NOT asserting that this f is multilinear as a Jordan polynomial, only as a function on $F((J))$. Now it follows from the additivity in all variables that there exist monomials $a_1 \hat{x}^{d_1}, \ldots, a_n \hat{x}^{d_n}$ [where $\hat{x}^{d_i} = x^{d_i}$ if $a_i \in F_0, \hat{x}^{d_i} = \bar{x}^{d_i}$ if $a_i \in F_1$], such that $f(a_1 \hat{x}^{d_i}, \ldots, a_n \hat{x}^{d_n}) \neq 0$. Suppose that q_i is the degree of the homogeneous polynomial $f(x_1, \ldots, x_n)$ with respect to x_i. Let $\sigma(a_i) = \begin{cases} 0 & \text{if } a_i \in F_0 \\ 1 & \text{if } a_i \in F_1 \end{cases}$. Then

$f(a_1\hat{x}^{t_1},\ldots,a_n\hat{x}^{t_n}) = h(t_1,\ldots,t_n)\hat{x}^{q_1t_1+\cdots+q_nt_n-q}$, where $q = \left[\left(\sum_{i=1}^{n}\sigma(a_i)q_it_i\right)/2\right]$ where $h(t_1,\ldots,t_n)$ is a polynomial in integers $t_1,\ldots,t_n$ with coefficients from F.

Since $h(d_1,\ldots,d_n)\neq 0$ it follows that there exist integers $t_1,\ldots,t_n$ $(t_i\geq m)$ such that $h(t_1,\ldots,t_n)\neq 0$, which finishes the proof of the lemma. ∎

LEMMA 2. *Let* $a = \sum_{i=0}^{\infty}a_i^{(0)}x^i + \sum_{i=0}^{\infty}a_i^{(1)}\bar{x}^i$, $b = \sum_{i=0}^{\infty}b_i^{(0)}x^i + \sum_{i=0}^{\infty}b_i^{(1)}\bar{x}^i$ *be arbitrary elements from* A. *If* $I_A(a)\circ I_A(b)=0$ *(*$\circ$ *the ordinary Jordan product in* A*) then* $a_i^{(\alpha)}\circ b_j^{(\beta)}=0$ *for all* $\alpha,\beta=0,1$ *and* $i,j=0,1,2,\ldots$.

PROOF: The elements a,b are Jordan polynomials in the free generators $X_1,X_2,\ldots$. Specializing the countably many free generators which are not involved in a,b to arbitrary elements of $F((J))$ (see p. 9), we see that

$$I_{F((J))}(a)\circ I_{F((J))}(b)=0. \tag{1}$$

Let $\tilde{a},\tilde{b}$ denote arbitrary elements $\Sigma\tilde{a}_i^{(0)}x^i+\Sigma\tilde{a}_i^{(1)}\bar{x}^i$, $\Sigma\tilde{b}_i^{(0)}x^i+\Sigma\tilde{b}_i^{(1)}\bar{x}^i$ of $F((J))$. As usual, for an arbitrary element c of a Jordan algebra we denote by $R(c)$ the multiplication operator $R(c): x\to xc$. For any $p,q\geq 1$ the operator

$$D_{p,q}=R(e_{op}^{(1)}\bar{x}^0)R(e_{oq}^{(1)}\bar{x}^0)$$

is an inner derivation of $F((J))$ (namely $e_{op}^{(1)}e_{oq}^{(1)}\left[\frac{d}{dx}+\frac{\bar{d}}{dx}\right]$). More precisely,

$$\tilde{a}D_{p,q}=e_{op}^{(1)}e_{oq}^{(1)}\{\sum_{i=1}^{\infty}i\tilde{a}_i^{(0)}x^{i-1}+\sum_{i=1}^{\infty}i\tilde{a}_i^{(1)}\overline{x^{i-1}}\}. \tag{2}$$

Then the formulas

$$aD_{p_1,q_1}\cdots D_{p_i,q_i}=e_{op_1}^{(1)}e_{oq_1}^{(1)}\cdots e_{op_i}^{(1)}e_{oq_i}^{(1)}[i!a_i^{(0)}x^0+i!a_i^{(1)}\bar{x}^0+\cdots], \tag{3}$$

$$aD_{p_1,q_1}\cdots D_{p_i,q_i}R(e_{0s}^{(1)}\bar{x})=-e_{op_1}^{(1)}e_{oq_1}^{(1)}\cdots e_{op_i}^{(1)}e_{oq_i}^{(1)}e_{os}^{(1)}[i!a_i^{(1)}x^0+i!a_i^{(0)}\bar{x}^1+\cdots] \tag{4}$$

show that we can find $\tilde{a}\in I_{F((J))}^{(a)}$ involving only finitely many Grassman variables $e_{or}^{(1)}$ whose constant term (coefficient of x^0) is $i!a_i^{(\alpha)}$ multiplied by a finite product of distinct Grassman variables $e_{or}^{(1)}$.

Thus $i!a_i^{(\alpha)}\circ j!a_j^{(\beta)}=0$ will follow if we can show that for any $\tilde{a}\in I_{F((J))}(a),\tilde{b}\in I_{F((J))}(b)$ involving only finitely many $e_{or}^{(1)}$ we have

$$\tilde{a}_0^{(0)}\circ\tilde{b}_0^{(0)}=0 \tag{5}$$

(and since the characteristic is zero this will imply $a_i^{(\alpha)} \circ b_j^{(\beta)} = 0$).

The coefficient of $\tilde{a} \circ \tilde{b}$ at the element x^0 is $\tilde{a}_0^{(0)} \circ \tilde{b}_0^{(0)} + \tilde{a}_1^{(1)} \circ \tilde{b}_0^{(1)} + \tilde{a}_0^{(1)} \circ \tilde{b}_1^{(1)}$. Hence (1) implies

(6) $$\tilde{a}_0^{(0)} \circ \tilde{b}_0^{(0)} + \tilde{a}_1^{(1)} \circ \tilde{b}_0^{(1)} + \tilde{a}_0^{(1)} \circ \tilde{b}_1^{(1)} = 0$$

We can find $e_{op}^{(1)}, e_{oq}^{(1)}$ not appearing in $\tilde{a}$ or $\tilde{b}$. The element $\tilde{\tilde{a}} = \tilde{a}D_{p,q} \in I_{F((J))}(a)$ has $\tilde{\tilde{a}}_0^{(\alpha)} = 0, \tilde{\tilde{a}}_1^{(\alpha)} = e_{op}^{(1)} e_{oq}^{(1)} a_1^{(\alpha)}$, so applying (6) to $\tilde{a}, \tilde{b}$ yields

$$\tilde{\tilde{a}}_1^{(1)} \circ b_0^{(1)} = [e_{op}^{(1)} e_{oq}^{(1)} \tilde{a}_1^{(1)}] \circ \tilde{b}_0^{(1)} = 0$$

and because we chose the Grassman variables $e_{op}^{(1)}$ not contained in $\tilde{a}, \tilde{b}$ we get

(7) $$\tilde{a}_1^{(1)} \circ \tilde{b}_0^{(1)} = 0.$$

Similarly

(8) $$\tilde{a}_0^{(1)} \circ \tilde{b}_1^{(1)} = 0.$$

Hence (6), (7), (8) imply

$$\tilde{a}_0^{(0)} \circ \tilde{b}_0^{(0)} = 0$$

as requested in (5). This finishes the proof. ∎

Let $<$ be an arbitrary linear order on the set $E^{(1)} = \{e_{ij}^{(1)}\}$ of all Grassman variables. For a finite subset π of $E^{(1)}$ let $e_\pi^{(1)}$ be the product $e_{i_1 j_1}^{(1)} \cdots e_{i_r j_r}^{(1)}$ of all elements in π such that $e_{i_1 j_1}^{(1)} < \cdots < e_{i_r j_r}^{(1)}$. Then every element a from F can be uniquely presented as $a = \Sigma_\pi a_\pi(e_{pq}^{(0)}) e_\pi^{(1)}$, where π are finite subsets of $E^{(1)}$ and $a_\pi(e_{pq}^{(0)})$ are polynomials in $e_{ij}^{(0)}$.

Let f be a nonzero homogeneous element from A. Then $f = \sum_{i=0}^{\infty} f_i^{(0)} x^i + \sum_{i=0}^{\infty} f_i^{(1)} \bar{x}^i$ where $f_i^{(\alpha)} \in F_\alpha$, so $f_i^{(\alpha)} = \Sigma f_{i\pi}^{(\alpha)} e_\pi$ for polynomials $f_{i\pi}^{(\alpha)}$ in the $e_{pq}^{(0)}$ for $p, q \geq 0$.

LEMMA 3. *For a given nonzero homogeneous element f in the free algebra A and for any finite subset τ of $E^{(1)}$ there exist integers $i \geq 0$, $\alpha = 0$ or 1, and a finite subset $\pi \subseteq E^{(1)}$ with $\pi \cap \tau = \phi$ such that the polynomial $f_{i\pi}^{(\alpha)}$ is nonzero.*

PROOF: Suppose on the contrary that for each i, α we have $f_{i\pi}^{(\alpha)} = 0$ whenever π does not contain an element from τ. Let $\tau = \{e_{i_1 j_1}^{(1)}, \ldots, e_{i_r j_r}^{(1)}\}, m = \max\{j_1, \ldots, j_r\}$. Then for any n elements $a_k = \sum_{i=0}^{\infty} a_{ki}^{(0)} x^i + \sum_{i=0}^{\infty} a_{ki}^{(1)} \bar{x}^i$ with $d(a_n) \geq m$, $[a_{ki}^{(\alpha)} \in F_\alpha$ $(1 \leq k \leq n)$ with $a_{ki}^{(1)} = 0$ for $i \leq m]$, we have $f(a_1, \ldots, a_n) = 0$. This contradicts the assertion of Lemma 1.

PROOF OF THEOREM 1: Let us suppose that the algebra A is not prime. Then there exist nonzero homogeneous elements $f, g \in A$ such that

$$\mathcal{I}_A(f) \circ \mathcal{I}_A(g) = 0.$$

Suppose that $f_i^{(\alpha)} \neq 0$ and let τ_0 be the finite subset of $E^{(1)}$ of minimal cardinality such that $f_{i\tau_0}^{(\alpha)} \neq 0$. Now let Σ be the union of all finite subsets τ of $E^{(1)}$ such that $f_{i\tau}^{(\alpha)} \neq 0$ (for this particular i). (Note that Σ is finite since $f_i^{(\alpha)}$ is a finite sum of $f_{i\pi}^{(\alpha)} e_\pi$'s). By Lemma 3 there exist an integer $j \geq 0, \alpha, \beta = 0$ or 1, and a finite subset $\pi_0 \subseteq E^{(1)}$ such that $\pi_0 \cap \Sigma = \phi$ and $g_{j\pi_0}^{(\beta)} \neq 0$. By Lemma 2 we have

$$(*) \qquad f_i^{(\alpha)} \circ g_j^{(\beta)} = \Sigma f_{i\tau}^{(\alpha)} g_{j\pi}^{(\beta)} e_\tau e_\pi = 0.$$

Let us show that

$$e_\tau e_\pi \neq \pm e_{\tau_0} e_{\pi_0}$$

whenever

$$(\tau, \pi) \neq (\tau_0, \pi_0), \; f_{i\tau}^{(\alpha)} \neq 0, \; g_{j\pi}^{(\beta)} \neq 0,$$

which would contradict (*). If $e_\tau e_\pi = \pm e_{\tau_0} e_{\pi_0}$ then $\tau \cup \pi = \tau_0 \cup \pi_0$, so from $\pi_0 \cap \Sigma = \phi$ for τ, $\tau_0 \subseteq \Sigma$ it follows that $\pi_0 \subseteq \pi$, $\tau \subseteq \tau_0$. By the minimality of τ_0 we have $\tau = \tau_0$, which yields $\pi = \pi_0$ by disjointness.

Thus we have proved that the algebra A is prime. Let us show that A is degenerate. It is straightforward to verify that any value of the Jordan polynomial

$$\{[x, y]\}^2 = 2x \circ \{y, x, y\} - \{x, y^2, x\} - \{y, x^2, y\}$$

on the Grassman envelope $G(J)$ of J lies in $G_0 \otimes J_0$. It is easy to verify also that for arbitrary elements $a \in G_0 \otimes J_0, z \in G(J)$ we have $(a, z, z) = 0$ where $(x, y, z) = (xy)z - x(yz)$ is the associator of the elements x, y, z. Hence A satisfies the polynomial identity

$$(\{[x, y]\}^2, z, z) = 0.$$

Since

$$[[e_1, u_{12}]^2, e_2 + u_{23}, e_2 + u_{23}] = \frac{1}{2}(e_3 - e_2) + \frac{1}{4} u_{23} \neq 0.$$

A cannot contain 3 connected idempotents. If A were nondegenerate, then by the classification theorem from [**12**] it would be isomorphic to a Jordan algebra of a symmetric bilinear form, and thus would satisfy the identity

$$(\{[x, y]\}^2, z_1, z_2) = 0.$$

However, one can check that the algebra $G(J)$ doesn't satisfy this identity for odd x, y, z_1, z_2, so the free algebra A in the variety of $G(J)$ doesn't either, and A must be degenerate. ∎

§3. The Pchelintsev Monster and the Jordan Superalgebra of Poisson Brackets.

In this chapter we shall construct Pchelintsev Monsters over a field K of prime characteristic $p \neq 2$. The idea will be the same as in §1 (and to a great extend as in [7]): we shall find a Jordan superalgebra $J = J_0 \oplus J_1$ such that the free algebra in the variety generated by $G(J)$ is prime and degenerate.

We remark that the superalgebra J will have to be infinite-dimensional or else A will be the free algebra in the variety generated by the ordinary Jordan algebra J_0. Indeed, if $\dim_K(J_1) = d < \infty$ then every element of $G_1 \otimes J_1$ generates a nilpotent ideal of index $\leq d+1$. If $f(x_1, \ldots, x_n)$ is a homogeneous identity of J_0 then it is an identity of $G_0 \otimes J_0$. For arbitrary elements $a_1, \ldots, a_n \in G(J)$ the ideal $I_{G(J)}(f(a_1, \ldots, a_n))$ is nilpotent of index $\leq (d+1)n$. Hence if A is the free algebra in the variety generated by $G(J)$ then for arbitrary elements $a_1, \ldots, a_n \in A$ the ideal $I_A(f(a_1, \ldots, a_n))$ is nilpotent of index $\leq (d+1)n$.

The superalgebra of §1, $K[x] \oplus \overline{K[x]}$ in one variable also doesn't work here, because x^p lies in the center of J, so J is finite-dimensional over its center.

This suggests we consider the algebra of truncated polynomials in an infinite number of variables

$$K_p[X, Y] = K[x_i, y_i \mid x_i^p = y_i^p = 0, i = 1, 2, \ldots]$$

with the Poisson brackets

$$[f, g] = \sum_i [\partial x_i(f) \partial y_i(g) - \partial y_i(f) \partial x_i(g)].$$

The superalgebra $J = J_0 \oplus J_1 = K_p[X, Y] \oplus \overline{K_p[X, Y]}$ with $f \circ \bar{g} = \bar{f} \circ g = \overline{fg}, \bar{f} \circ \bar{g} = [f, g]$ is a Jordan superalgebra.

THEOREM 2. *The free algebra in the variety generated by the Grassman envelope $G(J)$ of the Jordan superalgebra J of Poisson brackets is prime and degenerate over an infinite field K of characteristic p.*

We need a few preliminary lemmas. For an arbitrary element a of the Grassman envelope $G(J)$ consider the decomposition

$$a = \sum_i a_i \otimes w_i + \sum_j \bar{a}'_j \otimes w'_j \tag{9}$$

where $w_i \in G_0$, $\{a_i\}$ are linearly independent monomials from J_0; $w'_j \in G_1$, $\{a'_j\}$ are linearly independent monomials from J_1. We call (9) the canonical decomposition of a. Let $d_{\min}(a)$ be the minimum of degrees of all a_i, a'_j and let $d_{\max}(a)$ be the maximum of degrees of all a_i, a'_j.

LEMMA 4. *Let $f(x_1,\ldots,x_n)$ be a homogeneous polynomial which is not identically zero on $G(J)$. Then for any $m \geq 1$ there exist elements $u_1,\ldots,u_n \in G(J)$ such that $f(u_1,\ldots,u_n) \neq 0$ and each $d_{\min}(u_i)$ is greater than m.*

The proof of this lemma follows closely the proof of Lemma 1 and thus is omitted.

Let $A = A_0 + A_1$ be an arbitrary superalgebra. By a multiplication algebra $M(A)$ of A we mean the subalgebra of $\mathrm{End}_K(A)$ generated by all multiplication operators $R(a), a \in A_0 \cup A_1$. Clearly, $M(A) = M_0(A) + M_1(A)$ where $M_i(A)$ is spanned by all operators $R(a_1)\cdots R(a_m), m \geq 1, a_i \in A\alpha_i, \alpha_1 + \cdots + \alpha_m \equiv i \bmod 2$.

For an ordinary algebra its multiplication algebra is obtained from the above by ignoring the odd part.

LEMMA 5. *1) Let $a_1 \otimes w_1$ be a nonzero component in the canonical decomposition of an element $a \in G(J)$. Let E be an infinite set of Grassman variables. Then there exists an operator $W \in M_0(G(J))$ such that*

$$aW = 1 \otimes w_1 w,$$

where w is an even product of Grassman variables from E.

2) If $\bar{a}'_1 \otimes w'_1 \neq 0$ then there exists an operator $W \in M_1(G(J))$ such that

$$aW = 1 \otimes w'_1 w,$$

where w is an odd product of Grassman variables from E.

PROOF: 1) Let us assume that all $a_i, \bar{a}'_j$ lie in the sub-superalgebra

$$\begin{aligned} J^{(k)} &= J_0^{(k)} \oplus J_1^{(k)} \\ &= K_p[x_1,\ldots,x_k;y_1,\ldots,y_k] \oplus \overline{K_p[x_1,\ldots,x_k;y_1,\ldots,y_k]} \end{aligned}$$

on the first k variables. Since $J^{(k)}$ is simple as superalgebra it follows that $J_0^{(k)}$ and $J_1^{(k)}$ are irreducible modules over $M_0(J^{(k)})$. For both modules the centralizers of $M_0(J^{(k)})$ are just scalar multiplications. Thus by the Jacobson Density Theorem there exists an operator $W_0 \in M_0(J^{(k)})$ such that $a_1 W_0 = 1, a_i W_0 = 0$ for $i > 1$. Let us assume that

$$W_0 = \sum_S R(c_{s1})\cdots R(c_{sr_s}),$$

where $c_{st} \in J_0^{(k)} \cup J_1^{(k)}$ and for each s the number of odd c_{st} is even.

Now let r be an even number which is greater than all r_s and let u be the product of r variables from E, $u = e_1 \cdots e_r$. For each s we can represent u as $u = u_{s1} \cdots u_{sr_s}$, where u_{sq} is odd or even whenever c_{sq} is odd or even. Let

$$\tilde{W}_0 = \sum_s R(c_{s1} \otimes v_{s1}) \cdots R(c_{sr_s} \otimes v_{sr_s}) \in M_0(G(J)).$$

We have,

$$(a_i \otimes w_i)\tilde{W}_0 = a_i W_0 \otimes w_i u,$$

hence

$$a\tilde{W}_0 = 1 \otimes w_1 u + \Sigma \bar{b}_j \otimes \tilde{w}'_j,$$

where $\bar{b}_j \in J_1, \tilde{w}'_j \in G_1$. Let us assume that variables x_q, y_q are not involved in polynomials b_j; e_{r+1}, e_{r+2} are two additional variables from E. We have

$$[\bar{b}_j, \bar{x}_q] = 0,$$

hence

$$(\Sigma \bar{b}_j \otimes \tilde{w}'_j) R(\bar{x}_q \otimes e_{r+1}) = 0.$$

Let us consider the operator

$$W = \tilde{W}_0 R(\bar{x}_q \otimes e_{r+1}) R(\bar{y}_q \otimes e_{r+2}).$$

From what we showed above it follows that

$$aW = 1 \otimes w_1 u e_{r+1} e_{r+2}.$$

2) In the same way as we did in 1) one can prove the existence of an operator $\tilde{W}_0 \in M_0(G(J))$ such that

$$a\tilde{W}_0 = \Sigma b_i \otimes \tilde{w}_i + \bar{x}_1 \otimes w'_1 u,$$

where u is an even product of Grassman variables from E. Let e_{r+1} be a Grassman variable from E which is not involved in u. Then

$$a\tilde{W}_0 R(\bar{y}_1 \otimes e_{r+1}) \in 1 \otimes w'_1 u e_{r+1} + J_1 \otimes G_1$$

and we can use 1).

PROOF OF THEOREM 2: Let us suppose that the algebra A is not prime and thus there exist nonzero polynomials $f, g \in A$ (involving only finitely many variables $x_1, \ldots, x_n$) such that

$$I_A(f) \circ I_A(g) = 0.$$

Since A is free in the variety determined by $G(J)$ we can specialize $x_1, \ldots, x_n$ to get

$$I_{G(J)}(f(c_1, \ldots, c_n)) \circ I_{G(J)}(g(c_1, \ldots, c_n)) = 0$$

for any $c_1, \ldots, c_n \in G(J)$.

Since both polynomials f, g are not identically zero on $G(J)$ there exist elements $a_1, \ldots, a_n, b_1, \ldots, b_n \in G(J)$ such that $f(a_1, \ldots, a_n) \neq 0, g(b_1, \ldots, b_n) \neq 0$. By Lemma 4 we may assume that for any $1 \leq i, j \leq n$,

$$(d_{\max}(a_i) + 1)d \leq d_{\min}(b_j), \tag{10}$$

where d is the degree of f. Moreover, we may assume that the elements $a_1, \ldots, a_n$ have no Grassman variables in common with the elements $b_1, \ldots, b_n$. Now

$$f(a_1 + b_1, \ldots, a_n + b_n) = f(a_1, \ldots, a_n) + \Sigma \tilde{f}(\ldots a_i, \ldots, b_j, \ldots) + f(b_1, \ldots, b_n),$$

where the $\tilde{f}$ are partial linearizations of f containing both a_i and b_j. From (10) it follows that

$$d_{\max}(f(a_1, \ldots, a_n)) < d_{\min}(\Sigma \tilde{f}(\ldots a_i, \ldots, b_j, \ldots) + f(b_1, \ldots, b_n)).$$

Hence the canonical decomposition of $f(a_1, \ldots, a_n)$ is a part of the canonical decomposition of $f(a_1 + b_1, \ldots, a_n + b_n)$. Let

$$f(a_1 + b_1, \ldots, a_n + b_n) = \Sigma f_i \otimes w_i + \Sigma \bar{f}'_j \otimes w'_j,$$
$$g(a_1 + b_1, \ldots, a_n + b_n) = \Sigma g_k \otimes v_k + \Sigma \bar{g}'_e \otimes v'_e$$

be the canonical decompositions of elements $f(a_1+b_1, \ldots, a_n+b_n)$, $g(a_1+b_1, \ldots, a_n+b_n)$; f_i, g_k are monomials from J_0; $\bar{f}'_j, \bar{g}'_e$ are monomials from J_1.

Suppose that $w_i v_k \neq 0$ for some i, k (cases $w_i v'_e \neq 0$, $w'_j v_k \neq 0, w'_j v'_e \neq 0$ are treated similarly).

Let E_1, E_2 be two infinite sets of Grassman variables such that $E_1 \cap E_2 = \phi$ and E_1, E_2 don't have any Grassman variables in common with w_i, v_k. By Lemma 5 there exist products w, v of variables from E_1, E_2 respectively such that

$$I_{G(J)}(f(a_1 + b_1, \ldots, a_n + b_n)) \ni 1 \otimes w_i w,$$
$$I_{G(J)}(g(a_1 + b_1, \ldots, a_n + b_n)) \ni 1 \otimes v_k v.$$

Hence

$$w_i v_k w v = 0,$$

which contradicts our assumption.

Thus, the product of any members of the canonical decompositions of $f(a_1 + b_1, \ldots, a_n + b_n)$, $g(a_1 + b_1, \ldots, a_n + b_n)$ respectively is zero. In particular

$$f(a_1, \ldots, a_n) g(a_1 + b_1, \ldots, a_n + b_n) = 0.$$

Now since the ground field K is infinite and the polynomial g is homogeneous it follows that

$$f(a_1, \ldots, a_n) g(b_1, \ldots, b_n) = 0$$

for any elements $a_1, \ldots, a_n, b_1, \ldots, b_n$ satisfying (10). But since elements $a_1, \ldots, a_n$ have no Grassman variables in common with the elements $b_1, \ldots, b_n$ the product of Grassman parts of any two members of the canonical decompositions of $f(a_1, \ldots, a_n)$, $g(b_1, \ldots, b_n)$ respectively is different from zero and we are in a position to repeat the argument above. This establishes that A is prime.

Now let us suppose that A is a nondegenerate prime Jordan algebra. In [**13**] there was constructed a nonzero ideal T of the free Jordan algebra FJ and a certain integer $s \geq 1$ such that for any $m \geq s$ elements $t_1, \ldots, t_m \in T$ the operator $R(t_1) \cdots R(t_m)$ can be presented as a linear combination of operators $R(a_1) \cdots R(a_k)$ of length $k \leq s$ for some $a_i \in FJ$.

If $T(G(J)) \neq 0$ then as above $e_1 \cdots e_d \otimes 1 \in J_{\beta_i}$, $i = 1, 2, \ldots, m$, using independence of Grassman variables one sees that the operator $R(b_1) \cdots R(b_m)$ can be represented as a linear combination of operators $R(a_1) \cdots R(a_k)$ for $k \leq s$ and $a_j \in J_{\alpha_j}$ with the same parity $\Sigma\alpha_i = \Sigma\beta_j$. In particular, the operator

$$R(\bar{y}_1)R(\bar{1})R(\bar{y}_2)R(\bar{1}) \cdots R(\bar{y}_s)R(\bar{1})R(\bar{y}_{s+1})$$

is a linear combination of operators $R(a_1) \cdots R(a_k)$ for $k \leq s$, but

$$\begin{aligned}
&\overline{x_1 \cdots x_{s+1}} R(\bar{y}_1)R(\bar{1})R(\bar{y}_2)R(\bar{1}) \cdots R(\bar{y}_s)R(\bar{1})R(\bar{y}_{s+1}) \\
&= \partial_{x_s} \cdots \partial_{x_1}(x_1 \cdots x_{s+1})R(\bar{y}_{s+1}) = x_{s+1}R(\bar{y}_{s+1}) \\
&= \partial_{x_{s+1}}(x_{s+1}) = 1
\end{aligned}$$

while for any k elements $a_i \in J_{\alpha_i} (1 \leq i \leq k \leq s)$ with $\alpha_1 + \cdots + \alpha_k = 1$, the polynomial $\overline{x_1 \cdots x_{s+1}} R(a_1) \cdots R(a_k)$ lies in the ideal of J_0 generated by $x_1, \ldots, x_{s+1}$.

Thus $T(G(J)) = 0$, which implies $T(A) = 0$ for A in the variety generated by $G(J)$. Now from the classification of prime nondegenerate Jordan algebras and the nature of the ideal T (cf. [**12**], [**13**]) it follows that A is a central order in a Jordan algebra of symmetric bilinear form. However, again the algebra A doesn't satisfy the identity

$$(\{[x, y]\}^2, z_1, z_2) = 0,$$

a contradiction. Thus A must be degenerate. ∎

§4. The Radical of the Free Jordan Algebra.

Consider the 10-dimensional Kac superalgebra $J = J_0 \oplus J_1$ of Example 3, and let $G(J)$ be its Grassman envelope. Consider the elements

$$(*) \qquad \begin{aligned}
x &= e_1 \otimes b_1 + e_2 \otimes b_3 + e_3 \otimes b_4, \\
y &= e_4 \otimes b_1 + e_5 \otimes b_3 + e_6 \otimes b_4, \\
z &= e_7 \otimes b_1 + e \otimes b_2,
\end{aligned}$$

where $e_1, \ldots, e_8$ are different Grassman variables. The Glennie polynomial g_8 (cf. [**2**]) doesn't vanish on $G(J)$: tedious though routine computations show that for x, y, z as in (*)

$$g_8(x, y, z) = 4a_4 \otimes e_1 \cdots e_8 \neq 0.$$

Consider the polynomial

$$f(x_1, \ldots, x_{28}, y_1, z_1, z_2)$$
$$= \Sigma_{\sigma \in S_{28}} (-1)^\sigma p(x_{\sigma(1)}, \ldots, x_{\sigma(28)}, y_1, z_1, z_2)$$

for

$$p(x_1, \ldots, x_{28}, y_1, z_1, z_2)$$
$$= y_1 R(x_1)R(x_2)ZR(x_3)R(x_4)Z \cdots ZR(x_{27})R(x_{28})$$
$$\text{where} \qquad Z = R(z_1)R(z_2).$$

By construction, f is skew-symmetric in the 28 variables x_i and thus is identically zero on the 27-dimensional Albert algebra $H(C_3)$. Let

(**)
$$y_1 = a_2, \quad z_1 = a_5, \quad z_2 = a_2,$$
$$x_i = e_{i+8} \otimes b_3 \quad (1 \leq i \leq 28).$$

Then for these values in $G(J)$ f takes on the value

$$f = 28!\ e_8 \cdots e_{36} \otimes a_2.$$

Now the polynomial

$$h(x, y, z, t, y_1, z_1, z_2, x_1, \ldots, x_{28})$$
$$= [g_8(x, y, z) \circ t] \circ f(y_1, z_1, z_2, x_1, \ldots, x_{28})$$

is not identically zero on $G(J)$: for example, for the above values (*), (**) of $x, y, zy_1, z_1, z_2, x_i$ and the value

(***)
$$t = a_5$$

we have

(9)
$$h = 28!\ e_1 \cdots e_{36} \otimes a_2 \neq 0.$$

In [**11**] it was shown that all elements of the free Jordan algebra $FJ[X]$ which are identically zero on all special Jordan algebras and on the Albert algebra $\mathcal{H}(C_3)$ lie in the nil radical $Nil(FJ[X])$ of $FJ[X]$. Hence

$$h \in Nil(FJ[X]).$$

Now suppose that the factorial 28! is not equal to zero in the ground field K and the set X of free generators is infinite. Then $h \neq 0$ and $Nil(FJ[X]) \neq 0$ (which was first established in [**5**]).

THEOREM 3.. *If K has characteristic 0 or $p > 28$, and X is infinite, then the T-ideal I_h generated by h in $FJ[X]$ (the smallest T-ideal containing h) is nil but not solvable.*

PROOF: By (9) and simplicity of J we have

$$I_h(G(J)) \supseteq e_1 \cdots e_{36} \otimes J.$$

If the ideal I_h were solvable then so would the superalgebra J, whereas J is simple. ∎

We remark that in [8] it was shown that the radical of the free alternative algebra over a field of characteristic zero is nilpotent. The key part of the proof was the classification of prime alternative superalgebras. The more complicated structure of $Rad(FJ[X])$ is due to the diversity of simple Jordan superalgebras.

PROBLEM. *Suppose that char$(K) = 0$, and the element $f \in FJ[X]$ is identically zero on the Grassman envelopes of all finite-dimensional simple Jordan superalgebras over K. Is it true that the ideal I_f is solvable?*

Acknowledgement.

The second author wishes to thank Zaragoza University for its hospitality during his visit, supported by the Programa Europa of C.A.I. - C.O.N.A.I. in April 1989, during which the work on the present paper was completed. The authors also thank Kevin McCrimmon for his invaluable help and assistance in the preparation of this manuscript.

Bibliography

[1] N. Jacobson, *Structure and representations of Jordan algebras*, AMS, Providence, RI, 1969

[2] C.M. Glennie. *Some identities valid in special Jordan algebras but not valid in all Jordan algebras*, Pac. J. Math. **16** (1986).

[3] L. Hogben, V. Kac. *The correct multiplication table for the exceptional Jordan superalgebra F*, Comm. in Algebra, **11**(10) (1983), 1155-1156.

[4] V. Kac. *Classification of simple Z-graded superalgebras and simple Jordan superalgebras*, Comm. in Algebra, **5**(13) 1977, 1375-1400.

[5] Yu. Medvedev. *On nil elements of the free Jordan algebra*, Sib. Math. J., **26** no.2 1985, 140-148.

[6] Yu. Medvedev. *Radicals of free finitely generated Jordan algebras*, Sib. Math. J., **29** no. 4 1988, 139-148.

[7] S.V. Pchelintsev. *Prime Jordan algebras and absolute zero divisors*, Izv. Akad. Nauk SSSR, **50** no. 1 1986, 79-100.

[8] I.P. Shestakov, E. Zelmanov. *Prime alternative superalgebras and the nilpotency of the radical of the free alternative algebra*, submitted to Izv. Akad. Nauk SSSR.

[9] K.A. Zevlakov, A.M. Slinko, I.P. Shestakov, A.I. Shirshov. *Rings that are nearly associative*, Academic Press, New York, 1982.

[10] E. Zelmanov. *Absolute zero divisors and algebraic Jordan algebras*, Sib. Math. J., **23** no. 6 1982, 100-116.

[11] E. Zelmanov. *On prime Jordan algebras*, Alg. i Logika, **18** no. 2 1979, 286-310.

[12] E. Zelmanov. *On prime Jordan algebras 2*, Sib. Math. J., **24** no. 1 1983, 89-104.

[13] E. Zelmanov. *Birepresentations of infinite-dimensional Jordan algebras*, Sib. Math. J., **27** no. 6 1986, 79-94.

[14] E. Zelmanov. *On Lie algebras with Engel condition*, Doklady Akad. Nauk, **35** no. 1 1987, 44–47

[15] E. Zelmanov. *Jordan nil algebras of bounded degree*, to appear in Trudy Math. Inst. Akad. Nauk. SSSR (Sib. Branch), Novosibirsk, 1989.

[16] E. Zelmanov. *On the nilpotency of nil algebras*, Proceedings of the 5th National School in Algebra, Varna, 1986. Lecture Notes in Math. no. 1352, Springer-Verlag, Heidelberg, 1988.

Nonassociative
Algebraic
Models

ISBN 1-56072-050-6

Modular Jordan Nilpotent Algebras*

José Angel Anquela
Departamento de Matemáticas, Universidad de Zaragoza, 50009 Zaragoza, Spain

Abstract: We show that any Jordan nilpotent algebra, with modular lattice of subalgebras, has its cube trivial.

0. Introduction

The study of an algebra by means of the study of its lattice of subalgebras has been done by D. W.Barnes in [1], for Lie algebras, and [2], for associative algebras. A. C. Elduque studied this kind of things for Malcev algebras in his doctoral thesis [4] and J. A. Laliena for alternative algebras in [5].

For Jordan algebras J. A. Laliena recently completed "Lattice isomorphisms of Jordan algebras" .Following Laliena´s work, the author has just shown that over an algebraically closed field a Jordan algebra with the same (isomorphic) lattice of subalgebras as a given semisimple Jordan algebra must be isomorphic to it. The condition on the field in this type of work is very usual. One notes easily that the information provided by the lattice of subalgebras is too gross to differenciate between division algebras in many cases.

In the present paper we are going to see how the condition of modularity on the lattice of subalgebras determines some properties of the algebra. The easiest and most

* This paper has been written under the direction of Professor Santos González and it will be a part of the author's Doctoral Thesis. The author has been partially supported by the Ministerio de Educación y Ciencia (F.P.I. Grant) and the Diputación General de Aragón.

natural example of modular algebras with modular lattices of subalgebras are trivial algebras, and this makes one think that modularity could be a weaker condition than triviality but related with it. We will see the relationship for nilpotent Jordan algebras. This is not a very strong restriction: one can see that the only semisimple (finite dimensional) Jordan algebras which have a modular lattice of subalgebras are the base field or two copies of the base field and, thus, the (finite dimensional) Jordan algebras with modular lattice of subalgebras are at most two idempotents far from the nilpotent algebras.

1. Preliminaries

We will suppose that the reader knows the usual terms in lattice theory: sublattice, chain, lattice isomorphism, length of a lattice (the supremum of the lengths of all the chains, where the length of a chain is its cardinality minus one)...

A lattice $(L, \leq)=(L, \wedge, \vee)$ is said to be modular when one of the three following equivalent conditions holds:

(1) If $x \geq z$ then $x \wedge (y \vee z) = (x \wedge y) \vee z$.

(2) $x \wedge (y \vee z) = x \wedge [(y \wedge (x \vee z)) \vee z]$ ("shearing identity")

(3) L does not contain a pentagon (as sublattice).

In a lattice $(L, \leq)=(L, \wedge, \vee)$ we put $x -\!\!< y$ when $x \leq y$, $x \neq y$ and $\{z \in L / x \leq z \leq y \} = \{ x , y\}$.

A lattice $(L, \leq)=(L, \wedge, \vee)$ is said to be semimodular if :

$a -\!\!< b \Rightarrow a \vee c -\!\!< b \vee c$ or $a \vee c = b \vee c$ for any c in L.

It is well known that a modular lattice is always semimodular.We also have as a theorem the "Jordan-Hölder chain condition": In a semimodular lattice with finite length all the maximal chains have the same length.

(See [3] for these results on general lattice theory).

We will deal with algebras over a field F with characteristic not two.

Let J be an algebra. We will put L(J) for its lattice of subalgebras where $\leq$, $\wedge$, $\vee$ are naturally defined ($\leq$ is the inclusion relation). We define the length of J (and put l(J)) as the length of L(J).

It is very easy to prove that if J is a solvable algebra, then its length and dimension coincide.In general we have $l(J) \leq \dim_F J$.

DEFINITION: An algebra J is called modular (semimodular) if L(J) is modular (semimodular).

An inmediate consequence of the definition is the following:

PROPOSITION: *For any algebra J we have:*

(1) If J is modular then it is semimodular.

(2) If J is modular (semimodular), then all of its subalgebras and quotiens are modular (semimodular). •

In a power associative algebra we define the order of a nilpotent element x (and put o(x)) as the number n such that $x^n = 0$ and $x^{n-1} \neq 0$.

If X is a subset of an algebra J, we will put (X) for the subalgebra of J generated by X and $\langle X \rangle$ for the vectorial subspace of J spanned by X.

" $\leq$ " will mean "is a subalgebra of" when we deal with algebras.

" + " will mean sum of vectorial subspaces.

2. Modular Nilpotent Algebras

THEOREM : *Let N be a Jordan nilpotent algebra.Then the following conditions are equivalent:*

(1) N is semimodular.

(2) $S + T \leq J$ for all $S, T \leq J$.

(3) N is modular.

PROOF: (1) ⇒ (2) N is nilpotent, hence we only have to prove (2) for finite dimensional subalgebras. Let S, T be finite dimensional subalgebras of N, a semimodular Jordan algebra. Bearing in mind the Jordan-Hölder chain condition and the semimodularity condition, it is easy to show (see [3]):

$$l(S \vee T) - l(T) \leq l(S) - l(S \wedge T) = l(S) - l(S \cap T)$$

N is nilpotent and so are S, T, $S \vee T$, $S \cap T$. Hence they are solvable, and their lengths are their dimensions, showing that:

$$\dim_F(S \vee T) \leq \dim_F(S) + \dim_F(T) - \dim_F(S \cap T) = \dim_F(S+T)$$

and so, $S \vee T = S+T$ and S+T is a subalgebra of N.

(2) ⇒ (3) We only have to repeat the proof of the modularity of the lattice of submodules of a module. Indeed, with (2), the modularity condition has the form:

$T \subseteq S \Rightarrow S \cap (R + T) = (S \cap R) + T$, which is obviously true.

(3) ⇒ (1) is always true.●

PROPOSITION: *Let N be a modular Jordan nilpotent algebra, and suppose R, S are trivial subalgebras of N. Then R + S is trivial.*

PROOF: Let r be an element in R and s an element in S. We must show r.s = 0.

If r, s are not linearly independent, this is obvious. Hence let us suppose they are linearly independent. We have that $\langle r \rangle$ and $\langle s \rangle$ are subalgebras of N. Using the previous result we get that $\langle r \rangle + \langle s \rangle$ is a subalgebra of N. If $r.s = \alpha.r + \beta.s$ with, for example, $\alpha \neq 0$, then $((...(r.s).s)...^{n)}...s).s) \neq 0$ for all n, which contradicts the nilpotency of N. Thus r.s = 0, as we wanted to prove. ●

COROLLARY: *Let M be a modular Jordan nilpotent algebra.Then there exists a trivial subalgebra T(M) of M such that all trivial subalgebras are contained in it .Moreover* $T(M) = \{ x \in M \,/\, x^2 = 0 \}$. ●

PROPOSITION: *Let M be a modular Jordan nilpotent algebra. For any subalgebra S of M we have* $S \,.\, T(M) \subseteq S^2$.

PROOF: We will carry out an induction on the dimension of S.

If the dimension of S is one, the result is obvious because S is contained in T(M).

Let us suppose that $\dim_F(S) = n + 1$ with $n \geq 1$ and that the result is true for all modular Jordan nilpotent algebras and their subalgebras of dimension less than or equal to n. As M is nilpotent, so is S, and hence $S^2 \neq S$. If $S^2 = 0$ then S is contained in T(M) and everything is obvious.Let us suppose $S^2 \neq 0$. We have $\dim_F(S^2) < \dim_F(S)$ and using the induction assumption we obtain:

$$S^2 \triangleleft T(M) + S^2, \text{ hence, } S^2 \triangleleft T(M) + S.$$

The algebra $T(M) + S / S^2$ is modular and nilpotent and S/S^2 , $T(M)+S^2/S^2$ are trivial subalgebras; hence the sum is trivial and, thus, $T(M).S \subseteq S^2$. ●

PROPOSITION 1: *Let M be a modular Jordan nilpotent algebra.Let us suppose that y and a are elements in M such that* $y^2 = 0$ *and* $o(a)= 3$. *Then* $y.\,a = 0$.

PROOF: We know by the previous result that $T(M).(a) \subseteq (a)^2 = \langle a^2 \rangle$. Thus $y.a = \alpha.a^2$ .If $\alpha \neq 0$ we can change y and put $\frac{1}{\alpha}y$ and we have $y.a = a^2$ and

$$(a - \tfrac{1}{2}y)^2 = a^2 - a.y = 0$$

We have just seen that $a - \frac{1}{2}y$ is in T(M). As y is also in T(M) we know

$$(a - \tfrac{1}{2}y).y = 0, \text{ but } (a - \tfrac{1}{2}y).y = a.y = a^2 \neq 0$$

This contradiction comes from supposing $y.a \neq 0$. ●

PROPOSITION 2: *Let M be a modular Jordan nilpotent algebra.Let us suppose that y and a are elements in M such that* $y^2 = 0$ *and* $o(a) = 4$*.Then* $y \,.\, a$ *is contained in* $\langle a^3 \rangle$.

PROOF: We only have to use the previous result in the quotient $T(M)+(a)/\langle a^3\rangle$. •

PROPOSITION 3: *Let M be a modular Jordan nilpotent algebra.Let us suppose that y and a are elements in M such that* $y^2 = 0$ *and* $o(a) = 5$*.Then* $y \,.\, a$ *is contained in* $\langle a^3, a^4 \rangle$.

PROOF: Use Proposition 1 in $T(M)+(a)/\langle a^3, a^4\rangle$. •

PROPOSITION 4: *Let M be a modular Jordan nilpotent algebra. Then all the elements in M have orders less than or equal to 5.*

PROOF: Let a be an element in M such that $o(a) = 6$. Then $(a^2) = \langle a^2, a^4\rangle$, $(a^3) = \langle a^3\rangle$ and $(a^2) \vee (a^3) = \langle a^2, a^3, a^4, a^5\rangle \neq (a^2) + (a^3)$.Hence, we cannot find elements of order 6 in a modular nilpotent Jordan algebra. Moreover, we cannot find elements of order greater than 6: If a is such an element we have the quotient $(a) / \langle a^6, a^7, \ldots\rangle$ which is a modular nilpotent Jordan algebra with an element of order 6. •

REMARK: Proposition 4 cannot be improved because the algebras generated by a single element of order less than six are modular nilpotent Jordan algebras.(Modularity can be proved directly without many problems).

A consequence of the last four propositions is:

COROLLARY: *Let M be a modular Jordan nilpotent algebra. Then T(M) is an ideal of M.* •

The previous corollary can be used to construct a family of ideals of any modular Jordan nilpotent algebra in the following inductive way:

$T_1(M) = T(M)$.

As soon as we have $T_n(M)$ we define $T_{n+1}(M)$ as the ideal of M such that:

$T_{n+1}(M)/ T_n(M) = T(M/ T_n(M))$

By their very definition we have the following properties:

$T_n(M) \subset T_{n+1}(M)$ for all n´s and $T_n(M) \neq T_{n+1}(M)$ if $T_n(M) \neq M$.

$(T_{n+1}(M))^2 \subset T_n(M)$ for all n´s.

THEOREM: *Let M be a modular Jordan nilpotent algebra. Then $T_3(M) = M$.*

PROOF: Let a be an element in M.Its order can be (1), 2, 3, 4 or 5.

If $o(a) = 2$ we have a is in $T_1(M)$.

If $o(a) = 3$ or 4, then $o(a^2) = 2$. Hence a^2 is in $T_1(M)$ and a is in $T_2(M)$.

If $o(a) = 5$, then $o(a^2) = 3$. Hence a^2 is in $T_2(M)$ and a is in $T_3(M)$. •

COROLLARY: *Let M be a modular Jordan nilpotent algebra. Then $M^{(3)} = 0$.*

PROOF: $M = T_3(M)$ implies $M^2 = M^{(1)} \subseteq T_2(M)$.Hence $M^{(2)} \subseteq T_1(M)$ and $M^{(3)} = 0$. •

PROPOSITION: *Let M be a modular Jordan nilpotent algebra such that the elements in M have order less than or equal to 3. Then $M^3 = 0$.*

PROOF: We have to show that $(x.y).z = 0$ for all x, y, z in M. First we want to note that x.y is always in $T_1(M)$. Indeed, it is obvious if x or y is in $T_1(M)$. Hence let us suppose that x and y have order 3. We know that x.y is in $(x) \vee (y) = (x) + (y) = \langle x, y, x^2, y^2\rangle$, but it is easy to prove that the component in either x and y is zero because of the nilpotency (or factorizing $(x) + (y)$ over $\langle x^2, y^2\rangle$). Thus x.y is in $T_1(M)$. If $o(z) = 2$, z is in $T_1(M)$ and $(x.y).z = 0$. If $o(z) = 3$, $(x.y).z = 0$ by Proposition 1. •

THEOREM: *Let M be a modular Jordan nilpotent algebra. Then $(M^3)^2 = 0$.*

PROOF: We have to prove $M^3 \subseteq T_1(M)$, which is equivalent to $(M/T_1(M))^3 = 0$. This last statement is a corollary of the previous Proposition because in the algebra $M/T_1(M)$ all the elements have order less than or equal to three. •

FINAL REMARKS: (1) *The results obtained in this section are valid for nilpotent, power associative, commutative algebras, even of infinite dimension.*

(2) *Over an algebraically closed field the author has obtained a classification (up to isomorphism) of all the modular Jordan nilpotent algebras, even of infinite dimension. It can be seen that all these algebras are a direct sum of two ideals: one of them trivial, and as big as desired, and the other with dimension less than or equal to 6.*

(3) *Over an algebraically closed field, and for finite dimensional Jordan algebras, we can remove nilpotency in our first Theorem and it remains true. Moreover, the construction of the finite dimensional Jordan algebras over an algebraically closed field, by "glueing" idempotents to the algebras of (2), is easy because modularity acts strongly.*

REFERENCES

[1] D. W. BARNES. "Lattice isomorphisms of Lie algebras". J. Aust. Math. Soc. 4 (1964) 470-475.

[2] D. W. BARNES. "Lattice isomorphisms of associative algebras". J. Aust. Math. Soc. 6 (1966) 106-121.

[3] G. GRATZER. "General lattice theory". Math Reihe 52. Birkhäuser Verlag Basel und Stuttgart (1978).

[4] A. C. ELDUQUE. "Estructura de un algebra de Malcev a través de su retículo de subálgebras". Doctoral Thesis (1984).

[5] J. A. LALIENA. "Estructura reticular y cuasiideal en álgebras alternativas". Doctoral Thesis (1987).

Nonassociative
Algebraic
Models

ISBN 1-56072-050-6

Analogues du Théorème de Skornyakov-San Soucie

Habib Bouzraa
*Départment de Mathématiques, Faculté des Sciences II,
B.P. 6621, Sidi-Othman, Casablanca, Maroc*

El Amin Mokhtar Kaidi
*Département de Mathématiques et Informatique,
Faculté des Sciences de Rabat, B.P. 1014 Rabat, Maroc*

1. Introduction

Soit A une K-algèbre (non nécéssairement associative) unitaire sur un corps commutatif K. Pour $a \in A$, notons par $L_a : x \rightarrow ax$ (respectivement $R_a : x \rightarrow xa$) l'opérateur de multiplication à gauche (respectivement à droite) et par L(A) l'algèbre (associative) des K endomorphismes linéaires de A. On dira que A vérifie

i. La propriété d'inverse à gauche (PIG) si :

pour $a \in A$, L_a est inversible dans L(A) implique qu'il existe $b \in A$ tel que $L_a^{-1} = L_b$.

ii. La propriété d'inverse (PI) si :

pour $a \in A$, L_a et R_a sont inversibles dans L(A) entraîne $L_a^{-1} = L_b$ et $R_a^{-1} = R_b$ pour un $b \in A$.

iii. La propriété d'inverse quadratique (PIQ) si A est commutative et si :

pour $a \in A$, $U_a = 2R_a^2 - R_{a^2}$ est inversible dans $L(A)$ implique qu'il existe $b \in A$ tel que $U_a^{-1} = U_b$ et $R_b = U_a^{-1}R_a$.

On sait que toute algèbre de Moufang à gauche, une algèbre vérifiant les identités $L_x^2 = L_{x^2}$ et $L_xL_yL_x = L_{x(yx)}$, unitaire vérifie (PIG) (voir [4], [6]) et que toute algèbre alternative (respectivement de Jordan) unitaire possède (PI) (respectivement (PIQ)) [2].

Dans ce papier, on établit les résultats suivants :

1. Une K algèbre unitaire, vérifiant (PIG) (respectivement (PI)), est de Moufang à gauche (respectivement alternative) dans chacun des cas suivants :
   a. K infini et A de dimension finie
   b. $K = \mathbb{R}$ ou $\mathbb{C}$ et A normée complète.

2. Une K algèbre commutative unitaire, vérifiant (PIQ), est de Jordan dans chacun des cas suivants :
   c. K infini de caractéristique différente de 2 et 3 et A de dimension finie.
   d. $K = \mathbb{R}$ ou $\mathbb{C}$ et A normée complète.

Les résultats concernant (PIG) et (PI) sont inspirés du théorème de Skornyakov-San Soucie qui affirme que tout anneau non associatif unitaire de division, possédant (PIG), est alternatif. (voir [1], p.140).

En ce qui concerne (PIQ), à notre connaissance, elle n'a pas de précédents dans la littérature.

Dans la suite, K désigne un corps commutatif infini et une K algèbre est supposée unitaire, d'élément unité e.

2. Propriété d'inverse dans une algèbre

Dans une K algèbre non associative A de dimension finie (respectivement, normée complète réelle ou complexe, c'est à dire le K espace vectoriel sous-jaçant de A avec $K = \mathbb{R}$ ou $\mathbb{C}$, muni d'une norme $\| \ \|$ vérifiant $\|xy\| \leq \|x\| \ \|y\|$, est de Banach), l'ensemble $O = \{ x \in A : L_x$ est inversible dans $L(A) \}$, est un ouvert de Zariski (respectivement, $inv_g(A) = \{ x \in A$: L_x est inversible dans $BL(A) \}$, où $BL(A)$ est l'algèbre associative des opérateurs linéaires continus de A, est un ouvert de A. ([3], p.49)).

Considérons une K-algèbre A de dimension finie (respectivement, normée complète réelle ou complexe). On suppose que A possède (PIG).

Soit $x \in O$ (respectivement $x \in inv_g(A)$. Il existe $y \in A$ vérifiant $L_x^{-1} = L_y$. L'élement y est unique et déterminé par $y = L_x^{-1}(e)$. Posons $j(x) = L_x^{-1}(e)$. Dans le cas où A est de dimension finie, j est une inversion: une application birationnelle et homogène de degré -1 [5]. Dans le cas normé complet, j est une application continue, différentiable sur $inv_g(A)$ et sa différentielle en $x \in inv_g(A)$ est

$$- L_x^{-1} R_{L_x^{-1}(e)} \quad \text{([3], p. 49 et p. 51).}$$

Dans les deux situations précédentes, j vérifie $joj = id$ et $j(e) = e$.

Théorème 1. Une K algèbre A, vérifiant (PIG) (respectivement (PI)), est de Moufang à gauche (respectivement alternative) dans chacun des cas suivants :

a. A de dimension finie.

b. A normée complète réelle ou complexe.

Preuve. Soit A une K-algèbre vérifiant (PIG). $j(x) = L_x^{-1}(e)$ est différentiable dans chacun des cas a., b. et la derivée de j en x dans la direction de u est

$$- L_x^{-1} L_u L_x^{-1}(e) .$$

D'autre part, en différentiant l'identité $id = L_x L_{j(x)}$, on a :

$$0 = L_u L_{j(x)} - L_x L_{L_x^{-1}} L_u L_x^{-1}(e) \qquad (1)$$

En substituant dans (1), x par j(x), on obtient :

$$L_u L_x = L_x^{-1} L_{L_x L_u L_x(e)} \ .$$

D'où

$$L_x L_u L_x = L_{x(ux)} \qquad (2)$$

Faisons u = e dans (2), on aura

$$L_x^2 = L_{x^2} \qquad (3)$$

Les identités (2) et (3) ont lieu pour tout $x \in O$ (respectivement $x \in inv_g(A)$) et tout $u \in A$. Dans le cas a., O est dense dans A. A est donc une K algèbre de Moufang à gauche. Dans le cas b., soit $x \in A$, il existe $\alpha \in \mathbb{R}_+^*$ tel que $e-\alpha x \in inv_g(A)$ ([3], p.47).Les identités (2) et (3) ont lieu pour $e-\alpha x$ et u quelconque. Par identification, il découle que A est de Moufang à gauche.

3. Propriété d'inverse quadratique dans une algèbre

Etant donné une K algèbre A commutative vérifiant (PIQ) (K est supposé de caractéristique différente de 2). Soit $x \in A$ tel que U_x est inversible dans L(A). Il existe $y \in A$ vérifiant $U_x^{-1} = U_y$ et $R_y = U_x^{-1} R_x$. La dernière égalité, appliquée à e, donne $y = U_x^{-1}(x)$. En posant $J(x) = U_x^{-1}(x)$, il s'ensuit que

$$U_x^{-1} = U_{J(x)} \qquad (4)$$

Si A est de dimension finie (respectivement, normée complète réelle ou complexe), alors $J : A \to A$, $x \to U_x^{-1}(x)$ est une inversion (respectivement, une application continue différentiable) définie sur $O' = \{ x \in A : U_x$ est inversible dans $L(A) \}$ qui est un ouvert de Zariski de A (respectivement sur $\mathrm{inv}_q A = \{x \in A : U_x$ est inversible dans $BL(A) \}$ qui est un ouvert de A). De plus $JoJ = id$ car $J(J(x)) = U_{J(x)}^{-1}(J(x)) = U_x(J(x)) = x$, puisque $J(x) = U_x^{-1}(x)$.

Théorème 2. Une K algèbre commutative A, vérifiant (PIQ), est de Jordan dans chacun des cas suivants :

c. K de caractéristique différente de 2 et 3 et A de dimension finie.

d. $K = \mathbb{R}$ ou $\mathbb{C}$ et A normée complète.

Preuve. Soit A une K algèbre commutative de dimension finie vérifiant (PIQ). Notons par P_J le P-opérateur de J :

$$P_J(x) = - [(dJ)_x]^{-1}.$$

Par différentiation de l'identité $U_x(J(x)) = x$, on a

$$- U_x(P_J^{-1}(x)y) + 2U_{x,y}(J(x)) = y \text{ avec } U_{x,y} = 1/2(U_{x+y} - U_x - U_y) \quad (5)$$

Substituons dans (5), x par J(x) et utilisons le fait que $JoJ = id$ ainsi que $P_J^{-1}(J(x)) = P_J(x)$ ([5], Prop. 1.10), on obtient:

$$P_J(x)y = 2\, U_x\, U_{J(x),y}(x) - U_x(y) \quad (6)$$

Faison $y = e$ dans (6), on a

$$P_J(x)e = x^2 . \quad (7)$$

D'autre part, on différentie l'identité (4) , on trouve

$$U_{x,y} = U_x U_{J(x), P_J^{-1}(x)y)} U_x . \quad (8)$$

Remplaçons dans(8), x par J(x), y par e et utilisons (7), on aura $U_{x,x^2} = R_x U_x$, soit encore

$$R_{x^3} = 2R_x R_{x^2} + R_{x^2} R_x - 2R_x^3 \tag{9}$$

où $x^3 = xx^2 = x^2x$. En particulier, on a

$$xx^3 = x^2\, x^2 \tag{10}$$

Enfin, différentions l'identité (10) et identifions le avec (9), on aura $3R_{x^2}R_x = 3R_x R_{x^2}$ d'où

$$R_{x^2}R_x = R_x R_{x^2} \quad \text{pour tout} \quad x \in O' .$$

Comme O', ouvert de Zariski, est dense dans A, donc A est de Jordan.

Dans le cas normé complet, avec $K = \mathbb{R}$ ou $\mathbb{C}$, les mêmes étapes du calcul se font pour conclure que

$$R_{x^2}R_x = R_x R_{x^2} \quad \text{pour tout} \quad x \in \mathrm{inv}_q(A) ,$$

sauf qu'ici, on différentie relativement à la topologie induite par la norme.

Soit $x \in A$, il existe $\alpha \in \mathbb{R}_+^*$ tel que $e-\alpha x \in \mathrm{inv}_q(A)$ d'où

$$R_{(e-\alpha x)^2}R_{(e-\alpha x)} = R_{(e-\alpha x)}R_{(e-\alpha x)^2} ,$$

ce qui donne $R_{x^2}R_x = R_x R_{x^2}$. Donc A est de Jordan.

Remerciement

Les auteurs remercient vivement le réferée pour ses suggestions qui ont amélioré la rédaction du texte.

REFERENCES

[1] D.R. Hughes, F.C. Piper, *Projective planes* ; Springer-Verlag, New-York Heidelberg Berlin.

[2] N.Jacobson, *Structure and Representation of Jordan algebras*, Amer. Math. Soc. Coll. Publ. Vol XXXIX, 1969.

[3] A.M. Kaidi, Bases para una teoria de las algebras no associativas normadas, Tesis doctoral, Universidad de Granada Spain, 1977.

[4] K. McCrimmon, Finite-dimensional left Moufang algebras, Math. Ann. **224**, 179-187 (1976).

[5] K. McCrimmon, Axioms for inversion in Jordan algebras, J. Algebra. **47**, 201-222 (1977).

[6] A. Thedy, Right alternative rings, J. Algebra. **37**, 1-43 (1977).

Nonassociative
Algebraic
Models

ISBN 1-56072-050-6

Hilbert Modules Over H*-Algebras In Relation with Hilbert Ternary Rings

M. Cabrera García, J. Martínez Moreno and A. Rodríguez Palacios
Departamento de Análisis Matemático, Facultad de Ciencias,
Universidad de Granada, 18071-Granada, Spain

AMS 1980 Subject Classification $46K15, 46H25, 17A40$

Abstract

We prove that faithful Hilbert modules over H^*-algebras with zero annihilator are in one-to-one correspondence with some Hilbert ternary rings, namely the positive (associative) H^*-triples with zero annihilator. As a consequence we obtain a complete structure theory for H^*-triples.

Introduction

The study of Hilbert triple systems begins with the work by W. Kaup [11], where JH^*-triples are introduced and a complete structure theory for them is obtained (see also the book by E. Neher [12] for a "grid" approach). Related types of Hilbert triple systems have been considered also in [4] and [5]. In this paper, motivated by the observation that Hilbert modules over H^*-algebras with zero annihilator have a natural structure of a ternary ring in the sense of M. R. Hestenes [9], we introduce an abstract class of Hilbert ternary rings, called H^*-triples, so that Hilbert modules become actually "positive" H^*-triples with zero annihilator. Our main result states that, conversely, every positive H^*-triple with zero annihilator can be obtained from a suitable Hilbert module in the above way. The proof of this result uses intensively Rieffel's theory of induced representations of C^*-algebras [13]. Because the theory of Hilbert modules over H^*-algebras with zero annihilator is at this time widely developed (see

[14], [8], [18] and [3]) we derive from our main result a complete structure theory for H^*-triples, namely: essentially there are no other H^*-triples than the ones of the form $\mathcal{HS}(H, K)$ (Hilbert-Schmidt operators from H to K, where H and K are complex Hilbert spaces) with triple product $\{xyz\} = xy^*z$. In the development of this structure theory we also use some techniques close to the ones in the paper by H. Zettl [19], where a Gelfand-Naimark type theorem for ternary C^*-rings is obtained.

I. Positive H*-triples are Hilbert modules

Our definition of H^*-triple will become the associative counterpart of the Hilbert Jordan triple systems introduced and studied by W. Kaup [11] under the short name of JH^*-triples. Thus an *H^*-triple* will be a complex Hilbert space $(W, (.\mid .))$ endowed with a mapping $\{\ldots\} : W \times W \times W \to W$, called the *triple product* of W, which is linear in the outer variables, and conjugate linear in the middle variable, satisfying the associative identities

$$\{xy\{ztu\}\} = \{x\{tzy\}u\} = \{\{xyz\}tu\} \tag{1}$$

and the following H^*-conditions

$$(\{xyz\} \mid t) = (x \mid \{tzy\}) = (z \mid \{yxt\}) \tag{2}$$

for all x, y, z, t, u in W. If, for x and y in W, we denote by $x\Box y$ and $x\Delta y$ the operators on W given by $z \to \{xyz\}$ and $z \to \{zxy\}$, respectively, the H^*-conditions read as follows:

$$((x\Box y)(z) \mid t) = (z \mid (y\Box x)(t)) \text{ and } ((x\Delta y)(z) \mid t) = (z \mid (y\Delta x)(t)). \tag{3}$$

Therefore, by the closed graph theorem, $x\Box y$ and $x\Delta y$ lie in the C^*-algebra $BL(W)$ of all bounded linear operators on W, and we have

$$(x\Box y)^* = y\Box x \text{ and } (x\Delta y)^* = y\Delta x. \tag{4}$$

Moreover, if $P(x, y)$ denotes the conjugate-linear operator on W defined by $z \to \{xzy\}$, the equality

$$(P(x, y)(z) \mid t) = (P(y, x)(t) \mid z) \tag{5}$$

shows analogously that $P(x, y)$ is bounded. Now, from Corollary 2 to Theorem III.5.1 in [17], it follows easily that the triple product of W is continuous. We define the *annihilator* of W as the following set

$$Ann(W) := \{x \in W : \{xWW\} = 0\} =$$
$$\{x \in W : \{WxW\} = 0\} = \{x \in W : \{WWx\} = 0\}, \tag{6}$$

where the equalities follow from the H^*-conditions.

Following [1; Definition 9.11] a *left module* over an (associative) complex algebra $\mathcal{E}$ will mean a complex vector space W together with a bilinear mapping $(e, x) \to ex$

from $\mathcal{E} \times W$ into W satisfying $e(fx) = (ef)x$ for all e, f in $\mathcal{E}$ and x in W. W will be called a *faithful* left module if in addition for e in $\mathcal{E}$, $ex=0$ for all x in W implies $e = 0$. Similar definitions are stated for right modules. Saworotnow's original definition of Hilbert module over an H^*-algebra [14], with some remarks in [14] and [18], can be formulated as follows. Given an H^*-algebra $\mathcal{E}$ with zero annihilator (see for example [1; Section 34] for definitions), a ***Hilbert*** $\mathcal{E}$***-module*** is a faithful left module W over $\mathcal{E}$ provided with a mapping $[.\mid .] : W \times W \to \mathcal{E}$, called the $\mathcal{E}$*-valued inner product* of W, satisfying

i) $[\lambda x \mid y] = \lambda[x \mid y]$.
ii) $[x + x' \mid y] = [x \mid y] + [x' \mid y]$.
iii) $[ex \mid y] = e[x \mid y]$.
iv) $[x \mid y]^* = [y \mid x]$.
v) For each nonzero x in W there is a nonzero e in $\mathcal{E}$ such that $[x \mid x] = e^*e$.
vi) W is a Hilbert space under the inner product

$$(x \mid y) = tr([x \mid y]). \tag{7}$$

We note that, by the polarization law, property v) implies that, for x, y in W, $[x \mid y]$ actually lies in the "trace-class" of $\mathcal{E}$(see [15] and [16]), and so the **C**-*valued inner product* of W given in vi) is well defined. Also we note that for e in $\mathcal{E}$ and x, y in W we have

$$(e \mid [x \mid y]) = tr(e[x \mid y]^*) = tr(e[y \mid x]) = tr([ey \mid x]) = (ey \mid x), \tag{8}$$

so the linear hull of $[W \mid W]$ is dense in $\mathcal{E}$ thanks to the faithfulness of W.

From the above definitions, the following result can be easily verified.

Proposition 1. *Let W be a Hilbert module over an H^*-algebra $\mathcal{E}$ with zero annihilator. Then W is an H^*-triple with zero annihilator for the **C**-valued inner product and for the triple product defined by $\{xyz\} = [x \mid y]z$ for all x, y, z in W. Moreover, for all x in W, the operators $x\Box x$ and $x\Delta x$ are positive.*

We will say that an H^*-triple W is *positive* if, for all x in W, the operator $x\Delta x$ is positive. We will see later that this assumption implies that $x\Box x$ is also a positive operator. The fundamental result in this paper is the following theorem, which assures that the converse to the above proposition is true.

Theorem 1. *Every positive H^*-triple with zero annihilator arises from an essentially unique Hilbert module over a suitable H^*-algebra with zero annihilator in the form given by Proposition 1.*

The rest of this section is devoted to prove the above theorem. To this end, we begin by including some terminology from [13]. We recall that, given a pre$-C^*$-algebra B, a (right) B*-rigged space* [13; Definition 2.8] is a right B-module X together with a B-valued sesquilinear form (linear in the second variable and conjugate linear in the first one) $< , >_B$ such that

i) For all x in $X, < x, x >_B$ is a positive element in the completion of B.

ii) $< x, y >_B^* = < y, x >_B$.
iii) $< x, yb >_B = < x, y >_B b$.
iv) The range of $< \, , \, >_B$ generates a dense subalgebra of B.

B-module endomorphisms T of X satisfying that

1) $< Tx, Tx >_B \leq K_T^2 < x, x >_B$ for suitable nonnegative constant K_T and all x in X, and

2) There exist $T^* : X \to X$ such that satisfies 1) above and $< Tx, y >_B = < x, T^* y >_B$ for all x, y in X,

are called *bounded* operators on X [13; Definition 2.11]. The set of all "bounded" operators on X will be denoted by $L(X)$ and, for all T in $L(X)$, $||| T |||$ will denote the least constant K_T in 1). For the future application of Theorem 5.1 in [13], note that every (right) B-rigged space is in a natural way a "(left) pre-Hermitian B-rigged **C**-module" in the sense of [13; Definition 4.19].

Lemma 1. *Let W be a positive H^*-triple and let E denote the linear hull of $W \square W$. Then*

i) E is a self-adjoint subalgebra of the C^*-algebra $BL(W)$.

ii) E is a prehilbert space under an inner product $(. \mid .)$ determinated on the generators by $(x \square y \mid z \square t) = (\{xyt\} \mid z)$.

iii) If R, S, T *are in E, we have* $(RS \mid T) = (R \mid TS^*) = (S \mid R^* T)$, $\mid R^* \mid = \mid R \mid$, *and* $\mid ST \mid \leq \mid S \mid \| T \|$, *where* $\mid \cdot \mid$ *denotes the norm associated to the above inner product, while as usual* $\| \cdot \|$ *denotes the operator norm.*

Proof. The first assertion is easily verified from the H^*-conditions and the associative identities. Let B denote the linear hull of $W \Delta W$, which analogously is a self-adjoint subalgebra of $BL(W)$, so a pre$-C^*$-algebra. If V denotes the Hilbert space of W endowed with the action of B on W defined by $bv := b(v)$, obviously V is a "Hermitian (left) B-module" in the sense of [13; p. 183]. Let X denote the right B-module whose additive group is the same as the one of W, the multiplication by a complex number in X is the multiplication in W by the conjugate of the given number, and the right action of B on X is defined by $xb := b^*(x)$. Actually X becomes a (right) B-rigged space under the B-valued inner product defined by $< x, y >_B := y \Delta x$. Now, by regarding X as a pre-Hermitian B-rigged **C**-module, we can apply [13; Theorem 5.1] to obtain that there is a pre-inner product $(. \mid .)$ on $X \otimes_B V$ whose value on elementary tensors is given by

$$(x \otimes v \mid x' \otimes v') = (< x', x >_B v \mid v'). \tag{9}$$

Now, since V, X and W agree as additive groups, for n in $\mathbf{N}$, $i = 1, 2, \ldots n$, and x_i, y_i, z, t in W we have

$$(\sum_{i=1}^{n} y_i \otimes x_i \mid t \otimes z) = \sum_{i=1}^{n} (y_i \otimes x_i \mid t \otimes z) = \sum_{i=1}^{n} (< t, y_i >_B x_i \mid z) =$$

$$\sum_{i=1}^{n} ((y_i \Delta t)(x_i) \mid z) = \sum_{i=1}^{n} ((x_i \square y_i)(t) \mid z) = (\sum_{i=1}^{n} (x_i \square y_i)(t) \mid z). \tag{10}$$

Therefore we have that $\sum_{i=1}^{n} y_i \otimes x_i$ lies in the set

$$D := \{\alpha \in X \otimes_B V : (\alpha \mid \beta) = 0 \text{ for all } \beta \text{ in } X \otimes_B V\} \tag{11}$$

if and only if $\sum_{i=1}^{n} x_i \Box y_i = 0$, so

$$\sum_{i=1}^{n} y_i \otimes x_i + D \to \sum_{i=1}^{n} x_i \Box y_i \tag{12}$$

is a well defined linear isomorphism from the prehilbert space $(X \otimes_B V)/D$ onto E. By identifying in this way E with $(X \otimes_B V)/D$, the second assertion in our lemma follows. Concerning the last assertion in the lemma, note that for x, y in W $x \Box y$ lies in $L(X)$ [13; Lemma 6.2], so again by [13; Theorem 5.1] and the above identification $E \equiv (X \otimes_B V)/D$, for T in E, the correspondence $x \Box y \to x \Box T(y)$ determines a continuous operator on $(E, |\cdot|)$ with norm no greater than $||| T |||$. Since by the H^*-conditions and the associative identities we have that $x \Box T(y) = (x \Box y)T^*$, it follows that this operator is the right multiplication operator by T^* on E, so $| ST | \leq | S | \, ||| T^* ||| \; (S, T \in E)$. On the other hand it is easy to see that the B-rigged space X is in the most natural way a left module over the pre$-C^*$-algebra E satisfying that $< Tx, y >_B = < x, T^* y >_B$ for all x, y in X and T in E, and that for all x in X the mapping $T \to < x, Tx >_B$ from E into B is continuous. Therefore, by [13; Proposition 4.27], we have that $||| T ||| \leq \| T \| \; (T \in E)$. Now clearly $| ST | \leq | S | \, \| T \| \; (S, T \in E)$. The rest of the third assertion in the lemma is routine.

Lemma 2. *Let W be a positive H^*-triple. Then, with the notation in Lemma 1, for all S, T in E we have:*

i) $\| T \| \leq K \mid T \mid$ *for suitable positive number K.*

ii) $| (S \mid T) | \leq \| S \| \, p(T)$, where

$$p(T) := \inf\{\sum_{i=1}^{n} \| x_i \| \, \| y_i \| : T = \sum_{i=1}^{n} x_i \Box y_i\}.$$

Proof. By the definition of the inner product on E, for x, y in W and T in E, we have

$$| (T(x) \mid y) |^2 = | (T \mid y \Box x) |^2 \leq | T |^2 | y \Box x |^2 = | T |^2 \, (y \Box x \mid y \Box x) =$$

$$| T |^2 \, (\{yxx\} \mid y) \leq | T |^2 \| \{yxx\} \| \, \| y \| \leq M \mid T |^2 \| x \|^2 \| y \|^2, \tag{13}$$

where M denotes the norm of the triple product of W. Taking above $y = T(x)$, we obtain $\| T \| \leq K \mid T \mid$, with $K := M^{1/2}$, and *i)* is proved. If S and $T = \sum_{i=1}^{n} x_i \Box y_i$ are in E, again by definition of the inner product in E, we have

$$| (S \mid T) | = | \sum_{i=1}^{n} (S \mid x_i \Box y_i) | = | \sum_{i=1}^{n} (S(y_i) \mid x_i) | \leq$$

$$\sum_{i=1}^{n} |(S(y_i) \mid x_i)| \leq \sum_{i=1}^{n} \| S(y_i) \| \| x_i \| \leq \| S \| \sum_{i=1}^{n} \| x_i \| \| y_i \|. \tag{14}$$

and *ii)* follows.

Conclusion of the proof of Theorem 1. Let W be a positive H^*-triple with zero annihilator. Then, with the notation in Lemma 1, from Lemma 2.i) and a part of Lemma 1.iii) we have $| ST | \leq K | S | | T |$ for all S, T in E, so the product of E is continuous for the topology of the norm $| \cdot |$, and so, if $\mathcal{E}$ denotes the Hilbert space completion of $(E, | \cdot |)$, $\mathcal{E}$ is an associative algebra under the unique continuous extension of the product of E. Moreover, by the rest of Lemma 1.iii), $\mathcal{E}$ with the unique continuous extension of the involution of E is actually an H^*-algebra. The mapping $T \to T$ from $(E, | \cdot |)$ into $(BL(W), \| \cdot \|)$ is continuous (Lemma 2.i)), so it extends in a unique way to a continuous *-homomorphism φ from $(\mathcal{E}, | \cdot |)$ to $BL(W)$, which gives on W a natural structure of left $\mathcal{E}$-module. If $\varphi(T) = 0$ for some T in $\mathcal{E}$, then since $T = | \cdot | - \lim\{T_n\}$ with T_n in E, we have $\{T_n\} = \{\varphi(T_n)\} \xrightarrow{\|\cdot\|} \varphi(T) = 0$ and $\{(T_n \mid S)\} \to (T \mid S)$ for all S in E. Since $| (T_n \mid S) | \leq \| T_n \| p(S)$(Lemma 2.ii)), it follows that $(T \mid S) = 0$ for all S in E, so $T = 0$ by the density of E in $\mathcal{E}$, and W regarded as $\mathcal{E}$-module is a faithful module. Moreover, since now φ is a one-to-one *-homomorphism from $\mathcal{E}$ into $BL(W)$, clearly $\mathcal{E}$ must have zero annihilator. If x is in W and $Tx = 0$ for all T in $\mathcal{E}$, we have in particular $\{zyx\} = (z\Box y)(x) = 0$ for all y, z in W, so x lies in *Ann(W)*, and so $x = 0$ because by assumption W has zero annihilator. Also clearly $(Tx \mid y) = (x \mid T^*y)$ for all x, y in W and T in $\mathcal{E}$. Now we can apply the theorem in [8] to conclude that W regarded as $\mathcal{E}$-module is actually a Hilbert module over the H^*-algebra $\mathcal{E}$ with zero annihilator for the $\mathcal{E}$-valued inner product defined by

$$[x \mid y] = F_y^*(x), \tag{15}$$

where $F_y^* : W \to \mathcal{E}$ denotes the hilbertian adjoint of the operator $F_y : T \to Ty$ from $\mathcal{E}$ in W. But, by definition of the inner product on E, we have

$$(T \mid F_y^*(x)) = (F_y(T) \mid x) = (Ty \mid x) = (T \mid x\Box y), \tag{16}$$

so by the density of E in $\mathcal{E}$ we obtain $[x \mid y](= F_y^*(x)) = x\Box y$, and so $\{xyz\}(= (x\Box y)(z)) = [x \mid y]z$, which shows that the H^*-triple arising as in Proposition 1 from W regarded as Hilbert $\mathcal{E}$-module agrees with the original one. Concerning the essential uniqueness of the Hilbert module structure on W which gives via Proposition 1 the original H^*-triple, we only note that, for any such structure regarding the H^*-algebra over which W is a Hilbert module as an algebra of operators on W we have clearly $[x \mid y] = x\Box y$ $(x, y \in W)$ and also $([x \mid y] \mid [z \mid t]) = (\{xyt\} \mid z)$ (a not-difficult exercise which has inspired Lemma 1), which determines the H^*-algebra under consideration by the fact already noted that the linear hull of $[W \mid W]$ is dense in the H^*-algebra.

Remark 1. It follows from Theorem 1 and Proposition 1 that, for any x in a positive H^*-triple, we have $x\Box x \geq 0$.A better result will be obtained in the following section (see the proof of Theorem 2).

II. Structure theory for H*-triples

Our purpose in this section is to give a complete structure theory for H^*-triples. This theory will be achieved by reduction to the case of a topologically simple positive H^*-triple and then by using our main result (Theorem 1) and Smith's structure theory [18] for Hilbert modules.

A subspace I of an H^*-triple W satisfying

$$\{IWW\} + \{WIW\} + \{WWI\} \subset I, \tag{17}$$

will be called an *ideal* of W. It is clear that $Ann(W)$ is a closed ideal of W with zero triple product. From the H^*-conditions we have $Ann(W)=Lin\{WWW\}^{\perp}$, so $W = V \perp Ann(W)$ (orthogonal sum) where V is a closed ideal of W with zero annihilator in itself ("Wedderburn principal theorem for H^*-triples"). Thus the structure theory centres on the study of the H^*-triples with zero annihilator. The following characterization of the annihilator of an H^*-triple will be useful.

Proposition 2. *Let W be an H^*-triple. Then $Ann(W)= \{x \in W : \{xxx\} = 0\}$.*

Proof. Clearly $\{xxx\} = 0$ for all x in $Ann(W)$. To prove the converse, let x in W such that $\{xxx\} = 0$. Then the self-adjoint operator $x\Box x$ satisfies

$$(x\Box x)^2 = (x\Box x)(x)\Box x = \{xxx\}\Box x = 0, \tag{18}$$

and therefore $x\Box x = 0$. Now, for all y in W we have

$$(x\Delta y)(x\Delta y)^* = (x\Delta y)(y\Delta x) = y\Delta(x\Box x)(y) = 0, \tag{19}$$

hence $x\Delta y = 0$, and so x lies in $Ann(W)$.

Corollary 1. *Every closed subtriple of an H^*-triple with zero annihilator has also zero annihilator.*

In what follows, if W is an H^*-triple with zero annihilator we write

$$W_+ := \{x \in W : x\Delta x \geq 0\} \text{ and } W_- := \{x \in W : x\Delta x \leq 0\}. \tag{20}$$

Also, for x in W we denote by W_x the H^*-triple generated by x. We note that from the associative identities we have $x\Box x/W_x = x\Delta x/W_x$. Our first theorem shows that W_+ and W_- are orthogonal ideals which decompose W. This is an analogous result to Proposition 3.7 in [19].

Lemma 3. *Let W be an H^*-triple with zero annihilator. For each x in W there exist $u \in W_x \cap W_+$ and $v \in W_x \cap W_-$ such that $x = u+v$.*

Proof. We may assume that $\| x\Delta x \| \leq 1$. Put $a := x\Delta x$ and define

$$x_n := a_+^{1/n}(x) \text{ and } y_n := a_-^{1/n}(x) \tag{21}$$

for all n in $\mathbf{N}$. From [10; Lemma 5.1.5] the sequences $\{a_+^{1/n}\}, \{a_-^{1/n}\}$ are strong-operator convergent, so the sequences $\{x_n\}, \{y_n\}$ are norm-convergent in W_x. Let $u = \lim x_n$ and $v = \lim y_n$. It is clear that $u\Delta u = \lim x_n \Delta x_n = a_+$ and $v\Delta v = \lim y_n\Delta y_n = -a_-$, so $u \in W_+$ and $v \in W_-$. In a similar way to [19; Lemma 3.5] we have that $y\Delta(u+v) = y\Delta x$ for all y in W_x, equivalently $y\Delta[x-(u+v)] = 0$ for all y in W_x. In particular, since $x-(u+v)$ belongs to W_x, we have that $[x-(u+v)]\Delta[x-(u+v)] = 0$, so, by Proposition 2, $x = u+v$.

Lemma 4. *Let W be an H^*-triple with zero annihilator. Then $x\Box y = 0$ for all x in W_+ and y in W_-.*

Proof. Let $b := (x\Delta x)^{1/2}$. It is clear that

$$b(y)\Delta b(y) = b(y\Delta y)b \leq 0. \tag{22}$$

Thus

$$b(y)\Box b(y)/W_{b(y)} = b(y)\Delta b(y)/W_{b(y)} \leq 0. \tag{23}$$

But, on the other hand,

$$b(y)\Box b(y) = y\Box b^2(y) = y\Box(x\Delta x)(y) = (y\Box x)(x\Box y) = (x\Box y)^*(x\Box y) \geq 0. \tag{24}$$

Therefore $b(y)\Box b(y)/W_{b(y)} = 0$. By applying Corollary 1 and Proposition 2 to the H^*-triple $W_{b(y)}$, this implies $b(y) = 0$, so $(x\Box y)^*(x\Box y) = y\Box b^2(y) = 0$, and so $x\Box y = 0$.

Theorem 2. *For every H^*-triple W with zero annihilator, W_+ and W_- are closed ideals of W such that $W = W_+ \perp W_-$.*

Proof. Let x, y in W_+, by Lemma 3, there exist $u \in W_+$ and $v \in W_-$ such that $x+y = u+v$. Since $x\Box v = y\Box v = u\Box v = 0$ (Lemma 4), we have $v\Box v = (x+y-u)\Box v = 0$, so, by Proposition 2, $v = 0$, and therefore $x + y = u \in W_+$. This proves that W_+ is a subspace of W. Analogously one can prove that W_- is a subspace of W. From the continuity of the triple product it follows that W_+ and W_- are actually closed subspaces. By Proposition 2 $W_+ \cap W_- = 0$, and by Lemma 3 $W = W_+ + W_-$, so $W = W_+ \oplus W_-$. Now, we prove that W_+ and W_- are ideals in W. To this end we claim that $W_+ = \{x \in W : x\Box x \geq 0\}$ and $W_- = \{x \in W : x\Box x \leq 0\}$. Let x be in W such that $x\Box x \geq 0$. By Lemma 3, there exists $u \in W_x \cap W_+$ and $v \in W_x \cap W_-$ such that $x = u+v$. From the equality $x\Box x/W_x = x\Delta x/W_x$ it follows that $x \in (W_x)_+$. Now, from Lemma 4 applied to the H^*-triple W_x, we have $x\Box v/W_x = v\Box v/W_x = 0$. By Proposition 2 $v = 0$, so $x = u \in W_+$ and we have proved that $\{x \in W : x\Box x \geq 0\} \subset W_+$. The converse inclusion follows by applying the above inclusion to the H^*-triple obtained from W by changing its triple product by $\{xyz\}' := \{zyx\}$, and the equality for W_- follows by appliying the proved equality for W_+ to the H^*-triple obtained from W by changing its triple product by $<<xyz>>:= -\{xyz\}$. Once the claim is proved, the equalities

$$\{xyz\}\Delta\{xyz\} = (y\Delta z)(x\Delta x)(y\Delta z)^*, \{xyz\}\Box\{xyz\} = (x\Box y)(z\Box z)(x\Box y)^* \tag{25}$$

show that

$$\{W_+WW\} \subset W_+\,,\{W_-WW\} \subset W_- \text{ and } \{WWW_+\} \subset W_+\,,\{WWW_-\} \subset W_-. \quad (26)$$

By Lemma 4 $\{W_+W_-W\} = \{W_-W_+W\} = 0$, so from the above

$$\{WW_+W\} \subset \{W_+W_+W\} + \{W_-W_+W\} \subset W_+, \quad (27)$$

and so W_+ is an ideal of W. Similarly, W_- is an ideal of W. Finally, since by Corollary 1 W_+ has zero annihilator, it follows that W_+ is the closure of $Lin\{W_+W_+W_+\}$, so to prove that W_+ is orthogonal to W_- it is enough to show that $(W_- \mid \{W_+W_+W_+\}) = 0$, which now follows from the H^*-condition.

We say that an H^*-triple W is *topologically simple* if $\{WWW\} \neq 0$ and *0* and W are the only closed ideals of W. Also an H^*-triple W will be called *negative* if $x\Delta x \leq 0$ for all x in W, equivalently, if the H^*-triple obtained from W by defining the new triple product by $<<xyz>> := -\{xyz\}$ is positive. A direct consequence of Theorem 2 is the following.

Corollary 2. *Every topologically simple H^*-triple is either positive or negative.*

The Wedderburn principal theorem for H^*-triples reduces the theory of H^*-triples to those which have zero annihilator. At this time, a Wedderburn type second theorem for H^*-triples is almost folklore, so we omit the proof (see [5; Theorem 1.6], [6; Theorem 1], [7; Theorem 15] and [11; Satz 3.9] for related results whose proofs can be easily adapted to our context). This theorem reduces the knowledge of H^*-triples with zero annihilator to the one of topologically simple H^*-triples, namely: *every H^*-triple with zero annihilator is the closure of the orthogonal sum of its minimal closed ideals and these are topologically simple H^*-triples.* Note that this second Wedderburn theorem can be obtained also from our main result (Theorem 1) and [18; Theorem 2.10]. The above two Wedderburn theorems for H^*-triples, together with Corollary 2, allow us to conclude the structure theory for H^*-triples with the consideration of the topologically simple positive H^*-triples. If H and K are complex Hilbert spaces, the Hilbert space $\mathcal{HS}(H,K)$ of all Hilbert-Schmidt operators from H into K(see [10; p. 141]) has a natural structure of left module over the H^*-algebra $\mathcal{HS}(K)(:= \mathcal{HS}(K,K))$ under which $\mathcal{HS}(H,K)$ is actually a Hilbert module for the $\mathcal{HS}(K)$-valued inner product defined by $[x \mid y] := xy^*$. Also $\mathcal{HS}(H,K)$ has a natural structure of H^*-triple (a trivial application of Proposition 1) under the triple product defined by $\{xyz\} = xy^*z$. It can be easily verified that such an H^*-triple is topologically simple and positive. Conversely, we have

Theorem 3. *Up to a positive multiple of the inner product, every topologically simple positive H^*-triple is isomorphic and isometric to an H^*-triple of the form $\mathcal{HS}(H,K)$ for suitable complex Hilbert spaces H and K.*

Proof. Let W be a topologically simple positive H^*-triple. By Theorem 1 there are an H^*-algebra $\mathcal{E}$ with zero annihilator and a structure of Hilbert module over $\mathcal{E}$ on W such that $\{xyz\} = [x \mid y]z$ for all x, y, z in W. If $\mathcal{E}$ is not topologically simple, then

$\mathcal{E}$ equals the orthogonal sum of two proper nonzero closed ideals I and J and, by [18; Corollary 2.6 and Lemma 2.8], W equals the orthogonal sum of $M(I)$ and $M(J)$ where, for all closed ideal P of $\mathcal{E}$, $M(P)$ (defined as $\{x \in W : P^{\perp}x = 0\}$) is a Hilbert module over P under the restriction of the structure. Now it is easy to see that $M(I)$ and $M(J)$ are nonzero proper closed ideals of the H^*-triple W, which contradicts the assumption that W is topologically simple. Therefore the H^*-algebra $\mathcal{E}$ is topologically simple, so by [18; Theorem 3.1] (see also [3; Theorem 2.3]) there exists a complex Hilbert space K such that, up to a positive multiple of the inner product, $\mathcal{E}$ equals $\mathcal{HS}(K)$ and W (regarded as Hilbert module over $\mathcal{E}$) equals a suitable l^2-sum $\oplus_{\lambda\in\Lambda} K_\lambda$, where for all λ in Λ K_λ is a copy of K regarded as a Hilbert module over $\mathcal{HS}(K)$ (the $\mathcal{HS}(K)$-valued inner product of K is defined by $[k_1 \mid k_2] : k \to (k \mid k_2)k_1$). If H denotes the Hilbert space $l^2(\Lambda)$, each $x = \{x_\lambda\}$ in $\oplus_{\lambda\in\Lambda} K_\lambda$ defines a Hilbert-Schmidt operator $x\hat{}$ from H in K given by

$$x\hat{}(\{\mu_\lambda\}) = \sum_{\lambda\in\Lambda} \mu_\lambda x_\lambda \tag{28}$$

and the mapping $x \to x\hat{}$ is a total isomorphism from $\oplus_{\lambda\in\Lambda} K_\lambda$ onto $\mathcal{HS}(H, K)$ regarded as Hilbert modules over $\mathcal{HS}(K)$. Now clearly W and $\mathcal{HS}(H, K)$ agree also as H^*-triples.

Remark 2. Another concept of associative Hilbert triple system was introduced in [5]. We refer to these Hilbert triple systems as "involutive H^*-triples". An *involutive H^*-triple* is a complex Hilbert space W together with a conjugate-linear involutive mapping $\square : W \to W$ and a (complex) trilinear mapping $< \ >: W \times W \times W \to W$ satisfying for all x, y, z, t, u in W the following properties:

i) $<xyz>^\square = <x^\square y^\square z^\square>$.

ii) $<<xyz> tu> = <x <tzy> u> = <xy <ztu>>$.

iii) $(<xyz> \mid t) = (x \mid <tz^\square y^\square>) = (y \mid <z^\square t x^\square>) = (z \mid <y^\square x^\square t>)$.

It is clear that any involutive H^*-triple W is an H^*-triple for the triple product $\{xyz\} := <xy^\square z>$. Also a special type of Hilbert modules called involutive Hilbert modules were introduced by the authors in [2] in connection with the study of structurable H^*-algebras. If $\mathcal{E}$ is an H^*-algebra with zero annihilator and τ is an isometric linear involution on $\mathcal{E}$, we say that $(W, \square)$ is an *involutive Hilbert $(\mathcal{E}, \tau)$-module* if W is a Hilbert $\mathcal{E}$-module and $\square$ is a (conjugate-linear) vector space involution on W such that

i) $\tau([x \mid y]) = [y^\square \mid x^\square]$.

ii) $(ex)^\square = \tau(e*)x^\square$. It is straightforward that every involutive Hilbert module is an involutive H^*-triple under its **C**-valued inner product and the (complex) trilinear triple product $<xyz> := [x \mid y^\square]z$. It is easy to see, from our Theorem 1, that conversely all involutive H^*-triples appear in the above way. It follows that the structure theories for involutive Hilbert modules and for involutive H^*-triples, given in [2] and [5] respectively, can each be obtained from the other.

REFERENCES.

[1] **F. F. Bonsall and J. Duncan,** *Complete normed algebras.* Springer-Verlag. Berlin-Heidelberg-New York 1973.

[2] **M. Cabrera, J. Martínez and A. Rodríguez,** Structurable H^*-algebras. *J. Algebra* (to appear).

[3] **M. Cabrera, J. Martínez and A. Rodríguez,** Hilbert modules revised: orthonormal bases and Hilbert-Schmidt operators. Preprint. Universidad de Granada.

[4] **A. Castellón and J. A. Cuenca,** Compatibility in Jordan H^*-triple systems. *Bolletino U.M.I. (7)* 4-B (1990), 433-447.

[5] **A. Castellón and J. A. Cuenca,** Associative H^*-triple system. In *Workshop on nonassociative algebraic models*, Nova Science Publishers, New York (to appear).

[6] **J. A. Cuenca and A Rodríguez,** Structure theory for noncommutative Jordan H^*-algebras. *J. Algebra 106* (1987), 1-14 .

[7] **A. Fernández and A. Rodríguez,** A Wedderburn theorem for non-associative complete normed algebras. *J. London Math. Soc. 33* (1986), 328-338.

[8] **G. R. Giellis,** A characterization of Hilbert modules. *Proc. Amer. Math. Soc. 36* (1972), 440-442.

[9] **M. R. Hestenes,** A Ternary algebra with applications to matrices and linear transformations. *Arch. Rational Mech. Anal. 11* (1962), 138-194.

[10] **R. V. Kadison and J. R. Ringrose,** *Fundamentals of the theory of operator algebras.* Vol I. Academic Press. New York-London 1983.

[11] **W. Kaup,** Uber die klassifikation der symmetrischen hermiteschen mannigfaltigkeiten unendlicher dimension II. *Math. Ann. 262* (1983), 57-75.

[12] **E. Neher,** *Jordan triple systems by grid approach.* Lecture notes in Mathematics 1280. Springer-Verlag. Berlin-Heidelberg 1987.

[13] **M. A. Rieffel,** Induced representations of C^*-algebras. *Advan. in Math. 13* (1974), 176-257.

[14] **P. P. Saworotnow,** A generalized Hilbert space. *Duke Math. J. 35* (1968), 191-197.

[15] **P. P. Saworotnow,** Trace-class and centralizers of an H^*-algebra. *Proc. Amer. Math Soc. 26* (1970), 101-104.

[16] **P. P. Saworotnow and J. C. Friedell,** Trace-class for arbitrary H^*-algebras. *Proc. Amer. Math. Soc. 26* (1970), 95-100.

[17] **H. H. Schaefer,** *Topological vector spaces.* Springer-Verlag. New-York. 1971.

[18] **J. F. Smith,** The structure of Hilbert modules. *J. London Math. Soc. 8* (1974), 741-749.

[19] **H. Zettl,** A characterization of ternary rings of operators. *Advan. in Math. 48* (1983), 117-143.

Nonassociative
Algebraic
Models

ISBN 1-56072-050-6

Associative H*-Triple Systems[1]

Alberto Castellon Serrano and Jose Antonio Cuenca Mira
Departamento de Algebra, Geometría y Topología, Facultad de Ciencias, Universidad de Málaga, Apartado 59, 29080 Málaga, España

ABSTRACT

The structure theory of associative H^*-algebras was given by W.Ambrose [1] in the complex case and by I.Kaplansky [14] in the real case. H^*-triple systems are a ternary version of the H^*-algebras (see below for definition). They were introduced in [6] for the Jordan case in connection with a problem of compatibility for projections and they are a generalization of the Hilbert triples of [15] and [20]. In paragraph 1 of this paper we show a few elementary properties of the H^*-triple systems (not necessarily associative) and as in H^*-algebras [9] we prove that every nonzero H^*-triple system with zero annihilator is the closure of the direct orthogonal sum of its minimal closed ideals which are topologically simple H^*-triple systems. Paragraph 2 is devoted to developing projection techniques that will allow us in paragraph 3 to determine all the topologically simple associative H^*-triple systems. As an application we prove the structure theorem of Kaplansky [14] on real topologically simple associative H^*-algebras.

1 This work has been partially supported by the Plan Andaluz de Investigacion y Desarrollo Tecnologico.

1. H*-TRIPLE SYSTEMS.

Let A be a module over the commutative unital ring ϕ. We say that A is a *ϕ-triple system* if it is endowed with a trilinear map $< \ >$ of $A \times A \times A$ to A. If $\phi = \mathbb{R}$ or $\mathbb{C}$ a map $* : A \to A$ is called a *multiplicative involution* if it satisfies $<x\, y\, z>^* = <x^* y^* z^*>$ for any $x, y, z \in A$ and it is involutive linear if $\phi = \mathbb{R}$ and involutive antilinear if $\phi = \mathbb{C}$. The ϕ-triple system A is said to be an *H^*-triple system* if its underlying ϕ-module ($\phi = \mathbb{R}$ or $\mathbb{C}$) is a Hilbert space with inner product $(\mid)$ endowed with a multiplicative involution $x \to x^*$ satisfying

$$(<x\, y\, z> \mid t) = (x \mid <t\ z^* y^*>) = (y \mid <z^* t\ x^*>) = (z \mid <y^* x^* t>) \tag{1.1}$$

for any $x, y, z \in A$. We denote by L, M and R the bilinear maps from $A \times A$ to the space $End_\phi(A)$ of the ϕ-linear maps given by $L(x, y)\, z = M(x, z)\, y = R(z, y)x = <x\, y\, z>$ for any $x, y, z \in A$. As in [6] we obtain the continuity of the triple product. We recall that a triple system A is called *associative* if $<<x\, y\, z> t\, u> = <x <t\ z\, y> u> = <x\, y <t\ z\, u>>$ for any $x, y, z, t, u \in A$. If A is an associative H^*-triple system then the triple system A^+ with triple product $\{xyz\} = (<x\, y\, z> + <z\, y\, x>)$ and same Hilbert space and involution as A is a Jordan H^*-triple system in the sense of [6].

Let D be a unital associative H^*-algebra which has an isometric involutive $*$-antiautomorphism s. Let $\mathcal{A}$ and $\mathcal{B}$ be two non-vacuous sets, $T_1 = (\lambda_{ij})$ a $\mathcal{B} \times \mathcal{B}$-matrix and $T_2 = (\mu_{ij})$ an $\mathcal{A} \times \mathcal{A}$-matrix, both with entries in D such that the λ_{ii} and the μ_{ii} are either 1 or -1 with $\lambda_{ij} = 0 = \mu_{ij}$ if $i \neq j$. Let $\mathcal{M}_{\mathcal{A},\mathcal{B}}(D)$ be the set of $\mathcal{A} \times \mathcal{B}$-matrices (a_{ij}) with entries in D such that $\sum \|a_{ij}\|^2 < \infty$. This set is an associative H^*-triple system relative to the triple product, inner product and involution given by

$$<(a_{ij})\, (b_{ij})\, (c_{ij})> = (a_{ij})\, T_1\, (s\,(b_{ji}))\, T_2\, (c_{ij})$$
$$((a_{ij}) \mid (b_{ij})) = \sum (a_{ij} \mid b_{ij})$$
$$(a_{ij})^* = T_2\, (s\,(a_{ij}^*))T_1.$$

We shall denote it by $\mathcal{M}_{\mathcal{A},\mathcal{B}}(D, s, T_1, T_2)$.

Let A be an associative H^*-triple system and s an isometric involutive $*$-automorphism of A. We define the *s-isotope* $A^{(s)}$ of A to be the associative H^*-triple system with the same Hilbert space as A and triple product and involution given by $<xyz>^{(s)} = <x, s\,(y), z>$, $x^\nabla = s(x^*)$. We define the twin A^b of an H^*-triple system A as the H^*-triple system obtained from A replacing $*$ by $-*$.

A subset S of an H^*-triple system A is called *self-adjoint* if $S^* \subset S$. If S is a closed and *self-adjoint* subspace of A that satisfies $<S\ S\ S> \subset S$ we shall say S is an *H^*-subsystem* of A. The subspaces I of A satisfying $<I\ A\ A> + <A\ I\ A> + <A\ A\ I> \subset I$ are called *ideals* of A. An H^*-triple system A is *topologically simple* if $<A\ A\ A> \neq 0$ and its only closed ideals are 0 and A.

Let A be an H^*-triple system. From (1.1) we obtain

$$\{x \in A : <x\ A\ A> = 0\} = \{x \in A : <A\ x\ A> = 0\} =$$
$$= \{x \in A : <A\ A\ x> = 0\} = \overline{\mathcal{L}(<A\ A\ A>)}^{\perp} \qquad (1.2)$$

where $\mathcal{L}(<A\ A\ A>)$ denotes the linear span of set of the elements $<x\ y\ z>$ where $x,y,z \in A$. The set given by (1.2) is a self-adjoint closed ideal of A that we shall call the *annihilator* of A and we denote it by $Ann(A)$. If A is either Jordan or associative we have

$$Ann(A) = \{x \in A : <x\ A\ x> = 0\} \qquad (1.3)$$

(see [6, Lemma 1.7]). From the above for every H^*-triple system A we have a direct orthogonal sum

$$A = U \perp Ann(A) \qquad (1.4)$$

where U is a closed ideal of A that is an H^*-triple system with zero annihilator relative to the triple product and inner product induced by those of A and the involution $\square$ that assigns to each $u \in U$ the only element $u^{\square} \in U$ such that $u^* = u^{\square} + u'$ with $u' \in Ann(A)$. Moreover if the involution of A is continuous then U is an H^*-subsystem of A.

LEMMA 1.5. *Let A be an H^*-triple system with zero annihilator. Then every closed ideal I of A is an H^*-subsystem of A with zero annihilator. Moreover if I is minimal (i.e. $I \neq 0$ and I properly contains no nonzero closed ideals of A) then I is topologically simple.*

Proof. Since I is a closed ideal $I^{\perp}$ is also . From (1.4) it follows that $I = \overline{\mathcal{L}(<III>)}$. In a similar way to [9. Proposition 2] we can prove that $I^{\perp} = \{x \in A : <xIA> = 0\}$ and $I^* \subset I$. Hence I is an H^*-subsystem of A with zero annihilator. Let K be a closed ideal of I. We have $<K\ A\ A> = <K, I \perp I^{\perp}, I \perp I^{\perp}> = <K\ I\ I> \subset K$. An analogous argument shows that $<A\ K\ A> + <A\ A\ K> \subset K$, so K is a closed ideal of A. Therefore minimality of I implies its topological simplicity.

THEOREM 1.6. *An H^*-triple system $A \neq 0$ has zero annihilator iff* $A = \overline{\perp I_\alpha}$ *is the closure of the orthogonal sum of its minimal closed ideals (which are topologically simple H^*-triple systems).*

Proof. Assume A has zero annihilator. First we prove the existence of nonzero minimal closed ideals. For all $x \in A$ we define the continuous linear operator $L_x: A \to End(A)$, by $L_x(y) = L(x, y)$. By (1.2) we can define another norm in A given by $|x| = \| L_x \|$. Let I be a closed ideal of A and let $u \in I$, $v \in I^\perp$. We have

$$|u + v| = \| L_u + L_v \| = \sup_{\|y\|=1} \{ \| L(u, y) + L(v, y) \| \} . \qquad (1.7.)$$

It follows from [7, (1-3-4)] and the fact $[L(u, y) \circ L(v, y)]A = [L(u,y) \circ L(y^*, v^*)]A \subset I \cap I^\perp = 0$ that

$$\| L(u, y) + L(v, y) \| = \max \{ \| L(u, y) \|, \| L(v, y) \| \} . \qquad (1.8)$$

Taking into account (1.7) and (1.8) a routine calculation shows $|u + v| = \max \{ |u|, |v| \}$. As in [9, Proposition 3] or [7, (1-3-7)] we can prove the existence of a nonzero element $b \in A$ such that for every closed ideal I of A, either $b \in I$ or $b \in I^\perp$. Therefore the closed ideal generated by b is minimal. We conclude the proof as in [9; p. 5].

2. PROJECTIONS OF AN ASSOCIATIVE H*-TRIPLE SYSTEM.

For an arbitrary H^*-triple system A the element $e \in A$ is called a *projection* if $\langle eee \rangle = \varepsilon_e e$, $e^* = \delta_e e$; $\varepsilon_e, \delta_e \in \{1, -1\}$. As a consequence of (1.2) we have that if A is an associative H^*-triple system with zero annihilator then the Jordan H^*-triple system A^+ also has zero annihilator. From [6; Theorem 1.8] we obtain the existence of a nonzero projection e in every associative H^*-triple system A with zero annihilator. Similarly to [16, p.91] we define the Peirce subspaces $A_{ij}(e)$ relative to e in the following way

$$A_{ij}(e) = \{ x \in A : \varepsilon_e L(e, e)x = ix \text{ and } \varepsilon_e R(e, e)x = jx \}, \quad i, j \in \{0, 1\}$$

The relations with the Peirce subspaces of e in A^+ (see [6] for notations) are given by

$$(A^+)_2^{\varepsilon_e}(e) = A_{11}(e),\ (A^+)_1^{\varepsilon_e}(e) = A_{10}(e) \perp A_{01}(e),\quad (A^+)_0^{\varepsilon_e}(e) = A_{00}(e). \quad (2.1)$$

[16, p. 92]. As in the Jordan case we can speak of orthogonality and irreducibility of projections in associative H^*-triple system. By [6, Lemma 1.6], every nonzero projection of an associative H^*-triple system splits into a finite sum of pairwise orthogonal irreducible projections.

Let e be a nonzero projection of an associative H^*-triple system A. We denote by $A_{11}(e)^{(\varepsilon_e e)}$ the associative H^*-algebra with the same Hilbert space as $A_{11}(e)$ and whose product and involution are given by

$$x \circ y = \varepsilon_e \langle x\, e\, y\rangle, \qquad x^\nabla = \varepsilon_e\, \delta_e\, M(e, e)x^*.$$

Let e and f be two projections of an associative H^*-triple system A with $e \in A_{10}(f)$, $f \in A_{10}(e)$ (resp. $e \in A_{01}(f)$, $f \in A_{01}(e)$). Let ϕ_{ef} (resp. $\tilde{\phi}_{ef}$) be the map from $A_{11}(e)$ to $A_{11}(f)$ given by

$$\phi_{ef} = \varepsilon_e R(f, e) \qquad (\text{resp. } \tilde{\phi}_{ef} = \varepsilon_e L(f, e)). \qquad (2.2)$$

Taking into account that

$$\delta_e\, \phi_{ef}(x)^* = \delta_f\, \phi_{ef}(x^*) \quad (\text{resp. } \delta_e\, \tilde{\phi}_{ef}(x)^* = \delta_f\, \tilde{\phi}_{ef}(x^*)) \qquad (2.3)$$

for any $x \in A_{11}(e)$, it is easy to show that ϕ_{ef} (resp. $\tilde{\phi}_{ef}$) is an isometric $*$-isomorphism between the H^*-algebras $A_{11}(e)^{(\varepsilon_e e)}$ and $A_{11}(f)^{(\varepsilon_f f)}$ whose inverse is $\phi_{f,e}$ (resp. $\tilde{\phi}_{f,e}$). Moreover $\varepsilon_e\, \phi_{ef} \circ M(e, e)x = \varepsilon_f M(f, f) \circ \phi_{ef}(x)$ (resp. $\varepsilon_e\, \tilde{\phi}_{ef} \circ M(e, e)x = \varepsilon_e M(f, f) \circ \tilde{\phi}_{ef}(x)$). Now it is easy to show that $\varepsilon_e F(\langle x\, y\, z\rangle) = \varepsilon_f \langle F(x), F(y)\, F(z)\rangle$ for $F = \phi_{ef}$ (resp. $F = \tilde{\phi}_{ef}$). From [6, Lemma 2.4] $\varepsilon_e \delta_e = \varepsilon_f \delta_f$, thus F is an isometric $*$-isomorphism from $A_{11}(e)$ to either $A_{11}(f)$ or $[A_{11}(f)]^{(-Id)}$. By similarity with [18, (1.1)] we shall call to ϕ_{ef} and $\tilde{\phi}_{ef}$ the *exchange isomorphisms*. In this context the concept of reverse is useful to reduce the properties of the ϕ_{ef} to those of $\tilde{\phi}_{ef}$. We recall that the *reverse* A' of A is the triple system with a new triple product $\langle\,\rangle'$ given by $\langle x\, y\, z\rangle' = \langle z\, y\, x\rangle$, [16, p. 55]. This is an associative H^*-triple system with the same Hilbert space and involution as A.

Let A be an associative H^*-triple system and $\mathcal{A}, \mathcal{B}$ two non-vacuous sets. A family $\mathcal{E} = \{e_{ij} : i \in \mathcal{A}, j \in \mathcal{B}\}$ of projections of A with $< e_{ij}\, e_{ij}\, e_{ij}> = \varepsilon_{ij} e_{ij} = e_{ij}{}^*$ is said to be an $\mathcal{A} \times \mathcal{B}$-*net* if it satisfies the following conditions:

$$
\begin{gathered}
< e_{ik}\; e_{mk}\; e_{mj} > = \varepsilon_{mk}\, e_{ij} \\
\varepsilon_{ik}\;\varepsilon_{mk}\;\varepsilon_{mj} = \varepsilon_{ij}
\end{gathered}
\tag{2.4}
$$

for any $i, m \in \mathcal{A}$; $k, j \in \mathcal{B}$. By the Schröder-Bernstein theorem, the supposition $\mathcal{A} \subset \mathcal{B}$ or $\mathcal{B} \subset \mathcal{A}$ may be assumed without loss in generality. An $\mathcal{A} \times \mathcal{B}$-net of A is called an $\mathcal{A} \times \mathcal{B}$-*grid* if it *covers* A, that is

$$
A = \overline{\perp_{\substack{i \in \mathcal{A} \\ j \in \mathcal{B}}} A_{11}(e_{ij})}
$$

The family $\mathcal{E} = \{E_{ij} : i \in \mathcal{A}, j \in \mathcal{B}\}$ of unit matrices E_{ij} of the associative H^*-triple system $\mathcal{M}_{\mathcal{A},\mathcal{B}}(D, s, T_1, T_2)$ is an $\mathcal{A} \times \mathcal{B}$-grid with $\varepsilon_{ij} = \delta_{ij} = \lambda_{jj}\mu_{ii}$. Similarly to [18; 3.1], we define a *quadrangle* of projections to be an $\mathcal{A} \times \mathcal{B}$-net either of A or A^b, with $|\mathcal{A}| = |\mathcal{B}| = 2$.

LEMMA 2.5. *Let $\mathcal{E}$ be an $\mathcal{A} \times \mathcal{B}$-net of an associative H^*-triple system A. Then*

$$
< A_{11}(e_{ik})\, A_{11}(e_{mk})\, A_{11}(e_{mj}) > \subset A_{11}(e_{ij})
$$

for any $i, m \in A$; $k, j \in B$. Moreover products with index combinations other than above are zero.

Proof. Let $x_{\alpha,\beta} \in A_{11}(e_{\alpha,\beta})$ be for any $\alpha \in \mathcal{A}$, $\beta \in \mathcal{B}$. We have

$$
\begin{gathered}
< e_{ij}\, e_{ij} < x_{ik}\, x_{mk}\, x_{mj} >> = << e_{ij}\, e_{ij}\, x_{ik} > x_{mk} x_{mj} > = \\
\varepsilon_{ik} << e_{ij}\, e_{ij} < e_{ik} e_{ik} x_{ik} >> x_{mk} x_{mj} > = \varepsilon_{ik} <<< e_{ij}\, e_{ij}\, e_{ik} > e_{ik}\, x_{ik} > x_{mk} x_{mj} > = \\
\varepsilon_{ik}\varepsilon_{ij} << e_{ik}\, e_{ik}\; x_{ik} > x_{mk}\; x_{mj} > \;=\; \varepsilon_{ij} < x_{ik}\, x_{mk}\, x_{mj} >.
\end{gathered}
$$

Similar calculations complete the proof.

LEMMA 2.6. *Let e and f be two nonzero projections of an associative H^*-triple system A with $f \in A_{ij}(e)$ $(i \neq j)$ and e irreducible. Then the following assertions hold:*

i) $e \in A_{ij}(f)$
ii) f *is irreducible*
iii) $A_{11}(e) \subset A_{ij}(f)$ *and* $A_{11}(f) \subset A_{ij}(e)$
iv) $A_{ji}(e) \subset A_{00}(f)$ *and* $A_{ji}(f) \subset A_{00}(e)$
v) $A_{11}(e) \perp A_{ij}(e) = A_{11}(f) \perp A_{ij}(f)$
vi) $A = A_{11}(e) \perp A_{11}(f) \perp A_{ji}(e) \perp A_{ji}(f) \perp [A_{ij}(e) \cap A_{ij}(f)] \perp [A_{00}(e) \cap A_{00}(f)]$.

Proof. Since $f \in A_{ij}(e)$ we have $f \in (A^+)_1{}^{\varepsilon_e}(e)$. It follows from [6, Corollary 2.10] that either $e \in (A^+)_2{}^{\varepsilon_f}(f)$ or $e \in (A^+)_1{}^{\varepsilon_f}(f)$. Suppose $e \in (A^+)_2{}^{\varepsilon_f}(f) = A_{11}(f)$. We have $\varepsilon_f e/2 = \langle f\, e\, f \rangle \in \langle A_{ij}(e)\, A_{11}(e)\, A_{ij}(e) \rangle = 0$ as in [16; II, 9.3 and II, 9.8]. So $e = 0$, which is a contradiction. Therefore $e \in (A^+)_1(f) = A_{10}(f) \perp A_{01}(f)$ and the irreducibility of e and the Peirce relations prove *i)*. *ii)* follows by applying of the interchange isomorphisms given as in (2.2). Next we prove *iii)* and *iv)*. Let $x \in A_{11}(e)$, we have

$$\langle f f x \rangle = \varepsilon_e \langle f f \langle e\, e\, x \rangle\rangle = \varepsilon_e \langle\langle f f e \rangle e\, x \rangle = i\, \varepsilon_e\, \varepsilon_f \langle e\, e\, x \rangle = i\, \varepsilon_f x.$$

Similar calculations complete the proof of *iii)* and *iv)*. Let for example $i = 1, j = 0$. Let $x \in A_{10}(e)$, we can write $x = x_{11} + x_{10} + x_{01} + x_{00}$, $x_{kl} \in A_{kl}(f)$. By the Peirce relations we obtain $\varepsilon_e x = \langle e\, e\, x \rangle = \langle e\, e\, x_{11} \rangle + \langle e\, e\, x_{10} \rangle \in A_{11}(f) \perp A_{10}(f)$. Taking into account the above fact and *iii)* we obtain $A_{11}(e) \perp A_{10}(e) \subset A_{11}(f) \perp A_{10}(f)$. Therefore *v)* is satisfied. Finally *vi)* follows from *iii)*, *iv)* and the fact that the Peirce projections of e and f pairwise commute.

LEMMA 2.7. (Quadrangle Lemma).*(a) Let* e_{11}, e_{12}, e_{22} *(resp.* e_{11}, e_{21}, e_{22}*) be three projections of an associative* H^**-triple system* A *with* $e_{ij} \in A_{\Delta_{ik}, \Delta_{jl}}(e_{kl})$ ($\Delta_{\alpha\beta} = 1$ if $\alpha = \beta$, $\Delta_{\alpha\beta} = 0$ if $\alpha \neq \beta$)*, then we have a quadrangle of projections* $(e_{11}, e_{12}, e_{22}, \varepsilon_{12} \langle e_{22}\, e_{12}\, e_{11} \rangle)$ *(resp.* $(e_{11}, \varepsilon_{21} \langle e_{11}\, e_{21}\, e_{22} \rangle, e_{22}, e_{21})$*).*

(b) Let $(e_{11}, e_{12}, e_{22}, e_{21})$ *be a quadrangle of projections in* A*. Then we have*

$$\phi_{e_{21}, e_{22}}(x) = \varepsilon_{11} R(e_{11}, e_{12})\, x, \quad \tilde{\phi}_{e_{12}, e_{22}}(x) = \varepsilon_{11} L(e_{21}, e_{11})x$$

for each $x \in A_{11}(e_{22})$.

(c) Moreover if e_{11} *is irreducible and* $(e_{11}, e_{13}, e_{21}, e_{23})$ *(resp*$(e_{11}, e_{12}, e_{32}, e_{31})$*)is another quadrangle with* $e_{13} \in A_{10}(e_{12})$ *(resp* $e_{31} \in A_{01}(e_{21})$*), then* $(e_{12}, e_{13}, e_{23}, e_{22})$ *(resp*$(e_{21}, e_{22}, e_{32}, e_{31})$*) is also a quadrangle.*

Proof. Parts (a) and (b) follow from a straightforward calculation in a similar way to [17, 3.2] and [19, 1.3]. Next we prove (c).By the irreducibility of e_{11} and the quadrangularity of $(e_{11}, e_{13}, e_{21}, e_{23})$ it follows from Lemma 2.6 that

$$A_{11}(e_{12}) \perp A_{10}(e_{12}) = A_{11}(e_{13}) \perp A_{10}(e_{13}),$$
$$A_{11}(e_{21}) \perp A_{10}(e_{21}) = A_{11}(e_{23}) \perp A_{10}(e_{23}).$$

Therefore $<e_{22}\, e_{12}\, e_{13}> = \varepsilon_{21}<e_{22}<e_{11}e_{21}e_{22}>e_{13}> = \varepsilon_{21}<<e_{22}e_{22}e_{21}>e_{11}e_{13}> = \varepsilon_{21}\,\varepsilon_{22}\, <e_{21}\, e_{11}\, e_{13}> = \varepsilon_{21}\,\varepsilon_{22}\,\varepsilon_{11}\, e_{23} = \varepsilon_{12}e_{23} \in A_{00}(e_{12})$ which suffices for quadrangularity.The second case of (c) follows from the first argueing in the reverse.

The following Lemma is the associative H^*-version of the main theorem of [17].

LEMMA. 2.8. *Let e be a nonzero projection of an associative H^*-triple system A. Then every ideal K_1 of $A_{11}(e)$ is the projection of an ideal K of A given by*

$$K = K_1 \perp <K_1\, e\, A_{10}> \perp <A_{01}\, e\, K_1> \perp <A_{01}\, K_1\, A_{10}>.$$

In particular if A is topologically simple then $A_{11}(e)$ is simple.

Proof. Using the Peirce relations and the fact that $S = <e\, e\, S> = <S\, e\, e>$ for every $S \subset A_{11}(e)$, a direct calculation proves that K is an ideal of A. We suppose now that $A_{11}(e)$ is not simple and A is topologically simple. By Zornifying we obtain a nonzero maximal ideal K_1 of $A_{11}(e)$. This ideal is closed. By the first part of the proof, we can obtain a nonzero proper closed ideal of A, which is a contradiction.

LEMMA. 2.9. *Let e_{11}, e_{22} be two irreducible orthogonal projections of an associative H^*-triple system A with $Ann(A) = 0$ such that $A_{12} + A_{21} \neq 0$ ($A_{12} = A_{10}(e_{11}) \cap A_{01}(e_{22}), A_{21} = A_{01}(e_{11}) \cap A_{10}(e_{22})$). Then e_{11}, e_{22} can be imbedded in a quadrangle $(e_{11}, e_{12}, e_{22}, e_{21})$ of irreducible projections. In particular if A is topologically simple, then any pairwise irreducible orthogonal projections can be imbedded in a quadrangle.*

Proof. As in [16, II. 9.10] we denote by A_{ij} $(i, j \in \{0, 1, 2\})$ the H^*-subsystems of A given by

$$A_{ij} = \{x \in A \,/\, \varepsilon_{ll} <e_{ll}\, e_{ll}\, x> = \Delta_{il}\, x,\ \varepsilon_{ll} <x\, e_{ll}\, e_{ll}> = \Delta_{jl}\, x\}, \tag{2.10}$$

that is for $i, j \in \{1, 2\}$, $i \neq j$ we have

$A_{ii} = A_{11}(e_{ii}) \cap A_{00}(e_{jj}) = A_{11}(e_{ii})$
$A_{ij} = A_{10}(e_{ii}) \cap A_{01}(e_{jj})$
$A_{i0} = A_{10}(e_{ii}) \cap A_{00}(e_{jj})$
$A_{0i} = A_{01}(e_{ii}) \cap A_{00}(e_{jj})$
$A_{00} = A_{00}(e_{ii}) \cap A_{00}(e_{jj})$.

On the other hand $M(e_{22}, e_{11})$ is a bijection from A_{12} to A_{21} with inverse $\varepsilon_{11}\varepsilon_{12} M(e_{11}, e_{22})$. So $A_{12} \neq 0$ iff $A_{21} \neq 0$. It is easy to show that if S is an H^*-subsystem of A then $Ann(S) = 0$. Hence by [6, Lemma 1.2] there exists a nonzero projection $e_{12} \in A_{12}$. By Lemma 2.6, $e_{11} \in A_{10}(e_{12})$, $e_{22} \in A_{01}(e_{12})$. By Lemma 2.7, we have a quadrangle $(e_{11}, e_{12}, e_{22}, e_{21})$ of projections which are irreducible by Lemma 2.6. In particular if A is topologically simple, Lemma follows from [5, Lemma 4 and Proposition 3].

PROPOSITION 2.11. (Orthogonal decomposition). *Let* $\mathfrak{E} = \{e_i\}_{i \in \mathcal{A}}$ *be a maximal orthogonal family of irreducible projections of an associative* H^**-triple system* A *with* $Ann(A) = 0$. *Then* A *splits into the closure of the orthogonal sum*

$$A = \overline{\perp_{\substack{i,j \in \mathcal{A} \cup \{0\} \\ (i,j) \neq (0,0)}}} A_{ij}$$

where the A_{ij} *are the* H^**-triple subsystem of* A *given by*

$$A_{ii} = A_{11}(e_i) \qquad (i \in \mathcal{A})$$
$$A_{ij} = A_{10}(e_i) \cap A_{01}(e_j) \qquad (i,j \in \mathcal{A},\ i \neq j)$$
$$A_{i0} = A_{10}(e_i) \cap \bigcap_{j \in \mathcal{A} - \{i\}} A_{00}(e_j) \qquad (i \in \mathcal{A})$$
$$A_{0i} = A_{01}(e_i) \cap \bigcap_{j \in \mathcal{A} - \{i\}} A_{00}(e_j) \qquad (i \in \mathcal{A})$$

Moreover if every A_{ij} *is nonzero* $(i, j \in \mathcal{A},\ i \neq j)$, *then either*

$$\sum_{i \in \mathcal{A}} A_{i0} = 0 \quad \text{or} \quad \sum_{i \in \mathcal{A}} A_{0i} = 0.$$

In particular this holds when A *is topologically simple.*

Proof. The maximality of $\mathfrak{E}$ and the fact that $Ann\,(A) = 0$ prove that

$$\bigcap_{i \in \mathcal{A}} A_{00}(e_i) = 0 \tag{2.12}$$

By a straightforward verification we obtain

$$<<e_i\, e_i\; x> e_j\; e_j> = <e_i\; e_i <x\, e_j\; e_j>> \in A_{ij} \tag{2.13}$$

for any $x \in A$; $i, j \in \mathcal{A}$ with $i \neq j$. Let H be the orthogonal complement of

$$\overline{\bigperp_{\substack{i,j \in \mathcal{A} \cup \{0\} \\ (i,j) \neq (0,0)}} A_{ij}} \; .$$

We argue by contradiction. Assume $H \neq 0$. Let y be a nonzero element of H. By (2.13) we have $0 = (y \mid <e_i e_i <x e_j e_j>>) = (<<e_i e_i y> e_j e_j> \mid x) = (<e_i e_i <y e_j e_j>> \mid x)$. So

$$< e_i\; e_i\, y > \in A_{i0}, \qquad <y\; e_i\; e_i > \in A_{0i}\, . \tag{2.14}$$

For every $i \in \mathcal{A}$ we have $y = y_{11} + y_{10} + y_{01} + y_{00}$ with $y_{kl} \in A_{kl}(e_i)$. By (2.14) we have $\varepsilon_i <e_i\; e_i\, y> = y_{11} + y_{10} \in A_{10}(e_i)$ that is $y_{11} = 0$ and again it follows from (2.14) that $y_{10} \in A_{i0}, y_{01} \in A_{0i}$ and therefore $0 = (y \mid y_{10} + y_{01}) = \| y_{10} \|^2 + \| y_{01} \|^2$. So $y = y_{00} \in A_{00}(e_i)$ for all $i \in \mathcal{A}$ which is contrary to (2.12).

Suppose now that every $A_{ij} \neq 0$ and that there exists an $i \in \mathcal{A}$ such that $A_{i0} \neq 0 \neq A_{0i}$. Since A_{i0}, A_{0i} are of zero annihilator there are two nonzero projections e, f with $e \in A_{i0}, f \in A_{0i}$. From Lemmas 2.7 and 2.6 we deduce the existence of a quadrangle of irreducible projections

$$(e_i\, , e,\, \varepsilon_i \;<f\; e_i\; e\; >, f\;),\text{ where } 0 \neq \varepsilon_i \;<f\; e_i\; e\; > \in \bigcap_{j \in \mathcal{A}} A_{00}(e_j\,)$$

which contradicts (2.12). Suppose now that there are $i, j \in \mathcal{A}$, $i \neq j$ such that A_{i0} and A_{0j} are nonzero. By Lemma 2.9 there exists a quadrangle $(e_i, e_{ij}, e_j, e_{ji})$ of irreducible projections. Then for every nonzero projection $f \in A_{i0}$ we obtain the quadrangle $(e_i, f, \varepsilon_i <e_{ji}\; e_i\, f>, e_{ji})$ where $\varepsilon_i <e_{ji}\; e_i\, f> \in A_{j0}$. Therefore $A_{j0} \neq 0 \neq A_{0j}$ which is impossible as already shown. The last part of the theorem is a consequence of Lemma 2.9.

3. COORDINATIZATION THEOREM. TOPOLOGICALLY SIMPLE ASSOCIATIVE H^*-TRIPLE SYSTEMS.

Let D be one of the three complex associative composition algebras. There is a unique real associative composition algebra D_0 ($\mathbb{R}$, $\mathbb{C}$ or $\mathbb{H}$) of quadratic form Q_0 and canonical involutive antiautomorphism s_0 such that the algebras $D_0 \otimes \mathbb{C}$ and D agree. The involution ∇ and the inner product $(\mid)$ of D are determined by the equations

$$(x_0 \otimes \lambda)^\nabla = s_0(x_0) \otimes \bar{\lambda}, \quad (x_0 \otimes \lambda \mid y_0 \otimes \mu) = \lambda \bar{\mu} Q_0(x_0, y_0) \tag{3.1}$$

make D a complex H^*-algebra. Also we can make D a real H^*-algebra relative to the involution and the inner product determined by

$$(x_0 \otimes \lambda)^\nabla = s_0(x_0) \otimes \bar{\lambda}, \quad (x_0 \otimes \lambda \mid y_0 \otimes \mu) = Re\,(\lambda \bar{\mu} Q_0(x_0, y_0)) \tag{3.2}$$

Conversely, as in [9, p. 6] (resp. [10, Remark 2]) we can show that if $dim_{\mathbb{C}}(D) \neq 2$ each complex (resp. real) H^*-algebra structure of the associative composition algebra D is given as above from a real division algebra up to a positive factor of the inner product.

If D is $\mathbb{R}$, $\mathbb{C}$ or $\mathbb{H}$ and s_0 and Q_0 as above, then the real H^*-algebra structure of D is determined up to a positive factor of the inner product by the equations

$$x^\nabla = s_0(x), \qquad (x \mid y) = Q_0(x, y^\nabla) \tag{3.3}$$

(see [10, Proposition 5, Case 1 and Remark 1]). If D is a real associative split composition algebra whose quadratic form is Q then it is well known that there exist a composition division algebra D_1 of canonical involutive antiautomorphism s_1 such that the algebra D is obtained from (D_1, s_1) by the Cayley-Dickson process. So all $x \in D$ can be written in an unique way as $x = a + v\,b$ where $a, b \in D_1$, $Q(D_1, v) = 0$, $v^2 = 1$. With the following involution and inner product

$$x^\nabla = s(a) + v\,b, \qquad (x \mid y) = Q(x, s(y^\nabla)) \tag{3.4}$$

D is a real H^*-algebra and if $dim_{\mathbb{R}} D \neq 2$, then every real H^*-algebra structure is obtained, up to a positive factor of the inner product, from a division algebra D_1 in this way (see [10, Remark 1]).

Let A be an associative H^*-algebra and s an isometric $*$-antiautomorphism of A.

For $\sigma \in \{1, -1\}$ we denote by $A^{(\sigma s)}$ the associative triple system of same Hilbert space that A whose triple product $< >$ and involution ∇ are given by $<x\,y\,z> = \sigma x\, s\,(y)z$, $x^{\nabla} = s\,(x^*)$. We shall call it the $\sigma\, s$ *-isotope* of the H^*-algebra A. In the complex case $A^{(s)}$ and $A^{(-s)}$ are isometrically $*$-isomorphic.

LEMMA 3.5. *Let D be a real division composition H^*-algebra and let j be an isometric involutive automorphism of D. Then we have one of the following possibilities:*

i) $j = 1$

ii) There exists a real division composition H^*-subalgebra D_j such that D is obtained from D_j by the Cayley-Dickson process and $j(x + yv) = x - yu$ for any $x,y \in D_j$ where $v \in D_j^{\perp}$ is fixed with $v^2 = -1$.

Moreover if j_1 and j_2 are two isometric involutive automorphisms as in ii) then there exists an isometric $*$-automorphism φ of D such that $\varphi \circ j_1 = j_2 \circ \varphi$.

Proof. As in [7, (1, 2, 8)] we can prove that j is an $*$-automorphism. Assume $dimD \neq 2$. It follows from the fact that $\mathbb{R} = Z(D)$, the center of D, that j commutes with the Cayley antiautomorphism s. Let D_j and K_j be the parts of D given by

$$D_j = \{x \in D : j(x) = x\}$$
$$K_j = \{x \in D : j(x) = -x\}.$$

It follows from the isometric character of j that

$$D = D_j \perp K_j.$$

In the case $K_j = 0$ we obtain the first possibility of the Lemma. Suppose now $K_j \neq 0$. It is easy to show that D_j is a division H^*-subalgebra of D. It is well known ([13, p. 162]) that D is obtained from D_j by the Cayley-Dickson process. Choose v in $D_j^{\perp}$ such that $v^2 = -1$. Therefore j is given by $j(x + yv) = x - yv$ for any $x, y \in D_j$.

Finally let j_1 and j_2 be two isometric involutive automorphisms of D as in the second case. Since D_{j_1} and D_{j_2} are both division composition H^*-algebras, there exists an isometric $*$-isomorphism φ_1 from D_{j_1} to D_{j_2}. We define $\varphi : D \rightarrow D$ given by $\varphi(x + yv_1) = \varphi_1(x) + \varphi_1(y)v_2$ for any $x, y \in D_{j_1}$. A routine calculation proves that φ is an isometric $*$-automorphism of D such that $\varphi \circ j_1 = j_2 \circ \varphi$ and this finishes the proof.

COROLARY 3.6. *Let* D *be a real division composition* H^**-algebra and let* j *be an isometric involutive antiautomorphism of* D. *Then we have one of the following possibilities*:

i) $j = Id$, *if* $D = \mathbb{R}$

ii) $j = Id$ *or* $j = {}^-$, if $D = \mathbb{C}$

iii) j *is either the Cayley antiautomorphism* s *or the antiautomorphism* t *given by*

$$t(a + bv) = \bar{a} + bv \tag{3.7}$$

with a, b *belonging to an* H^**-subalgebra* $D' \equiv \mathbb{C}$ *such that* D *is obtained from* D' *by means of the Cayley-Dickson process where* $v \in D'^{\perp}$ *is fixed with* $v^2 = -1$.

Proof. We note that the map $f \to s \circ f$ is a bijection between the set of the isometric involutive automorphisms and the set of the isometric involutive antiautomorphisms.

Now we can characterize the Peirce H^*-subsystem $A_{11}(e)$ for e an irreducible projection of an associative H^*-triple system.

PROPOSITION 3.8. *Let* A *be an associative* H^**-triple system over* $\mathbb{K}$ *and* $e \in A$ *an irreducible projection*.

a) *If* $\mathbb{K} = \mathbb{C}$ *then up to twins and a positive factor of the inner product the Peirce* H^**-subsystem* $A_{11}(e)$ *is the s-isotope with* s *the Cayley antiautomorphism of a complex associative composition* H^**-algebra of involution and inner product given by (3.1).*

b) *If* $\mathbb{K} = \mathbb{R}$ *up to twins and a positive factor of the inner product the Peirce* H^**-subsystem* $A_{11}(e)$ *is one of the following*:

b-1) *The* s*-isotope of the* H^**-algebra* D, *being* (D, s) *either a complex associative composition algebra with real* H^**-algebra structure given as in* (3.2) *or the 2-dimensional real split composition* H^**-algebra of involution and inner product as in* (3.4).

b-2) *The* t*-isotope of* $\mathbb{H}$ *where* t *is the involutive* $*$*-antiautomorphism given as in* (3.7).

b-3) $\mathbb{H} \oplus \mathbb{H}$ *with the triple product, inner product and involution given by*

$$<(x, y)\,(z, t)\,(u, v)> = (x\,t\,u,\ v\,z\,y)$$
$$((x, y) \mid (z, t)) = (x \mid z) + (y \mid t)$$

$$(x, y)^* = (s(y),\ s(x))$$

for any $x, y, z, t, u, v \in \mathbb{H}$.

b-4) The ε_e *s-isotope of the* H^**-algebra* D, *being* (D, s) *either the real split quaternions with inner product and involution given by* (3.4) *or* $\mathbb{R}, \mathbb{C}$ *or* $\mathbb{H}$ *with the* H^**-algebra structure determined by* (3.3).

Proof. Let $A^1{}_{11}(e) = \{x \in A: \langle e\ x\ e\rangle = \varepsilon_e\, x\}$. By [6, Lemma 2.2], $A^1{}_{11}(e)$ is a division Jordan H^*-subalgebra of $[A_{11}(e)^{(\varepsilon_e e)}]^+$. If we denote by

$$j : A_{11}(e)^{(\varepsilon_e e)} \rightarrow A_{11}(e)^{(\varepsilon_e e)}$$

the map given by $j(x) = \varepsilon_e M(e, e)x$ then by a well known theorem on ring theory (see [11, Theorem 2.1.7] or [13 , p. 170]) we have that $A_{11}(e)^{(\varepsilon_e e)}$ is: *i*) a division H^*-algebra, or *ii*) the direct sum $\Delta \oplus \Delta^{op}$ with Δ a division algebra and j the exchange involution, or *iii*) the algebra of the split quaternions over a field $\hat{\mathbb{K}}$ which extends $\mathbb{K}$ with the standard antiautomorphism. From the Arens-Gelfand-Mazur theorem the above division algebras are $\mathbb{C}$ in the complex case and $\mathbb{R}$, $\mathbb{C}$ or $\mathbb{H}$ in the real case. Also $\hat{\mathbb{K}}=\mathbb{C}$ in the complex case and $\hat{\mathbb{K}}=\mathbb{R}$ or $\mathbb{C}$ if $\mathbb{K}=\mathbb{R}$. Since $\Delta x\{0\}$ and $\{0\}x\Delta^{op}$ are closed ideals of an H^*-algebra these are H^*-subalgebras. Since δ_e* agrees with the composition of j with the involution $\varepsilon_e \delta_e M(e, e) \circ *$ of $A_{11}(e)^{(\varepsilon_e e)}$ we obtain that the involution of the H^*-triple system $A_{11}(e)$ is given by $(a, b)^* = \delta_e(b^\nabla, a^\nabla)$ when $A_{11}(e)^{(\varepsilon_e e)} = \Delta \oplus \Delta^{op}$ and j are as in *ii*), being ∇ the involution of the H^*-algebra Δ given either by (3.1) or (3.2) (depending if $\mathbb{K}=\mathbb{C}$ or $\mathbb{K}=\mathbb{R}$). Hence $A_{11}(e)$ is the triple system $\Delta \oplus \Delta$ with triple product

$$<(x, y)\ (z, t)\ (u, v)> = \varepsilon_e(x\,t\,u,\ v\,z\,y),$$

and from the isometric character of j its inner product is given by $((x, y) \mid (z, t)) = (x|z) + (y|t)$. The proof can be completed using Corollary 3.6 and the following remarks:

1) $D^{(j)}$ and $D^{(-j)}$ are isometrically $*$-isomorphic for an H^*-algebra D over $\mathbb{K}$ ($\mathbb{K}=\mathbb{R}$ or $\mathbb{C}$) which is a composition algebra over $\mathbb{C}$.

2) If $\Delta \neq \mathbb{H}$ then the above H^*-triple system structure on $\Delta \oplus \Delta$ agrees with the $\varepsilon_e s$-isotope of a split two-dimensional composition H^*-algebra over $\mathbb{K}$ (whose inner product and involution are given as in (3.1), (3.2), (3.4)).

3) In the above H^*-triple system structure on $\Delta \oplus \Delta$ the two choices for ε_e $(=\pm 1)$ are isometrically $*$-isomorphic.

4) If $\mathbb{H}$ is obtained from an H^*-*subalgebra* $D_1 \equiv \mathbb{C}$ by the Cayley-Dickson process then every $x \in \mathbb{H}$ can be writen in a unique way as $x = a + b\,v$ with $a, b \in D_1$ and v fixed in $D_1^{\perp}$ such that $v^2 = -1$. If $t : \mathbb{H} \to \mathbb{H}$ is the isometric antiautomorphism given by Corollary 3.6 then the map $f : \mathbb{H}^{(t)} \to \mathbb{H}^{(-t)}$ given by $f(a + bv) = b + av$ is an isometric $*$-isomorphism.

THEOREM 3.9. *(Coordinatization Theorem). Let* A *be an associative* H^**-triple system endowed with an* $\mathcal{A} \times \mathcal{B}$*-grid* $\{e_{ij}\}_{(i,j) \in \mathcal{A} \times \mathcal{B}}$ *and* ω *an arbitrary element of* $\mathcal{A} \cap \mathcal{B}$. *If* $D = A_{11}(e_{\omega\omega})^{(\varepsilon_{\omega\omega}\, e_{\omega\omega})}$ *and* $s : D \to D$ *is given by* $s(x) = \varepsilon_{\omega\omega} M(e_{\omega\omega}, e_{\omega\omega})x$ *for any* $x \in D$ *then there is an isometric* $*$*-isomorphism*

$$\phi : A \to \mathcal{M}_{\mathcal{A},\mathcal{B}}\,(D, s, T_1, T_2)$$

for T_1 *(resp.* T_2*) the diagonal* $\mathcal{B} \times \mathcal{B}$*-matrix* (λ_{jl}) *(resp. diagonal* $\mathcal{A} \times \mathcal{A}$*-matrix* (μ_{il})*) such that* $\lambda_{jj} = \varepsilon_{j\omega}$ *(resp.* $\mu_{ii} = \varepsilon_{\omega i}\,\varepsilon_{\omega\omega}$*). Moreover* $\phi(e_{ij}) = E_{ij}$ *for any* $(i,j) \in \mathcal{A} \times \mathcal{B}$. *Conversely every associative matrix* H^**-algebra* $\mathcal{M}_{\mathcal{A},\mathcal{B}}(D, s, T_1, T_2)$ *has an* $\mathcal{A} \times \mathcal{B}$*-grid.*

Proof. As we have noted the family $\mathcal{E} = \{E_{ij} : i \in \mathcal{A}, j \in \mathcal{B}\}$ is an $\mathcal{A} \times \mathcal{B}$-grid of $\mathcal{M}_{\mathcal{A},\mathcal{B}}\,(D, s, T_1, T_2)$ and the converse is obvious. Now we assume A an associative H^*triple system with an $\mathcal{A} \times \mathcal{B}$-grid $\mathcal{E} = \{e_{ij} : (i,j) \in \mathcal{A} \times \mathcal{B}\}$. First we suppose $\varepsilon_{ij} = 1$ for any $i \in \mathcal{A}, j \in \mathcal{B}$. Thus $\delta_{ij} = 1$. Let ϕ_{ij} be the map from $A_{11}(e_{ij})$ to $A_{11}(e_{\omega\omega})$ given by

$$\phi_{ij}(x_{ij}) = \langle\langle e_{\omega j}\, e_{ij}\, x_{ij}\rangle\, e_{\omega j}\ e_{\omega\omega}\rangle$$

Note that ϕ_{ij} is given in some of the following ways

$$\phi_{ij} = \begin{cases} \mathrm{Id} & \text{for } i = j = \omega \\ \phi_{e_{\omega j}, e_{\omega\omega}} & \text{for } i = \omega,\ j \neq \omega \\ \tilde{\phi}_{e_{i\omega}, e_{\omega\omega}} & \text{for } i \neq \omega,\ j = \omega \\ \phi_{e_{\omega j}, e_{\omega\omega}} \circ \tilde{\phi}_{e_{ij}, e_{\omega j}} & \text{for } i \neq \omega,\ j \neq \omega \end{cases}$$

where the ϕ and $\tilde{\phi}$ are as in (2.2). In any cases we have a family of isometric $*$-isomorphisms between the associative unital involutive H^*- algebras $A_{11}(e_{ij})^{(\varepsilon_{ij}e_{ij})}$ and $D = A_{11}(e_{\omega\omega})^{(\varepsilon_{\omega\omega}e_{\omega\omega})}$ which are isometric $*$-isomorphisms of H^*-subsystems of A. We define the map ϕ from A to $\mathcal{M}_{\mathcal{A},\mathcal{B}}(D, s, Id, Id)$ by

$$\phi(x) = \phi(\Sigma x_{ij}) = \Sigma \phi_{ij}(x_{ij}) E_{ij},$$

where $x_{ij} \in A_{11}(e_{ij})$ for any $i \in \mathcal{A}, j \in \mathcal{B}$. It is obvious that ϕ is an isomorphism of Hilbert spaces satisfying

$$(\phi(x) \mid \phi(y)) = (x \mid y).$$

To prove that ϕ is an isomorphism of triple systems it suffices to show that

$$\phi_{ij} \langle x_{ik}\, y_{mk}\, z_{mj} \rangle = \langle \phi_{ik}(x_{in})\, \phi_{mk}(y_{mk})\, \phi_{mj}(z_{mj}) \rangle$$

for any $i, m \in \mathcal{A}; j, k \in \mathcal{B}$, which is obtained by a direct calculation using the Peirce relations, Lemma 2.5 and (2. 4). The fact that ϕ is a $*$-isomorphism follows from (2.3).

Suppose now that $\mathcal{E} = \{e_{ij} : i \in \mathcal{A},\ j \in \mathcal{B}\}$ is an arbitrary $\mathcal{A} \times \mathcal{B}$-grid of A. Thus

$$\varepsilon_{ij}\, \delta_{ij} = 1 \tag{3.10}$$

for any $i \in \mathcal{A},\ j \in \mathcal{B}$. Let $S_{\mathcal{E}}$ be the map from A to A given by

$$S_{\mathcal{E}}\Big(\sum_{\substack{i \in \mathcal{A} \\ j \in \mathcal{B}}} x_{ij}\Big) = \sum_{\substack{i \in \mathcal{A} \\ j \in \mathcal{B}}} \varepsilon_{ij}\, x_{ij}.$$

Next we prove that $S_{\mathcal{E}}$ is an automorphism. Let $x_{ik} \in A_{11}(e_{ik})$, $y_{mk} \in A_{11}(e_{mk})$, $z_{mj} \in A_{11}(e_{mj})$. Using Lemma 2.5 we obtain

$$\langle S_{\mathcal{E}}(x_{ik})\, S_{\mathcal{E}}(y_{mk})\, S_{\mathcal{E}}(z_{mj}) \rangle = \varepsilon_{ik}\, \varepsilon_{mk}\, \varepsilon_{mj} \langle x_{ik}\, y_{mk}\, z_{mj} \rangle = S_{\mathcal{E}} \langle x_{ik}\, y_{mk}\, z_{mj} \rangle.$$

Therefore $S_{\mathcal{E}}$ is an isometric involutive $*$-automorphism. We have

$$\langle e_{ij}\, e_{ij}\, e_{ij} \rangle^{(S_{\mathcal{E}})} = \varepsilon_{ij} \langle e_{ij}\, e_{ij}\, e_{ij} \rangle = e_{ij},$$

for every $e_{ij} \in \mathfrak{E}$. So by (3.10) every e_{ij} is a self-adjoint tripotent of the $S_{\mathfrak{E}}$-isotope $A^{(S_{\mathfrak{E}})}$. By what we already shown there exists an isometric $*$-isomorphism $\phi: A^{(S_{\mathfrak{E}})} \to \mathcal{M}_{\mathcal{A},\mathfrak{B}}(D,s,Id,Id)$. It is easy to see that the following diagram commutes

$$\begin{array}{ccc} A^{(S_{\mathfrak{E}})} & \xrightarrow{S_{\mathfrak{E}}} & A^{(S_{\mathfrak{E}})} \\ \phi\downarrow & & \downarrow\phi \\ \mathcal{M}_{\mathcal{A},\mathfrak{B}}(D,s,Id,Id) & \xrightarrow{S_{T_1,T_2}} & \mathcal{M}_{\mathcal{A},\mathfrak{B}}(D,s,Id,Id) \end{array}$$

where $S_{T_1,T_2}(M) = T_2 M T_1$ for any $M \in \mathcal{M}_{\mathcal{A},\mathfrak{B}}(D,s,Id,Id)$. Hence A is isometrically $*$-isomorphic with $\mathcal{M}_{\mathcal{A},\mathfrak{B}}(D,s,T_1,T_2)$. This completes the proof.

The coordinatization methods have been used by N. Jacobson [13], K. McCrimmon and K. Meyberg [19], J.A. Cuenca and A. Rodríguez [9],and E. Neher [20] among others.

THEOREM 3.11. *Let A be a topologically simple associative H^*-triple system over $\mathbb{K}$ ($\mathbb{R}$ or $\mathbb{C}$).*

a) If $\mathbb{K} = \mathbb{C}$ then up to twins and a positive factor of the inner product A is $*$-isometrically isomorphic with the H^*-triple system $\mathcal{M}_{\mathcal{A},\mathfrak{B}}(D,s,Id,Id)$ for $\mathcal{A}$ and $\mathfrak{B}$ two non-vacuous sets and (D,s) the complex associative composition H^*-algebra whose involution and inner product are given as in (3.1).

b) If $\mathbb{K} = \mathbb{R}$ then up to twins and a positive factor of he inner product A is $*$-isometrically isomorphic with one of the following H^*-triple systems:

b-1) $\mathcal{M}_{\mathcal{A},\mathfrak{B}}(D,s,Id,Id)$ for (D,s) either a complex associative composition algebra with real H^*-algebra structure given as in (3.2) or the 2-dimensional real split composition H^*-algebra of involution and inner product as in (3.4).

b-2) $\mathcal{M}_{\mathcal{A},\mathfrak{B}}(\mathbb{H},t,Id,Id)$ where t is the involutive antiautomorphism given by (3.7) and $\mathbb{H}$ is endowed with one H^*-algebra structure as in (3.3).

b-3) $\mathcal{M}_{\mathcal{A},\mathfrak{B}}(\mathbb{H} \oplus \mathbb{H}^{op}, ex, Id, Id)$ where ex is the exchange antiautomorphism and $\mathbb{H} \oplus \mathbb{H}^{op}$ is the H^*-algebra with inner product and involution given in the following way

$$((x, y) \mid (z, t)) = (x \mid z) + (y \mid t),$$
$$(x, y)^* = (s(x), s(y)),$$

s being the Cayley antiautomorphism.

b-4) $\mathcal{M}_{\mathcal{A},\mathcal{B}}(D, s, T_1, T_2)$ *where* (D, s) *is either the real split quaternions with inner product and involution given by (3.4) or* $\mathbb{R}$ $\mathbb{C}$ *or* $\mathbb{H}$ *with the* H^**-algebra structure determined by (3.3).*

Conversely every associative H^**-triple system described above is topologically simple.*

Proof. It is easy to show that the H^*-triple systems of a), b-1), b-2), b-3), b-4) are topologically simple. As we have noted, if A is a topologically simple H^*-triple system the Zorn Lemma and the existence of irreducible projections give the existence of a maximal orthogonal family $\mathcal{E} = \{e_{ii}\}_{i\in\mathcal{A}}$ of irreducible projections. Let ω be a fixed but arbitrary element of $\mathcal{A}$. Changing $*$ to $-*$ if necessary, we can suppose $\varepsilon_{\omega\omega} = \delta_{\omega\omega}$. By Lemma 2.9, for all $i\in\mathcal{A}$, $i \neq \omega$, we have the quadrangle $(e_{\omega\omega}, e_{\omega i}, e_{ii}, e_{i\omega})$. Using Lemma 2.6 we can prove that for every quadrangle $(e_{11}, e_{12}, e_{22}, e_{21})$ of irreducible projections we have

$$A_{11}(e_{12}) = A_{10}(e_{11}) \cap A_{01}(e_{22}), \quad A_{11}(e_{21}) = A_{01}(e_{11}) \cap A_{10}(e_{22}) \tag{3.12}$$

and by [6; Lemma 2.4]

$$\varepsilon_{mn}\,\delta_{mn} = \varepsilon_{rs}\,\delta_{rs} \tag{3.13}$$

for any $m, n, r, s \in \{1, 2\}$. In particular $\varepsilon_{ii} = \delta_{ii}$, $\varepsilon_{\omega i} = \delta_{\omega i}$, $\varepsilon_{i\omega} = \delta_{i\omega}$ for all $i\in\mathcal{A}-\{\omega\}$. As in [18; 2.9] it is easy to show that $e_{i\omega}, e_{\omega i} \in A_{00}(e_{jj})$ for any $j\in\mathcal{A}$, $\omega\neq j\neq i$. From Lemma 2.6 we have $e_{\omega j}\in A_{11}(e_{\omega i}) \perp A_{10}(e_{\omega i})$ and taking into account $e_{\omega j}\in A_{00}(e_{ii})$, (3.12) yields $e_{\omega j}\in A_{10}(e_{\omega i})$. Similarly we can prove that $e_{i\omega}\in A_{01}(e_{\omega j})$. By Lemma 2.7, for any $i, j \in \mathcal{A}$, $i \neq j \neq \omega \neq i$ the element $e_{ij} = \varepsilon_{\omega\omega}\langle e_{i\omega}\, e_{\omega\omega}\, e_{\omega j}\rangle$ satisfies that $(e_{\omega\omega}, e_{\omega j}, e_{ij}, e_{i\omega})$ is a quadrangle. Next we prove that the family $\mathcal{E}_{\mathcal{A}} = \{e_{ij} : i, j \in \mathcal{A}\}$ is an $\mathcal{A} \times \mathcal{A}$-net. By Lemma 2.7, $(e_{\omega i}, e_{\omega j}, e_{ij}, e_{ii})$ and $(e_{i\omega}, e_{ii}, e_{ji}, e_{j\omega})$ are quadrangles. Again by Lemma 2.7 from the quadrangularities of $(e_{\omega\omega}, e_{\omega j}, e_{ij}, e_{i\omega})$, $(e_{\omega\omega}, e_{\omega j}, e_{jj}, e_{j\omega})$ follows that of $(e_{i\omega}, e_{ij}, e_{jj}, e_{j\omega})$ and from those of $(e_{\omega\omega}, e_{\omega j}, e_{ij}, e_{i\omega})$ and $(e_{\omega\omega}, e_{\omega j}, e_{jj}, e_{j\omega})$ that of

$(e_{i\omega}, e_{ij}, e_{jj}, e_{j\omega})$. Other quadrangles obtained in this way are $(e_{ii}, e_{ij}, e_{jj}, e_{ji})$ and $(e_{\omega i}, e_{\omega j}, e_{jj}, e_{ji})$. We have proved that every quadruple $(e_{ik}, e_{lk}, e_{lj}, e_{ij})$ containing some element of the form $e_{mm}, e_{m\omega}$ or $e_{\omega m}$ is a quadrangle. In particular, if $|\mathcal{A}| = 3$ then $\mathfrak{E}_{\mathcal{A}}$ is an $\mathcal{A} \times \mathcal{A}$-net. Now we assume that i, j, k, l are pairwise different elements of $\mathcal{A}$-$\{\omega\}$. The quadrangularity of $(e_{ij}, e_{ik}, e_{lk}, e_{lj})$ is a consequence of those of $(e_{\omega j}, e_{\omega k}, e_{lk}, e_{lj})$ and $(e_{\omega j}, e_{\omega k}, e_{ik}, e_{ij})$. Hence every quadruple $(e_{ik}, e_{lk}, e_{lj}, e_{ij})$ with $i, j, k, l \in \mathcal{A}$ is a quadrangle. By (3.13), $\mathfrak{E}_{\mathcal{A}}$ is an $\mathcal{A} \times \mathcal{A}$-net and from (3.12) we can write

$$\overline{\perp_{i,j \in \mathcal{A}} A_{ij}} = \overline{\perp_{i,j \in \mathcal{A}} A_{11}(e_{ij})} \,. \tag{3.14}$$

The Peirce subspaces A_{i0} and A_{0i} relative to the maximal orthogonal family $\mathfrak{E}_{\mathcal{A}}$ introduced as in Proposition 2.11 satisfie either $\Sigma A_{0i} = 0$ or $\Sigma A_{i0} = 0$. We assume $\Sigma A_{0i} = 0$, since the case $\Sigma A_{i0} = 0$ can be argued in a similar way. We shall extend the net $\mathfrak{E}_{\mathcal{A}}$ to a grid of A. By (3.14) we can assume $\Sigma A_{i0} \neq 0$. The properties of the Peirce decomposition yields $\langle e_{ii}\, e_{\omega i}\, x\rangle \in A_{i0}$ for any $x \in A_{\omega 0}$, $i \in \mathcal{A}$. So for every $i \in \mathcal{A}$-$\{\omega\}$ we have a linear map $\Psi_i : A_{\omega 0} \to A_{i0}$ given by $\Psi_i(x) = \varepsilon_{\omega i} \langle e_{ii} e_{\omega i}\, x\rangle$. The map θ_i given by $\theta_i(y) = \varepsilon_{ii} \langle e_{\omega i} e_{ii}\, y\rangle$ is the inverse map of Ψ_i. So $\Sigma A_{i0} \neq 0$ implies $A_{\omega 0} \neq 0$. Since the H^*-subsystems of associative H^*-triple systems of zero annihilator are of zero annihilator, there is some nonzero projection in $A_{\omega 0}$ which is irreducible by Lemma 2.6. By the Zorn Lemma there exists a family of irreducible projections $\mathfrak{E}_{\omega} = \{e_{\omega j} : j \in \mathfrak{C}\}$ of $A_{\omega 0}$ such that $e_{\omega j} \in A_{10}(e_{\omega k})$ for any $j, k \in \mathfrak{C}$, $j \neq k$, and maximal for this property. Let H be the orthogonal complement of

$$\overline{\perp_{j \in \mathfrak{C}} A_{11}(e_{\omega j})}$$

in $A_{\omega 0}$. Assume $H \neq 0$ and let g be a nonzero projection of H. By Lemma 2.6, for every $j \in \mathfrak{C}$ we have

$$A_{11}(e_{\omega\omega}) \perp A_{10}(e_{\omega\omega}) = A_{11}(e_{\omega j}) \perp A_{10}(e_{\omega j}),$$
$$[A_{01}(e_{\omega j}) \perp A_{00}(e_{\omega j})] \cap A_{\omega 0} = 0.$$

Therefore $g = g_{11}(j) + g_{10}(j)$ with $g_{rs}(j) \in A_{rs}(e_{\omega j}) \cap A_{\omega 0}$ for all $j \in \mathfrak{C}$. Hence $0 = (g_{11}(j) \mid g) = \| g_{11}(j) \|^2$ and

$$0 \neq g \in \bigcap_{j \in \mathbf{C}} A_{10}(e_{\omega j}),$$

which contradict the maximality of $\mathcal{E}_\omega$. Therefore

$$A_{\omega 0} = \overline{\perp_{j \in \mathbf{C}} A_{11}(e_{\omega j})}.$$

It is easy to show that $\Psi_i(\langle x\, u\, v\rangle) = \varepsilon_{\iota\iota}\,\varepsilon_{\omega\iota} \langle \Psi_i(x), \Psi_i(u), \Psi_i(v)\rangle$ for any $i \in \mathcal{A} - \{\omega\}$ and $x, u, v \in A_{\omega 0}$. Now the isometric character of Ψ_i yields

$$A_{i0} = \overline{\perp_{j \in \mathbf{C}} A_{11}(e_{ij})}.$$

where $e_{ij} = \Psi_i(e_{\omega j})$ for all $i \in \mathcal{A} - \{\omega\}, j \in \mathbf{C}$. By Lemma 2.7 when $i \in \mathcal{A} - \{\omega\}, j \in \mathbf{C}$ the quadruples $(e_{\omega i}, e_{\omega j}, e_{ij}, e_{ii})$ are quadrangles. Let $l \in \mathbf{C}$, $l \neq j$. From the quadrangularities of $(e_{\omega i}, e_{\omega j}, e_{ij}, e_{jj})$ and $(e_{\omega i}, e_{\omega l}, e_{il}, e_{ii})$ we obtain that of $(e_{\omega j}, e_{\omega l}, e_{il}, e_{ij})$. Therefore by (3.13), (3.14), Proposition 2. 11 and what was proved above we have that $\mathcal{E}_{\mathcal{A}} \cup (\cup_{j \in \mathcal{A}} \mathcal{E}_j)$ where $\mathcal{E}_j = \Psi_i(\mathcal{E}_\omega)$, is a grid of irreducible projections of A. If $A_{11}(e_{\omega\omega})$ is an H^*-triple system as in a), b-1), b-2) or b-3) of Proposition 3.8, then there are nonzero self-adjoint tripotents and nonzero skew-adjoint antitripotents in $A_{11}(e_{\omega\omega})$.These projections have the same Peirce subspaces (see [6; Corollary 2.10] and beginning of the proof of Lemma 2.6). Using the exchange isomorphisms we see that this holds for any $A_{11}(e_{ij})$, so in this case the elements of $\mathcal{E}$ and of $\{e_{\omega j} : j \in \mathcal{A} \cup \mathbf{C}\} \cup \{e_{i\omega} : i \in \mathcal{A}\}$ can be taken self-adjoint tripotents without loss of generality. As in proof of Lemma 2.9, all the elements of the grid

$$\mathcal{E}_{\mathcal{A}} \cup \Big(\bigcup_{j \in \mathcal{A}} \mathcal{E}_j\Big)$$

are self-adjoint tripotents in this case. We conclude the proof using the Coordinatization Theorem and Proposition 3.8.

Several applications can be obtained from our results, for instance the structure theory for H^*-triples [4], or the following result of Kaplansky [14] which we prove here , (see [2;

Theorem 4] and [3; Theorem 4] for other proofs).

PROPOSITION 3.15. *Let* A *be a real associative topologically simple* H^**-algebra. Then there exist a non-vacuous set* $\mathcal{A}$, *a real* H^**-algebra* $D = \mathbb{R}, \mathbb{C}$ *or* $\mathbb{H}$ *(described as in (3.1)) such that up to a positive factor of the inner product* A *is isometrically* $*$*-isomorphic with the* H^**-algebra* $\mathcal{M}_{\mathcal{A}}(D)$ *of the* $\mathcal{A} \times \mathcal{A}$*-matrices* $(a_{ij})_{(i,j) \in \mathcal{A} \times \mathcal{A}}$ *such that* $\sum \|a_{ij}\|^2 < \infty$ *whose product, inner product and involution are given in the following way:*

$$(a_{ij})(b_{ij}) = (c_{ij}), \qquad c_{ij} = \sum_{k \in \mathcal{A}} a_{ik} b_{kj}$$
$$((a_{ij}) \mid (b_{ij})) = \sum_{ij} (a_{ij} \mid b_{ij})$$
$$(a_{ij})^* = (a_{ji}^{\nabla}) .$$

Proof. Taking into account that A agrees with the closure of the linear envelope of $\{xy : x, y \in A\}$, the $*$-isotope $A^{(*)}$ is a topologically simple H^*-triple system of triple product $<x\,y\,z> = x\,y^* z$ and identity involution. Let $\{e_i\}_{i \in \mathcal{A}}$ be a maximal family of pairwise orthogonal irreducible self-adjoint idempotents [10; proof of Theorem 2]. For $k, l \in \{0, 1\}$ and $i \in \mathcal{A}$ we have $A_{kl}^{(*)}(e_i) = A_{kl}(e_i)$. In particular $\{e_i : i \in \mathcal{A}\}$ is a maximal family of orthogonal tripotents of $A^{(*)}$. By Proposition 2.11 either $\sum A_{0i}^{(*)} = 0$ or $\sum A_{i0}^{(*)} = 0$. Thus from $(\sum A_{0i}^{(*)})^* = \sum A_{i0}^{(*)}$ results $\sum A_{0i}^{(*)} = \sum A_{i0}^{(*)} = 0$. By Theorem 3.8 and its proof there are a non-vacuous set $\mathcal{A}$, a grid $\{e_{ij}\}_{(i,j) \in \mathcal{A} \times \mathcal{A}}$ such that $e_{ii} = e_i$ for any $i \in \mathcal{A}$ and a $*$-isomorphism $\eta : A^{(*)} \to \mathcal{M}_{\mathcal{A},\mathcal{A}}(D, s, T_1, T_2)$ that multiplies all the norms by the same positive real factor, with $\mathcal{M}_{\mathcal{A},\mathcal{A}}(D, s, T_1, T_2)$ as in b-1), b-2), b-3), b-4) of Theorem 3.11. Moreover $\eta(e_{ij}) = E_{ij}$ for any $i, j \in \mathcal{A}$ (see Theorem 3.9). Since $A^{(*)}$ has identity involution and the e_i are tripotents, we have $T_1 = T_2 = Id$ and (D, s) is either $(\mathbb{R}, Id)$, or $(\mathbb{C}, -)$ or $(\mathbb{H}, -)$. Let $\dagger$ be the unique linear map making the following diagram commutative:

$$\begin{array}{ccc} A^{(*)} & \xrightarrow{\ *\ } & A^{(*)} \\ \eta \downarrow & & \downarrow \eta \\ \mathcal{M}_{\mathcal{A},\mathcal{A}}(D, s, Id, Id) & \xrightarrow{\ \dagger\ } & \mathcal{M}_{\mathcal{A},\mathcal{A}}(D, s, Id, Id) \end{array} \qquad (3.16)$$

For any non-vacuous finite subset F of $\mathcal{A}$ the involution $*$ induces an involutive antiautomorphism of every $(A_{11}^{(*)}(e_F))^{(e_F)}$, where $e_F = \sum_{i \in F} e_i$. Since $E_{ii}^{\dagger} = E_{ii}$

for each $i \in \mathcal{A}$, either from a direct calculation or from a well known theorem [12; p. 79] we obtain the existence of an isometric involutive $*$-antiautomorphism t of the H^*-algebra D such that

$$\Big(\sum_{i,j\in\mathcal{A}} a_{ij}\, \mathrm{E}_{ij}\Big)^{\dagger} = \sum_{i,j\in\mathcal{A}} t\,(a_{ij})\,\mathrm{E}_{j\,i} \tag{3.17}$$

for every arbitrary matrix $\sum a_{ij}\mathrm{E}_{ij}$ of $\mathcal{M}_{\mathcal{A},\mathcal{A}}(D, s, Id, Id)$. Since $*$ induces the involution of the H^*-algebra D, by (3.17) this agrees with t. Thus $\dagger$ is the involution of the H^*-algebra $\mathcal{M}_{\mathcal{A}}(D)$. Since the product of two elements x and y of the subalgebra $A_{11}(e_F)$ agrees with $<x\ e_F\, y> = x\ e_F\, y$ for any finite subset F of $\mathcal{A}$, from the density of

$$\bigcup_{\substack{F\subset\mathcal{A}\\ |F|<\infty}} A_{11}(e_F) = \bigcup_{\substack{F\subset\mathcal{A}\\ |F|<\infty}} \overset{(*)}{A_{11}}(e_F)$$

we obtain that η is an isomorphism between the H^*-algebras A and $\mathcal{M}_{\mathcal{A}}(D)$. Its $*$-isomorphic character is equivalent to the commutativity of (3.16).

REFERENCES

[1] W. AMBROSE. Structure theorems for a special class of Banach algebras, *Trans. Amer.Math. Soc.* 57 (1945), 364-386.

[2] V. K. BALACHANDRAN and N. SWAMINATHAN. Real H*-algebras, *J. Funt. Anal.* 65 (1986), 64-75.

[3] M. CABRERA, J. MARTINEZ and A. RODRIGUEZ. Nonassociative real H*-algebras, Publications Matemáttiques, Vol **32** (1988), 267-274.

[4] M. CABRERA, J. MARTINEZ and A. RODRIGUEZ., *Hilbert modules over H*-algebras in relation with Hilbert Ternary Rings*, Proceedings of the Workshop on Nonassociative Algebraic Models, Nova Science Publishers, New York.

[5] A. CASTELLON and J. A. CUENCA. Alternative H*-triple systems, Preprint, Universidad de Málaga.

[6] A. CASTELLON and J. A. CUENCA. Compatibility in Jordan H*-Triple Systems, Bolletino U.M.I. (7) 4-B (1990), 433-447.

[7] J. A. CUENCA.*Sobre H^*-álgebras no asociativas. Teoría de estructura de las H^*-álgebras de Jordan no conmutativas semisimples.* Secretariado de Publicaciones de la Universidad de Málaga, Málaga, 1984.

[8] J. A. CUENCA. Sur la thèorie de structure des H*-algèbres de Jordan non commutatives, Colloque sur les algèbres de Jordan, Montpellier, 1985.

[9] J. A. CUENCA and A. RODRIGUEZ. Structure theory for noncommutative Jordan H^*-algebras, *J. Algebra*, 106 No. 1, (1987), 1-14.

[10] J. A. CUENCA and A. SANCHEZ. Structure theory for real noncommutative Jordan H^*-algebras, Preprint, Universidad de Málaga.

[11] I. N. HERSTEIN. *Rings with involution.* The University Chicago Press, Lectures in Mathematics, 1976.

[12] N. JACOBSON. *Structure of rings*, Amer. Math. Soc. Coll. Publ. Vol. XXXVII, Providence R. I., 1968.

[13] N. JACOBSON. *Structure and representations of Jordan algebras*, Amer. Math. Soc. Coll. Publ. Vol. XXXIX, Providence R. I., 1968.

[14] I. KAPLANSKY. Dual rings, *Ann. Math.* 49, No. 3 (1948), 689-701.

[15] W. KAUP. Uber die klassification der symmetrichen hermiteschen Mannigfaltigkeiten unerdlicher dimension II, *Math. Ann.*, 262 (1983), 57-75.

[16] O. LOOS. *Jordan pairs*, Lecture Notes in Mathematics, Vol. 460, Springer-Verlag, Berlin-Heidelberg-New York, 1975.

[17] K. McCRIMMON. Peirce ideals in Jordan triple systems, *Pacific J. Math.* 83 (1979), 415-439.

[18] K. McCRIMMON. Compatible Peirce decomposition of Jordan triple systems, *Pacific J.Math.* 103 (1982).

[19] K. McCRIMMON and K. MEYBERG. Coordinatization of Jordan triple systems, *Com. in Alg.* 9 (14), (1981), 1495-1542.

[20] E. NEHER. *Jordan triple systems by the grid approach*, Lecture Notes in Mathematics, Springer-Verlag, Berlin-Heidelberg-New York, 1987.

Nonassociative
Algebraic
Models

ISBN 1-56072-050-6

Bernstein Algebras: Lattice Isomorphisms and Isomorphisms*

Teresa Cortés
Departamento de Matemáticas, Universidad de Zaragoza, 50009, Zaragoza, Spain

Abstract: We present here some cases of Bernstein algebras which are determined up to isomorphisms by their lattices of subalgebras.

0. Introduction

The origin of Bernstein algebras lies in genetics and in the study of the stationary evolution operators, see Lyubich [1]. Holgate [2] was the first to give a formulation of Bernstein's problem into the language of nonassociative algebras. For a summary of known results see Wörz-Busekros [3], Ch 9. Further investigations on Bernstein algebras have been taken up by Alcalde, Burgueño, Labra, Micali, (see [5]).

On the other hand, the study of the relationship between lattice isomorphisms and isomorphisms has been done by Barnes ([7] and [8]) for associative and Lie algebras and by J. A. Laliena for alternative algebras [9]. For Jordan algebras similar studies has been done by J. A. Laliena (presented in "Jornadas sobre Modelos Algebraicos no Asociativos y sus Aplicaciones", celebrated in Zaragoza, April 1989) and completed by J. A. Anquela (personal communication, still not submitted).

* This paper has been written under the direction of Professor Santos Gonzalez and it will be a part of the author's Doctoral Thesis. The author has been partially supported by the Ministerio de Educación y Ciencia (F.P.I. Grant) and the Diputación General de Aragón.

1. Preliminaries

A finite-dimensional commutative algebra A over a field K is called *baric* if there exists a nontrivial homomorphism $\omega : A \to K$, called *weight homomorphism.*

A baric algebra is called a *Bernstein algebra* if:

$$(x^2)^2 = \omega(x)^2 x^2 \text{ for all } x \text{ in } A.$$

In the following, let K be a commutative infinite field of characteristic different from 2.

Let us list several results on Bernstein algebras which can be found in [3].

For every Bernstein algebra the nontrivial homomorphism $\omega : A \to K$ is uniquely determined.

Every Bernstein algebra A possesses at least one non-zero idempotent.

Every Bernstein algebra A with non-zero idempotent e can be decomposed into the internal direct sum of subspaces:

$$A = Ke \oplus \operatorname{Ker} \omega,$$

with: $\operatorname{Ker} \omega = U_e \oplus V_e,$

where: $U_e = \{ex \mid x \in \operatorname{Ker} \omega\} = \{x \in A \mid ex = \frac{1}{2}x\},$

$$V_e = \{x \in A \mid ex = 0\}.$$

The subspaces U_e and V_e of A satisfy:

$$U_e V_e \subseteq U_e, \quad V_e^2 \subseteq U_e, \quad U_e^2 \subseteq V_e, \quad U_e V_e^2 = \langle 0 \rangle.$$

The set of idempotent elements of A is given by:

$$\mathcal{J} = \{e + u + u^2 \mid u \in U_e\} \text{ for any idempotent } e \text{ in } A.$$

If $e_1 = e + \sigma + \sigma^2$ with σ in U_e is another idempotent of A we have the following relations between the corresponding subspaces:

$$U_{e_1} = \{ u + 2\sigma u \mid u \in U_e \},$$

$$V_{e_1} = \{ v - 2(\sigma + \sigma^2) v \mid v \in V_e \}.$$

It follows that, although the decomposition of a Bernstein algebra depends on the choice of the idempotent e, the dimension of the subspace U_e of A is an invariant of A. If $\dim_K A = n+1$, then one can associate to $A = Ke \oplus U_e \oplus V_e$ a pair of integers $(r+1, s)$, called *the type* of A, where:

$$r = \dim_K U_e, \qquad s = \dim_K V_e, \qquad \text{hence } r + s = n.$$

In the same way Wörz-Busekros shows in [3] that $\dim_K U_e^2$ and $\dim_K (U_eV_e+V_e^2)$ are also invariants of the algebra A.

Other useful identities can be found in [3] and [5] and will be cited if necessary.

A classification of all the Bernstein algebras of dimensions 2 and 3 is given by Wörz-Busekros in [3]. We reproduce it here to get a classification of these algebras up to isomorphisms.

In dimension 2 there exist up to isomorphisms exactly 2 Bernstein algebras, one algebra of type (1,1): $Ke \oplus Kv$, with $e^2 = e$, $ev = 0$, $v^2 = 0$, and one algebra of type (2,0) : $Ke \oplus Ku$, with $e^2 = e$, $eu = \frac{1}{2}u$, $u^2 = 0$.

In dimension 3 there exists up to isomorphisms exactly one Bernstein algebra of type (1,2): $Ke \oplus Kv_1 \oplus Kv_2$, whose multiplication table is $e^2 = e$, $ev_i = 0$, $v_i v_k = 0$, and one algebra of type (3,0): $Ke \oplus Ku_1 \oplus Ku_2$ whose multiplication table is $e^2 = e$, $eu_i = \frac{1}{2}u_i$, $u_i u_k = 0$. We will denote these algebras by $A_{(1)}$ and $A_{(2)}$ respectively. The 3-dimensional Bernstein algebras of type (2,1) are:

$Ke \oplus Ku \oplus Kv$ with $e^2 = e$, $eu = \frac{1}{2}u$, $ev = 0$ and the remaining products given by the table 1, depending on $\dim_K U_e^2$ and $\dim_K (U_eV_e+V_e^2)$.

TABLE 1.

$\dim_K(U_eV_e+V_e^2)$	$\dim_K U_e^2$	u^2	uv	v^2	
0	0	0	0	0	(a)
0	1	αv	0	0	(b)
1	0	0	βu	γu	(c)

with $\alpha \neq 0$ and $(\beta, \gamma) \neq (0,0)$.

The algebra (a) is the trivial algebra of type (2,1), see [6], and will be called $A_{(3)}$. The algebra (b) is isomorphic to another algebra with the same multiplication table and $\alpha = 1$ and will be called $A_{(4)}$. In (c), if $\beta \neq 0$ then we can put $\beta = 1$ and $\gamma = 0$ to obtain an isomorphic algebra which will be called $A_{(5)}$. If $\beta = 0$, the resulting algebra is isomorphic to another algebra of (c) with $\beta = 0$ and $\gamma = 1$ which will be called $A_{(6)}$. On the other hand one can see that the algebras $A_{(1)} \dots A_{(6)}$ are nonisomorphic.

In the same way Wörz-Busekros shows in [6] that for every decomposition n=r+s, there exists, up to isomorphism, exactly one (n+1)-dimensional, trivial (with $(\operatorname{Ker}\omega)^2 = 0$), Bernstein algebra of type (r+1, s).

2. On the Length of a Bernstein Algebra

Let A be an algebra over a commutative field K. We denote by $\mathfrak{L}(A)$ the lattice of all subalgebras of A. By an $\mathfrak{L}$-isomorphism (lattice isomorphism) of the algebra A onto an algebra B over the same field, we mean an isomorphism:

$$\mathfrak{L}(A) \to \mathfrak{L}(B) \quad \text{of } \mathfrak{L}(A) \text{ onto } \mathfrak{L}(B).$$

We put $\ell(A)$, the *length* of A, for the supremum of the lengths of all the chains in $\mathfrak{L}(A)$ (by the length of a chain we mean its cardinality minus one). Clearly we have $\dim_K A \geq \ell(A)$ and if the algebra A is finite-dimensional then $\ell(A)$ is the maximum, not only the supremum. We remark that, for a solvable algebra A, we have $\ell(A) = \dim_K A$.

THEOREM 1: *Let A be a Bernstein algebra over a commutative field K of characteristic different from 2. Then $\mathcal{L}(A) = dim_K A$.*

PROOF: Let $\omega : A \rightarrow K$ be the weight homomorphism of the algebra A, $e \in A$ be an idempotent and $A = Ke \oplus U_e \oplus V_e$ be the decomposition of A with respect to e. In [3] we can see $A^2 = Ke \oplus U_e \oplus U_e^2$, and this algebra is a Bernstein algebra with weight homomorphism the restriction $\tilde{\omega}$: $A^2 \rightarrow K$ of ω to the algebra A^2. Then we have that the corresponding subspaces of the decomposition of the algebra A^2 with respect to e are:

$$\tilde{U}_e = \{ x \in A^2 \mid ex = \tfrac{1}{2}x \} = U_e,$$

$$\tilde{V}_e = \{ x \in A^2 \mid ex = 0 \} = U_e^2 = \tilde{U}_e^2.$$

Then $(A^2)^2 = A^2$ and, in this situation, we can apply the theorem 2 of [4] and conclude that $\operatorname{Ker}\tilde{\omega} = U_e \oplus U_e^2 \supseteq (\operatorname{Ker}\omega)^2$ is right nilpotent, hence nilpotent, (see proposition 4.1. on page 82 of [10]), and then solvable. But now $\operatorname{Ker}\omega$ is solvable, too, and hence $\mathcal{L}(\operatorname{Ker}\omega) = \dim_K \operatorname{Ker}\omega$.

So we have:

$\mathcal{L}(A) \geq \mathcal{L}(\operatorname{Ker}\omega) + 1 = \dim_K \operatorname{Ker}\omega + 1 = \dim_K A$ and this completes the proof.•

In what follows, by "subalgebra" we mean a proper subalgebra.

COROLLARY: *The dimensions of the subalgebras of a Bernstein algebra are invariant by any $\mathcal{L}$-isomorphism.*

PROOF: If A_1 is a subalgebra of a Bernstein algebra A such that $A_1 \subseteq \operatorname{Ker}\omega$, then A_1 is solvable and hence $\mathcal{L}(A_1) = \dim_K A_1$. In the other case the algebra A_1 is a Bernstein algebra and we have $\mathcal{L}(A_1) = \dim_K A_1$. Clearly the length of an algebra is invariant by any $\mathcal{L}$-isomorphism.•

REMARK: If K is a field of characteristic 2 the proofs of the previous results are not valid because $\operatorname{Ker}\omega$ is not, in general, solvable. We can consider, for example, the commutative algebra $A = Ke \oplus Kx \oplus Ky \oplus Kz$ whose multiplication table is given by $e^2 = e$, $xy = z$, $yz = x$, $xz = y$, where the missing products are zero, over any field K of characteristic 2. Clearly A is a Bernstein algebra over K with weight homomorphism

$\omega : A \rightarrow K$ given by $\omega(e) = 1$, $\omega(x) = \omega(y) = \omega(z) = 0$ linearly extended, but $(\mathrm{Ker}\,\omega)^2 = \mathrm{Ker}\,\omega$.

3. Subalgebras of Dimension Two

We are going to study the 2-dimensional algebras which can be found in a Bernstein algebra.

If A_1 is a 2-dimensional subalgebra of a Bernstein algebra A such that $A_1 \not\subset \mathrm{Ker}\,\omega$, then A_1 is a 2-dimensional Bernstein algebra and hence it is either $Ke \oplus Ku$ or $Ke \oplus Kv$, with the notation of §1 .

In the other case we have $A_1 \subseteq \mathrm{Ker}\,\omega$ and then A_1 is a commutative solvable algebra such that $(x^2)^2 = 0$ holds in A_1. If A_1 is not trivial we have $A_1{}^2 = Kz$ with $z \neq 0$ and one can see that A_1 is isomorphic to another algebra $B = Kz \oplus Kw$ whose multiplication table is given by:

$z^2 = 0$, $zw = 0$, $w^2 = z$, which will be called $B_{(2)}$, or:

$z^2 = 0$, $zw = z$, $w^2 = 0$, which will be called $B_{(3)}$.

By $B_{(1)}$ we mean A_1 when is trivial. Clearly these algebras are nonisomorphic.If we look at the lattice of subalgebras of these algebras we can conclude the following table 2:

TABLE 2.

ALGEBRA	Multiplication table	Subalgebras	Number of subalgebras
$Ke \oplus Ku$	$e^2 = e$ $eu = \frac{1}{2}u$ $u^2 = 0$	Ku, $K(e + \alpha u)$	$\lvert K \rvert + 1$
$Ke \oplus Kv$	$e^2 = e$ $ev = 0$ $v^2 = 0$	Ke, Kv	2
$B_{(1)} = Kz \oplus Kw$	$z^2 = 0$ $zw = 0$ $w^2 = 0$	Kw, $K(z + \alpha w)$	$\lvert K \rvert + 1$
$B_{(2)} = Kz \oplus Kw$	$z^2 = 0$ $zw = 0$ $w^2 = z$	Kz	1
$B_{(3)} = Kz \oplus Kw$	$z^2 = 0$ $zw = z$ $w^2 = 0$	Kz, Kw	2

where α is in K and we write $|K|$ for the cardinality of the field K, which is infinite, in this case.

4. Lattice Isomorphism and Isomorphism

We begin with the low-dimensional cases.

THEOREM 2: *Two Bernstein algebras of dimension less than or equal to 3 over an infinite field of characteristic different from 2 are $\mathfrak{L}$-isomorphic if and only if they are isomorphic.*

PROOF: The "if" is obvious. Let us prove the "only if".

Clearly they have the same dimension (see §2), and the result is valid in dimension 2 (see table 2). Let (A, ω) and $(\widetilde{A}, \widetilde{\omega})$ be two 3-dimensional Bernstein algebras over K, and $\phi: \mathfrak{L}(A) \rightarrow \mathfrak{L}(\widetilde{A})$ be an $\mathfrak{L}$-isomorphism.

(I) Let us suppose $A \cong A_{(1)}$. Then there exits only one idempotent e and $A = Ke \oplus Kv_1 \oplus Kv_2$ with $e^2=e$, $ev_i=0$, $v_iv_k=0$, $V_e = \mathrm{Ker}\,\omega = Kv_1 \oplus Kv_2 \cong B_{(1)}$.

The 1-dimensional subalgebras of A are Kx with $x^2=x$ or $x^2=0$. So it is easy to see that, in this case, they are Ke or Kv with v in V_e.

We consider the 2-dimensional subalgebras of A: if A_1 is a 2-dimensional subalgebra of A and $A_1 \subseteq \mathrm{Ker}\,\omega$ then $A_1 = \mathrm{Ker}\,\omega \cong B_{(1)}$ which has $|K| + 1$ subalgebras. In the other case A_1 possesses an idempotent, i.e., $Ke \subseteq A_1$, and $A_1 = Ke \oplus Kv$ with v in V_e, which has 2 subalgebras.

Hence Ke is the only 1-dimensional subalgebra of A not contained in 2-dimensional subalgebras with $|K| + 1$ subalgebras. On the other hand, Ke is contained in an infinite number of 2-dimensional subalgebras with only 2 subalgebras.

If $\widetilde{A}$ is a Bernstein algebra of type (1, 2), then clearly $\widetilde{A} \cong A$. Let us suppose this is false and we will get a contradiction.

We denote $Kx = \phi(Ke)$ with $0 \neq x$ in $\widetilde{A}$ and let $\widetilde{A}_1$ be a 2-dimensional sulalgebra of $\widetilde{A}$ such that $Kx \subseteq \widetilde{A}_1$, then $\widetilde{A}_1$ has exactly 2 subalgebras. It follows from table 2:

$$\widetilde{A}_1 = K\tilde{e} \oplus K\tilde{v} \quad \text{with } \tilde{e}^2 = \tilde{e}, \quad \tilde{e}\tilde{v} = 0, \quad \tilde{v}^2 = 0 \quad \text{, or:}$$

$$\widetilde{A}_1 = Kz \oplus Kw \quad \text{with } z^2 = 0, \quad zw = z, \quad w^2 = 0.$$

In the second case we have $\widetilde{A}_1 = \operatorname{Ker}\tilde{\omega}$; we take $\widetilde{A}_2$ and $\widetilde{A}_3$, two different subalgebras of $\widetilde{A}$ satisfying the same conditions as $\widetilde{A}_1$. Then, for the same reasons we have:

$$\widetilde{A}_2 = K\tilde{e} \oplus K\tilde{v} \quad \text{with } \tilde{e}^2 = \tilde{e}, \quad \tilde{e}\tilde{v} = 0, \quad \tilde{v}^2 = 0 \quad , \quad \text{and } \tilde{\omega}(\tilde{e}) = 1$$

$\widetilde{A}_3 = K\tilde{e}_1 \oplus K\tilde{v}_1$ with $\tilde{e}_1^2 = \tilde{e}_1$, $\tilde{e}_1\tilde{v}_1 = 0$, $\tilde{v}_1^2 = 0$, and $\tilde{\omega}(\tilde{e}_1) = 1$, since $\widetilde{A}_2$ and $\widetilde{A}_3$ cannot be $\operatorname{Ker}\tilde{\omega}$.

But now $\tilde{v}$ is in $\widetilde{V}_{\tilde{e}}$, hence $\dim_K \widetilde{V}_{\tilde{e}} \geq 1$ and $\dim_K \widetilde{V}_{\tilde{e}} \leq 1$, because the type of $\widetilde{A}$ is not (1, 2). It follows $\widetilde{V}_{\tilde{e}} = K\tilde{v}$. In the same way $\widetilde{V}_{\tilde{e}_1} = K\tilde{v}_1$. Besides that, $\widetilde{A}$ is a Bernstein algebra of type (2, 1).

We have: $0 \neq x \in \operatorname{Ker}\tilde{\omega} \cap \widetilde{A}_2 \cap \widetilde{A}_3$, hence $Kx = K\tilde{v} = K\tilde{v}_1$.

On the other hand, the idempotents $\tilde{e}$ and $\tilde{e}_1$ are connected by:

$\tilde{e}_1 = \tilde{e} + \sigma + \sigma^2$ with σ in $\widetilde{U}_{\tilde{e}}$, and $\sigma \neq 0$, since $\widetilde{A}_2$ and $\widetilde{A}_3$ are different subalgebras. Therefore $\widetilde{U}_{\tilde{e}} = K\sigma$.

Since $0 \neq \tilde{v} \in K\tilde{v} = K\tilde{v}_1 = \widetilde{V}_{\tilde{e}_1}$, we have $\tilde{v} = \lambda\{\tilde{v} - 2(\sigma+\sigma^2)\tilde{v}\}$, with λ in K. Since $(\sigma+\sigma^2)\tilde{v}$ is in $\widetilde{U}_{\tilde{e}}$, and the sum $\widetilde{U}_{\tilde{e}} \oplus \widetilde{V}_{\tilde{e}}$ is direct, we can conclude:

$$\lambda = 1 \quad \text{and} \quad (\sigma+\sigma^2)\tilde{v} = 0.$$

On the other hand σ^2 is in $\widetilde{V}_{\tilde{e}} = K\tilde{v}$, and $\tilde{v}^2 = 0$, hence $\sigma^2\tilde{v} = 0$, and thus $\sigma\tilde{v} = 0$, i.e., $\widetilde{U}_{\tilde{e}}\widetilde{V}_{\tilde{e}} = 0$.

If $\sigma^2 \neq 0$, we can consider $K\sigma \oplus K\sigma^2 = \tilde{U}_{\tilde{e}} \oplus \tilde{V}_{\tilde{e}} \cong B_{(2)}$, which has only one subalgebra. As A has not any subalgebras of this type, we get a contradiction.Then $\sigma^2 = 0$. Therefore we take $K\sigma \oplus K\tilde{v} = \tilde{U}_{\tilde{e}} \oplus \tilde{V}_{\tilde{e}} \cong B_{(1)}$, which has $|K|+1$ subalgebras and contains $Kx = K\tilde{v}$ and this is a contradiction.

Then, the second posibility for $\tilde{A}_1$ is not valid, i.e., $x \notin \mathrm{Ker}\,\tilde{\omega}$. Hence $Kx = K\tilde{e}_0$, where $\tilde{e}_0$ is an idempotent of $\tilde{A}$. As above, we take $\tilde{A}_2$ and $\tilde{A}_3$, which are Bernstein algebras with only one idempotent. Thus $\tilde{e}_0 = \tilde{e} = \tilde{e}_1$. Now, $\tilde{v}$ and $\tilde{v}_1$ cannot be linearly independent, because $\dim_K \tilde{V}_{\tilde{e}} \leq 1$, and cannot be linearly dependent, since $\tilde{A}_2$ and $\tilde{A}_3$ are different subalgebras, which is a contradiction.

(II) Let us suppose $A \cong A_{(2)}$. Then $A = Ke \oplus Ku_1 \oplus Ku_2$ with $e^2 = e$, any idempotent of A, $eu_i = \frac{1}{2}u_i$, $u_i u_k = 0$, $U_e = \mathrm{Ker}\,\omega = Ku_1 \oplus Ku_2 \cong B_{(1)}$.

The 1-dimensional subalgebras of A are exactly the 1-dimensional subspaces of A, Kx, since $x^2 = \omega(x)\,x$ holds for all x in A.

Concerning the 2-dimensional subalgebras of A, if A_1 is a 2-dimensional subalgebra of A and $A_1 \subseteq \mathrm{Ker}\,\omega$ then $A_1 = \mathrm{Ker}\,\omega \cong B_{(1)}$ which has $|K| + 1$ subalgebras. In the other case A_1 possesses an idempotent e, and $A_1 = Ke \oplus Ku$ with u in U_e, which has $|K| + 1$ subalgebras. Thus, none of the 2-dimensional subalgebras of A has a finite number of subalgebras.

From (I), $\tilde{A}$ cannot be isomorphic to $A_{(1)}$. If it were isomorphic to $A_{(3)}$, $A_{(4)}$, or $A_{(5)}$, it would have a 2-dimensional subalgebra with exactly 2 subalgebras ($Ke \oplus Kv$, with the notation of Table 1). If it were isomorphic to $A_{(6)}$, it would have a 2-dimensional subalgebra with only one subalgebra ($Ku \oplus Kv$, with the same notation). Both situations lead us to a contradiction. Then it must be that $\tilde{A} \cong A_{(2)} \cong A$.

(III) Finally, let us suppose A is a Bernstein algebra of type (2, 1). Then we can write $A = Ke \oplus Ku \oplus Kv$, with $e^2 = e$, $eu = \frac{1}{2}u$, $ev = 0$, for any idempotent e of A. From (I) and (II) we can conclude that $\tilde{A}$ is a Bernstein algebra of the same type.

If $A \cong A_{(3)}$, we can write $u^2 = 0$, $uv = 0$, $v^2 = 0$ for any decomposition of A. Besides that, either a 2-dimensional subalgebra A_1 of A is $A_1 = \mathrm{Ker}\,\omega \cong B_{(1)}$, which has $|K|+1$

subalgebras, or A_1 possesses an idempotent. In the second case $A_1 = Ke \oplus Kx$ with $0 \neq x$ in $\operatorname{Ker}\omega = Ku \oplus Kv$. We put $x = \alpha u + \beta v$, with α, β in K. If $\alpha\beta \neq 0$, we have $u = (2\alpha)^{-1} ex \in A_1$, hence $v \in A_1$ and then $A_1 = A$, which is a contradiction. Therefore we have $Ke \oplus Kx = Ke \oplus Ku$, with an infinite number of subalgebras, or $Ke \oplus Kx = Ke \oplus Kv$, which has 2 subalgebras. Either way A has only 2-dimensional subalgebras with 2 or more subalgebras.

If $A \cong A_{(5)}$, we can put $u^2=0$, $uv=u$, $v^2=0$. Then, a 2-dimensional subalgebra A_1 of A is $A_1 = \operatorname{Ker}\omega \cong B_{(3)}$, which has 2 subalgebras, or A_1 possesses an idempotent $e + \lambda u$, λ in K. In the last case we put $A_1 = K(e + \lambda u) \oplus Kx$ with $0 \neq x$ in $\operatorname{Ker}\omega = Ku \oplus Kv$. As above, $x = \alpha u + \beta v$, and, if $\alpha\beta \neq 0$, $x^2 = 2\alpha\beta u \in A_1$ and hence $A_1 = A$, which is absurd. Thus we conclude $\alpha=0$ or $\beta=0$. If we had $\alpha=0$, then we would have $\beta \neq 0$ and $\lambda\beta u = (e+\lambda u)\beta v \in A_1$; hence $\lambda=0$ and $A_1 = Ke \oplus Kv$, which has 2 subalgebras. If we had $\beta=0$, then we would have $\alpha \neq 0$ and $A_1 = Ke \oplus Ku$, which has $|K|+1$ subalgebras. Again, A has only 2-dimensional subalgebras with 2 or more subalgebras.

If $A \cong A_{(4)}$ or $A \cong A_{(6)}$, it has a 2-dimensional subalgebra, $\operatorname{Ker}\omega \cong B_{(2)}$ with a unique proper subalgebra.

Moreover, we have shown that the algebras $A_{(3)}$ and $A_{(5)}$ are determined by their lattices. They can be $\mathfrak{L}$-isomorphic neither to $A_{(4)}$ nor to $A_{(6)}$ because of what we said above. And the algebra $A_{(3)}$ cannot be $\mathfrak{L}$-isomorphic to $A_{(5)}$ because $A_{(5)}$ has only one 2-dimensional subalgebra with an infinite number of subalgebras, $Ke \oplus Ku$, and $A_{(3)}$ has two of them, $\operatorname{Ker}\omega$ and $Ke \oplus Ku$, using the previous notation.

It remains to show why the algebras $A_{(4)}$ and $A_{(6)}$ cannot be $\mathfrak{L}$-isomorphic.

As above, if we put $A_{(6)} = Ke \oplus Ku \oplus Kv$, with $e^2 = e$, $eu = \frac{1}{2}u$, $ev = 0$, $u^2 = 0$, $uv = 0$, $v^2 = u$, a 2-dimensional subalgebra A_1 of $A_{(6)}$ such that $A_1 \not\subset \operatorname{Ker}\omega$ possesses an idempotent $e + \lambda u$, λ in K. We write $A_1 = K(e + \lambda u) \oplus Kx$, $x = \alpha u + \beta v$, with α, β in K. If $\beta \neq 0$, we have $x^2 = \beta^2 u$ is in A_1, and hence $A_1 = A$, which is a contradiction. Thus it must be $\beta=0, \alpha \neq 0$, and $A_1 = Ke \oplus Ku$, which has an infinite number of subalgebras. Then we can conclude that the 2-dimensional subalgebras of $A_{(6)}$ have only one subalgebra or an infinite number of subalgebras. Now if we put $A_{(4)} = Ke \oplus Ku \oplus Kv$, with $e^2 = e$, $eu = \frac{1}{2}u$, $ev = 0$, $u^2 = v$, $uv = 0$, $v^2 = 0$, we can consider $Ke \oplus Kv$, which has exactly 2 subalgebras. It follows that $A_{(4)}$ and $A_{(6)}$ cannot be $\mathfrak{L}$-isomorphic.

In our situation this means that the algebra $\widetilde{A}$ must be isomorphic to the algebra A.•

We are going to generalize the ideas of the previous theorem to get similar results for some extreme cases of Bernstein algebras in any dimension.

THEOREM 3: *If two Bernstein algebras are $\mathfrak{L}$-isomorphic and one of them is an $(n+1)$-dimensional Bernstein algebra of type $(n+1, 0)$, then they are isomorphic.*

PROOF: Let $(\widetilde{A}, \widetilde{\omega})$ be an (n+1)-dimensional Bernstein algebra of type (n+1, 0). Let (A, ω) be a Bernstein algebra and $\phi : \mathfrak{L}(\widetilde{A}) \to \mathfrak{L}(A)$ be an $\mathfrak{L}$-isomorphism.

We can write: $\widetilde{A} = K\widetilde{e} \oplus \mathrm{Ker}\,\widetilde{\omega}$ with $\widetilde{U}_{\widetilde{e}} = \mathrm{Ker}\,\widetilde{\omega}$, $\widetilde{V}_{\widetilde{e}} = 0$, for any idempotent $\widetilde{e}$ in $\widetilde{A}$ and suppose $n \geq 2$.

If $\widetilde{A}_1$ is a 2-dimensional subalgebra of $\widetilde{A}$, either $\widetilde{A}_1 \subseteq \mathrm{Ker}\,\widetilde{\omega}$ or $\widetilde{A}_1$ possesses an idempotent $\widetilde{e}$.In the first case, $\widetilde{A}_1$ is trivial and it has $|K| + 1$ subalgebras. In the second case we have $\widetilde{A}_1 = K\widetilde{e} \oplus K\widetilde{u}$, with $\widetilde{u}$ in $\widetilde{U}_{\widetilde{e}}$, which also has $|K| + 1$ subalgebras.

Besides that, we can see that:

-Every pair of 1-dimensional subalgebras of $\widetilde{A}$ generates a 2-dimensional subalgebra of $\widetilde{A}$.

-The n-dimensional subalgebras of $\widetilde{A}$ are $\mathrm{Ker}\,\widetilde{\omega}$ or $K\widetilde{e} \oplus W$ where W is (n-1)-dimensional and $W \subseteq \widetilde{U}_{\widetilde{e}}$. In both cases we can see that a n-dimensional subalgebra of $\widetilde{A}$ contains n 1-dimensional subalgebras generated by elements linearly independent, let $\widetilde{x}_1, \ldots, \widetilde{x}_n$ be those elements. Moreover any subset of m elements of $\{\widetilde{x}_1, \ldots, \widetilde{x}_n\}$ generates a m-dimensional subalgebra of $\widetilde{A}$.

Now, let us write $A = Ke \oplus U_e \oplus V_e$ for any idempotent e in A. Then, since $\mathrm{Ker}\,\omega = U_e \oplus V_e$ is a n-dimensional subalgebra of A, we have $\mathrm{Ker}\,\omega = \phi(\widetilde{A}_2)$, where

$\widetilde{A}_2$ is a n-dimensional subalgebra of $\widetilde{A}$. Therefore we can write $\widetilde{A}_2 = K\widetilde{x}_1 \oplus \ldots \oplus K\widetilde{x}_n$ for the preceding elements.

We take $Kx_i = \phi(K\widetilde{x}_i)$, $0 \neq x_i \in A$, for i=1, ..., n. If the elements $x_1, \ldots, x_n$ were not linearly independent, there would exist j such that:

$$x_j \in Kx_1 + \overset{\wedge}{\underset{\ldots}{j}} + Kx_n$$

Then we would have:

$$\phi(K\widetilde{x}_j) = Kx_j \subseteq Kx_1 \vee \overset{\wedge}{\underset{\ldots}{j}} \vee Kx_n = \phi(K\widetilde{x}_1) \vee \overset{\wedge}{\underset{\ldots}{j}} \vee \phi(K\widetilde{x}_n) =$$

$$= \phi(K\widetilde{x}_1 \vee \overset{\wedge}{\underset{\ldots}{j}} \vee K\widetilde{x}_n) = \phi(K\widetilde{x}_1 \oplus \overset{\wedge}{\underset{\ldots}{j}} \oplus K\widetilde{x}_n)$$

Thus: $K\widetilde{x}_j \subseteq K\widetilde{x}_1 \oplus \overset{\wedge}{\underset{\ldots}{j}} \oplus K\widetilde{x}_n$, which is a contradiction. Hence the elements $x_1, \ldots, x_n$ are linearly independent and we have:

$$\mathrm{Ker}\,\omega = U_e \oplus V_e = Kx_1 \oplus \ldots \oplus Kx_n .$$

If $x_i \in U_e$ for all i = 1, ..., n we have $\mathrm{Ker}\,\omega = U_e$ and A is a Bernstein algebra of type (n+1, 0) isomorphic to $\widetilde{A}$, as we wanted to prove.

In the other case there would exist j such that:

$$x_j = u + v \text{ with } u \text{ in } U_e \text{ and } 0 \neq v \text{ in } V_e.$$

If $u^2 = w \neq 0$ we would have $w^2 = 0$ and we could consider the subalgebra $Ke \oplus Kw$ which has only 2 subalgebras, and this is impossible because A and $\widetilde{A}$ are $\mathcal{L}$-isomorphic. Hence, $u^2 = 0$ and Ku is a subalgebra of A. Now we take the subalgebras $Kx_j = K(u+v)$ and Ku; they generate a 2-dimensional subalgebra which contains u and v, and thus it is $Ku \oplus Kv$. This algebra must have $|K| + 1$ subalgebras and, since it is contained in $\mathrm{Ker}\,\omega$, it must be trivial. But then $v^2 = 0$ and we get a contradiction (as above with w).•

THEOREM 4: *If two Bernstein algebras are $\mathcal{L}$-isomorphic and one of them is a Bernstein algebra (n+1)-dimensional of type (1, n), then they are isomorphic.*

PROOF: We will carry out an induction on n.

For $n = 0,1$ the result is obvious. Let us suppose the result valid for all $0 \leq k \leq n-1$, with $n \geq 2$ and we will prove it for n.

Let $(\widetilde{A}, \widetilde{\omega})$ be a (n+1)-dimensional Bernstein algebra of type (1, n), (A, ω) be a Bernstein algebra and $\phi : \mathfrak{L}(\widetilde{A}) \rightarrow \mathfrak{L}(A)$ be an $\mathfrak{L}$-isomorphism.

We can write: $\widetilde{A} = K\widetilde{e} \oplus \operatorname{Ker} \widetilde{\omega}$ with $\widetilde{V}_{\widetilde{e}} = \operatorname{Ker} \widetilde{\omega}$, $\widetilde{U}_{\widetilde{e}} = 0$, for the unique idempotent $\widetilde{e}$ in $\widetilde{A}$.

If $\widetilde{A}_1$ is a 2-dimensional subalgebra of $\widetilde{A}$, either $\widetilde{A}_1 \subseteq \operatorname{Ker} \widetilde{\omega}$ or $\widetilde{A}_1$ possesses an idempotent $\widetilde{e}$. In the first case, $\widetilde{A}_1$ is trivial and it has $|K| + 1$ subalgebras. In the second case we have $\widetilde{A}_1 = K\widetilde{e} \oplus K\widetilde{v}$, with $\widetilde{v}$ in $\widetilde{V}_{\widetilde{e}}$, which has 2 subalgebras.

Besides that, $K\widetilde{e}$ is the only 1-dimensional subalgebra of $\widetilde{A}$ not contained in 2-dimensional subalgebras with an infinite number of subalgebras. Moreover, a 2-dimensional subalgebra of $\widetilde{A}$ contains $K\widetilde{e}$ if and only if it has exactly 2 subalgebras.

On the other hand, $\widetilde{A}$ has at least four different n-dimensional subalgebras: $\operatorname{Ker} \widetilde{\omega}$ and at least three Bernstein algebras with decomposition $K\widetilde{e} \oplus \widetilde{W}$ where $\widetilde{W}$ is a (n-1)-dimensional subspace of $\widetilde{V}_{\widetilde{e}}$.

Therefore A has at least four different n-dimensional subalgebras, and we can take two different n-dimensional subalgebras which are different from $\operatorname{Ker} \omega$ and $\phi(\operatorname{Ker} \widetilde{\omega})$. Then, each one of them is a Bernstein algebra $\mathfrak{L}$-isomorphic to a Bernstein algebra of type (1, n-1), hence isomorphic to a Bernstein algebra of that type, by the induction assumption.

Let us write these algebras:

$$Ke \oplus W, \text{ with } W \subseteq V_e,\ e^2 = e,$$

$$Ke_1 \oplus W_1, \text{ with } W_1 \subseteq V_{e_1},\ e_1^2 = e_1.$$

If V_e were n-dimensional, we would have the type of A is (1, n) and hence it is isomorphic to $\widetilde{A}$.

If V_e is (n-1)-dimensional, we have $V_e=W$, $V_{e_1}=W_1$, and A is of type (2, n-1) with $V_e^2=0$. We put $U_e=Ku$. If $0\neq v=u^2$, we would have $v^2=0$, $uv=0$, since $u^3=0$ (see [3]), and $Ku\oplus Kv$ would be a 2-dimensional subalgebra of A with only one subalgebra, but that is not possible in A. Thus, it must be $u^2=0$.

Hence the idempotents are connected by $e_1= e + \lambda u$, $0\neq\lambda$ in K, since these algebras are different.

Now, we consider their sum as vectorial subspaces:

$$(Ke\oplus V_e) + (Ke_1\oplus V_{e_1}) = A\,,$$

hence the subalgebra $A_1=(Ke\oplus V_e)\cap(Ke_1\oplus V_{e_1})$ is (n-1)-dimensional.

If $A_1\not\subset \mathrm{Ker}\,\omega$, A_1 possesses an idempotent, but since the algebras $Ke\oplus V_e$ and $Ke_1\oplus V_{e_1}$ have only one idempotent, it follows $e = e_1$ and we have a contradiction. Hence $A_1\subseteq \mathrm{Ker}\,\omega$, and then $A_1\subseteq V_e\cap V_{e_1}$. Because of the dimensions of the subspaces, we have shown:

$$V_e\cap V_{e_1}= V_e=V_{e_1}$$

Now, any w in V_e is in $V_{e_1}=\{\, v-2\lambda uv \mid v\in V_e \,\}$ and we can write:

$$w = v-2\lambda uv \text{ with } w, v \text{ in } V_e.$$

Since $U_e\oplus V_e$ is direct we conclude $uv=0$, then $uw=0$, hence $U_eV_e=0$.

Therefore A is a trivial Bernstein algebra of type (2, n-1).

If we take $Kx = \phi(K\tilde{e})$, it is contained in a 2-dimensional subalgebra of A with 2 subalgebras, say A_2. If A_2 were contained in $\mathrm{Ker}\,\omega$, it would be isomorphic to $B_{(3)}$, but this impossible because the latter is not trivial. Then $A_2= Ke\oplus Kv$ with v in V_e for an idempotent e of A and $Kx = Ke$ or $Kx = Kv$. In both cases Kx would be contained in a 2-dimensional subalgebra of A with an infinite number of subalgebras, which is a contradiction.•

THEOREM 5: *If two Bernstein algebras are $\mathfrak{L}$-isomorphic and one of them is a trivial Bernstein algebra which is (n+1)-dimensional of type (r+1, s), then they are isomorphic.*

PROOF: We will carry out an induction on n. For $n = 0,1$ the result is obvious, and we know it for $n = 2$. Let us suppose the result valid for all $0 \leq k \leq n-1$, with $n \geq 3$ and we will prove it for n. The cases $r = 0$ and $s = 0$ are the theorems 3 and 4 and we will suppose $r\,s \neq 0$.

Let $(\widetilde{A}, \widetilde{\omega})$ be an (n+1)-dimensional Bernstein algebra of type (r+1, s), (A, ω) be a Bernstein algebra and $\phi : \mathfrak{L}(\widetilde{A}) \rightarrow \mathfrak{L}(A)$ be an $\mathfrak{L}$-isomorphism.

We can write: $\widetilde{A} = K\widetilde{e} \oplus \operatorname{Ker}\widetilde{\omega}$ with $\operatorname{Ker}\widetilde{\omega} = \widetilde{U}_{\widetilde{e}} \oplus \widetilde{V}_{\widetilde{e}}$, for any idempotent $\widetilde{e}$ in $\widetilde{A}$ and it is easy to see that any 2-dimensional subalgebra of $\widetilde{A}$ must have at least 2 subalgebras.

(I) We will show here that if $A \not\cong \widetilde{A}$, then A has two n-dimensional subalgebras which are trivial Bernstein algebras, one of type (r, s), and the other of type (r+1, s-1).For any idempotent $\widetilde{e}$ in $\widetilde{A}$, let

$\widetilde{U}_{\widetilde{e}} = K\langle \widetilde{u}_1, \ldots, \widetilde{u}_r \rangle$, with $\{\widetilde{u}_1, \ldots, \widetilde{u}_r\}$ a basis of $\widetilde{U}_{\widetilde{e}}$ over K

$\widetilde{V}_{\widetilde{e}} = K\langle \widetilde{v}_1, \ldots, \widetilde{v}_s \rangle$, with $\{\widetilde{v}_1, \ldots, \widetilde{v}_s\}$ a basis of $\widetilde{V}_{\widetilde{e}}$ over K.

(a) If $r \geq 2$ and $s \geq 2$, we take for example the following subalgebras of A:

$A_1 = \phi(K\widetilde{e} \oplus K\langle \widetilde{u}_2, \ldots, \widetilde{u}_r \rangle \oplus \widetilde{V}_{\widetilde{e}})$, $\quad A_2 = \phi(K\widetilde{e} \oplus \widetilde{U}_{\widetilde{e}} \oplus K\langle \widetilde{v}_2, \ldots, \widetilde{v}_s \rangle)$,

$A_3 = \phi(K\widetilde{e} \oplus K\langle \widetilde{u}_1, \ldots, \widetilde{u}_{r-1} \rangle \oplus \widetilde{V}_{\widetilde{e}})$, $\quad A_4 = \phi(K\widetilde{e} \oplus \widetilde{U}_{\widetilde{e}} \oplus K\langle \widetilde{v}_1, \ldots, \widetilde{v}_{s-1} \rangle)$.

At most one of them can be contained in $\operatorname{Ker}\omega$. Hence among them there are at least 3 Bernstein algebras and therefore, at least one $\mathfrak{L}$-isomorphic to a trivial Bernstein algebra of type (r, s) and one $\mathfrak{L}$-isomorphic to a trivial Bernstein algebra of type (r+1, s-1).We apply the induction assumption for these algebras.

(b) If $r = 1$, we have $s = n - r = n - 1 \geq 2$ and we can take the algebras:

$A_1 = \phi(K\tilde{e} \oplus \tilde{V}_{\tilde{e}})$, $\quad A_2 = \phi(K\tilde{e} \oplus \tilde{U}_{\tilde{e}} \oplus K\langle \tilde{v}_2, \ldots, \tilde{v}_s \rangle)$,

$A_3 = \phi(K(\tilde{e}+\tilde{u}_1) \oplus \tilde{V}_{\tilde{e}})$, $\quad A_4 = \phi(K\tilde{e} \oplus \tilde{U}_{\tilde{e}} \oplus K\langle \tilde{v}_1, \ldots, \tilde{v}_{s-1} \rangle)$,

and work as in (a).

(c) Finally, if $s = 1$, we have $r = n - s = n - 1 \geq 2$. Now we take, for example:

$$\phi(K\tilde{e} \oplus K\langle \tilde{u}_2, \ldots, \tilde{u}_r \rangle \oplus \tilde{V}_{\tilde{e}}),$$

$$\phi(K\tilde{e} \oplus K\langle \tilde{u}_1, \ldots, \tilde{u}_{r-1} \rangle \oplus \tilde{V}_{\tilde{e}}),$$

$$\phi(K(\tilde{e}+\tilde{u}_1) \oplus K\langle \tilde{u}_2, \ldots, \tilde{u}_r \rangle \oplus \tilde{V}_{\tilde{e}}),$$

and we conclude that there are in A two trivial Bernstein algebras of type (r, s).

Put $A_1 = \phi(K\tilde{e} \oplus \tilde{U}_{\tilde{e}})$. If it is not contained in $\mathrm{Ker}\,\omega$, it is a Bernstein algebra, and hence it is a trivial Bernstein algebra of type (r+1, s-1) by the induction assumption. Otherwise $A_1 = \mathrm{Ker}\,\omega$ and each of its 2-dimensional subalgebras has an infinite number of subalgebras.

We consider: $Kx_0 = \phi(K\tilde{u}_0)$, where we put $\tilde{u}_0 = \tilde{e}$, $Kx_i = \phi(K\tilde{u}_i)$, for $i=1, \ldots, r$. If $x_0, \ldots, x_r$ are not linearly independent, there must exist i such that:

$$0 \neq Kx_i \cap (Kx_0 \vee \overset{\wedge}{\underset{\ldots}{i}} \vee Kx_r) = \phi(K\tilde{u}_i) \cap (\phi(K\tilde{u}_0) \vee \overset{\wedge}{\underset{\ldots}{i}} \vee \phi(K\tilde{u}_r)) =$$

$$= \phi(K\tilde{u}_i \cap (K\tilde{u}_0 \vee \overset{\wedge}{\underset{\ldots}{i}} \vee K\tilde{u}_r)) = \phi(K\tilde{u}_i \cap (K\tilde{u}_0 + \overset{\wedge}{\underset{\ldots}{i}} + K\tilde{u}_r)).$$

Then $0 \neq K\tilde{u}_i \cap (K\tilde{u}_0 + \overset{\wedge}{\underset{\ldots}{i}} + K\tilde{u}_r)$, which is a contradiction. Hence $\{x_0, \ldots, x_r\}$ is a basis of $A_1 = \mathrm{Ker}\,\omega$ over K.

Besides that, for all $i \neq j$ the algebra $Kx_i \vee Kx_j$ is a 2-dimensional subalgebra of $\mathrm{Ker}\,\omega$ and it has an infinite number of subalgebras, hence it is trivial. Thus we have $\mathrm{Ker}\,\omega$ is also trivial.

On the other hand, we can take the two trivial Bernstein algebras of type (r, s=1) which are contained in A:

$B = Ke \oplus W \oplus Kv$, with $0 \neq v$ in V_e, $W \subsetneq U_e$

$B_1 = Ke_1 \oplus W_1 \oplus Kv_1$, with $0 \neq v_1$ in V_{e_1}, $W_1 \subsetneq U_{e_1}$.

If U_e is r-dimensional, A is of type (r+1, s) and trivial, hence isomorphic to $\widetilde{A}$, but we are supposing this is false. Therefore U_e is (r-1)-dimensional and $W = U_e$, $W_1 = U_{e_1}$, V_e and V_{e_1} are 2-dimensional. But since A is a trivial Bernstein algebra, we can write $V_e = V_{e_1}$, $U_e = U_{e_1}$ and so we can put:

$B = Ke \oplus U_e \oplus Kv$, $B_1 = Ke_1 \oplus U_e \oplus Kv_1 = Ke \oplus U_e \oplus Kv_1$, with v, v_1 in V_e.

Since $B \neq B_1$, the elements v and v_1 must be linearly independent and besides that, we know the multiplication table of A. If we consider U, an (r - 2)-dimensional subspace of U_e (we include the case r - 2 = 0 !), and the subalgebra $A_2 = Ke \oplus U \oplus V_e$ of A, which is n-dimensional, we can take $\phi^{-1}(A_2)$ and it cannot be Ker $\widetilde{\omega}$ since it has a 2-dimensional subalgebra, $\phi^{-1}(Ke \oplus Kv)$, with only 2 subalgebras, and that is not possible in a trivial algebra. Hence $\phi^{-1}(A_2)$ is a trivial Bernstein algebra of type (r -1, 2) by the induction assumption, which is impossible since $\widetilde{A}$ has type (r+1, 1).

(II) For any e, idempotent in A, it is easy to see that $U_e^2 = 0$:

For if there exists u in U_e such that $u^2 \neq 0$, then the subalgebra $Ku \oplus Ku^2$ would have only one subalgebra, which is not possible in A.

(III) Finally, we will show that A must be isomorphic to $\widetilde{A}$.

We consider in A the two subalgebras of (I):

$B = Ke \oplus W \oplus H$, a trivial Bernstein algebra with $e^2 = e$, W an (r -1)-dimensional subspace of U_e, H a s-dimensional subspace of V_e, and:

$B_1 = Ke_1 \oplus W_1 \oplus H_1$, a trivial Bernstein algebra with $e_1^2 = e_1$, W_1 an r-dimensional subspace of U_{e_1}, H_1 a (s -1)-dimensional subspace of V_{e_1}.

We can conclude from these decompositions that the type of A is (r+1, s) and as a consequence of (II) $U_{e_1} = U_e$ and then we have:

$B = Ke \oplus W \oplus H = Ke \oplus W \oplus V_e$, $B_1 = Ke_1 \oplus W_1 \oplus H_1 = Ke_1 \oplus U_e \oplus H_1$

$U_e^2 = 0$, $V_e^2 = 0$, $W\, V_e = 0$, $U_e\, H_1 = 0$.

Now, we must show $U_eV_e = 0$ and this will complete the proof.

It is easy to see that two different 1-dimensional subalgebras of $\widetilde{A}$ generate a subalgebra of dimension 2 or 3, hence the same must happen in A. Besides that, if the algebra is 3-dimensional, it has a 2-dimensional subalgebra with only 2 subalgebras.

Let us suppose there exist u in U_e and v in V_e with $0 \neq uv$ in U_e. Then $u \notin W$ and $v \notin H_1$ and we can write $U_c = Ku \oplus W$, $V_e = Kv \oplus H_1$.

Now if $Ku \vee Kv$ is 2-dimensional, it is equal to $Ku \oplus Kv$, hence $uv \in Ku$ and $Ke \oplus Ku \oplus Kv \cong A_{(5)}$. As in the last part of the proof of (I), $\phi^{-1}(Ke \oplus Ku \oplus Kv)$ cannot be in Ker $\tilde{\omega}$, and hence it is a Bernstein algebra isomorphic to $A_{(5)}$, which cannot be in $\widetilde{A}$.

If $Ku \vee Kv$ is 3-dimensional, the work is a bit harder.

Clearly $uv \notin Ku$; we put $uv = \lambda u + w$, w in W. If we had $\lambda \neq 0$, by taking $u_1 = uv$, we would have $u_1 v = \lambda u_1$, and we could work as above. Therefore, we can suppose $uv \in W$ and then $Ku \vee Kv = Ku \oplus Kv \oplus K(uv)$ has a 2-dimensional subalgebra which has exactly 2 subalgebras. But since we are working in Ker ω, this latter subalgebra is isomorphic to $B_{(3)}$, say $Ka \oplus Kb$, $a^2 = b^2 = 0$, $ab = b$. Then if one writes a and b as a linear combination of u, v, and uv, from $ab = b$, one obtains $b = 0$, which is a contradiction.●

Finally we include here another extreme case of Bernstein algebra of any dimension which is also determined by its lattice of subalgebras.

LEMMA: *Any non trivial (n+1)-dimensional Bernstein algebra of type (2, n-1) has the property that for any idempotent e, either* $U_eV_e + V_e^2 = 0$ *or* $U_e^2 = 0$.

PROOF :We write $A = Ke \oplus Ku \oplus V_e$, the decomposition of A with respect to an idempotent e of A, and let us suppose $u^2 \neq 0$. We define the bilinear form:

$$F: \; V_e \times V_e \to K$$

by $F(v, w) \in K$ such that $vw = F(v, w)\, u$.

From the elementary theory of bilinear forms, we can get a basis of V_e, $\{v_1, \ldots, v_{n-1}\}$ such that $v_i v_j = 0$ for $i \neq j$. We put $v_i^2 = a_i u$, $u v_i = b_i u$.

As one can see in [3], $u (u v_i) = 0$ and $(v_i^2)^2 = 0$. From this we have $a_i = b_i = 0$, which concludes the proof.•

If a Bernstein algebra satisfies the condition $U_e V_e + V_e^2 = 0$ for any idempotent e it is called in [3] a *normal Bernstein algebra.*

As a consequence of this lemma, we have:

COROLLARY: *Over a commutative field of characteristic different from 2 there exist, up to isomorphism, exactly one non trivial, normal Bernstein algebra of type (2, n-1) :* $A = Ke \oplus Ku \oplus V$, *where* $e^2 = 0$, $eu = \frac{1}{2} u$, $eV = 0$, $uV = 0$, $V^2 = 0$, $0 \neq u^2$ *is in V, and* $V = V_e$ *does not depend on the choice of e.* •

Now we can prove the following theorem.

THEOREM 6: *If two Bernstein algebras are $\mathfrak{L}$-isomorphic and one of them is a normal Bernstein algebra (n+1)-dimensional of type (2, n-1), then they are isomorphic.*

PROOF: Let (A, ω) be a normal Bernstein algebra (n+1)-dimensional of type (2, n-1); we can suppose it is not trivial by theorem 5. We write as above $A = Ke \oplus Ku \oplus V$ and put $\{ u^2, v_2, \ldots, v_{n-1} \}$, a basis of V. Let $(\tilde{A}, \tilde{\omega})$ be a Bernstein algebra $\mathfrak{L}$-isomorphic to A. As we have already proved the result for n = 1, 2, we can take $n \geq 3$.

It is easy to see some properties of the lattice of subalgebras of A:

-the 1-dimensional subalgebras of A are Ke_1, with $e_1^2 = e_1$, and Kv for any v in V.

-the 2-dimensional subalgebras of A are:

$Kv \oplus Kw$, v, w in V, a trivial algebra isomorphic to $B_{(1)}$, with an infinite number of subalgebras

$Ku^2 \oplus K(u+v)$, v in V, which is isomorphic to $B_{(2)}$, with Ku^2 as its only subalgebra.

$Ke_1 \oplus Kv$, $e_1^2 = e_1$, v in V, i.e., the Bernstein algebra of type (1, 1).

Hence the subalgebras of the form Ke_1 are contained only in these latter subalgebras, and as we are supposing $n \geq 3$, the dimension of V is at least 2 and there are an infinite number of subalgebras containing Ke_1.

We can also take $Ke \oplus V$ and $K(e+u) \oplus V$, two different subalgebras which are trivial Bernstein algebras of type (1, n-1). In this way, if we consider their images under ϕ, at most one of them coincides with Ker $\tilde{\omega}$ and so, at least one of them is a trivial Bernstein algebra of type (1, n-1) by using theorem 4. We put for this algebra $K\tilde{e} \oplus W$, W a (n-1)-dimensional subspace of $\tilde{V}_{\tilde{e}}$. If $\tilde{V}_{\tilde{e}}$ were n-dimensional, $\tilde{A}$ should be trivial and so should be A, again by theorem 4, but it is not. Then $W = \tilde{V}_{\tilde{e}}$ and $\tilde{A}$ is of type (2, n-1), with $\tilde{V}_{\tilde{e}}^2 = 0$.

If we proof $\tilde{U}_{\tilde{e}} \tilde{V}_{\tilde{e}} = 0$, $\tilde{A}$ will be normal, and since it cannot be trivial by theorem 5, we would conclude the result by using the preceding corollary.

Let us suppose the contrary and we will get a contradiction. By the previous lemma $\tilde{U}_{\tilde{e}}^2$ must be 0.

We can take $\tilde{v}$ in $\tilde{V}_{\tilde{e}}$ such that, putting $\tilde{U}_{\tilde{e}} = K\tilde{u}$, we have $\widetilde{uv} = \tilde{u}$. Hence $K\tilde{u} \oplus K\tilde{v} \cong B_{(3)}$ and $K\tilde{u}$, $K\tilde{v}$ are its subalgebras. Hence $K\tilde{u} \oplus K\tilde{v} = \phi(Ke_1 \oplus Kv)$, the image by ϕ of one 2-dimensional subalgebra of A with only 2 subalgebras, and $\phi(Ke_1) = K\tilde{u}$ or $\phi(Ke_1) = K\tilde{v}$. In either case this means that $\phi(Ke_1)$ is in a 2-dimensional subalgebra with an infinite number of subalgebras, and that is not possible.•

REMARK: The results we have obtained are also valid if the field K is finite but has at least 5 elements, in order to have the identities and properties of the decomposition respect an idempotent (see [3]).

This paper is just a first approach to lattice theory applied to the study of Bernstein algebras. The "high degree of nilpotency" of these algebras makes one think that the lattice of subalgebras determines the structure of the algebra to a "high degree" since the techniques of this paper are simple in general.

REFERENCES

[1] YU. I. LYUBICH, *"Basic concepts and theorems of the evolutionary genetics of free populations"* , Russian Math. surveys, 26, (1971), 51-123.

[2] P. HOLGATE, *"Genetic algebras satisfying Bernstein's stationarity principle"* , J. London Math. Soc., 9, (1975), 613-623.

[3] A. WÖRZ-BUSEKROS, *"Algebras in genetics"* , Lecture notes in Biomathematics 36 (Springer-Verlag, Berlin Heidelberg, 1980).

[4] A. N. GRISHKOV,*"On the genetic property of Bernstein algebras"* , Soviet Math. Dokl., 35, (1987), 489-492.

[5] M.T. ALCALDE, C. BURGUEÑO, A. LABRA, A. MICALI, *"Sur les algèbres de Bernstein"* , Proc. London Math. Soc., 58, (1989), 51-68.

[6] A. WÖRZ-BUSEKROS, *"Further remarks on Bernstein algebras"* , Proc. London Math. Soc., 58, (1989), 69-73.

[7] D. W. BARNES, *"Lattice isomorphisms of associative algebras"* , J. Aust. Math. Soc., 6, (1966), 106-121.

[8] D. W. BARNES, *"Lattice isomorphisms of Lie algebras"* , J. Aust. Math. Soc., 4, (1964), 470-475.

[9] J. A. LALIENA,*"Estructura reticular y cuasiideal en álgebras alternativas"*, Doctoral Thesis, (1987).

[10] K. A. ZHEVLAKOV, A. M. SLIN'KO, I. P. SHESTAKOV, A I. SHIRSHOV, *"Rings that are nearly associative"*, Pure and Applied Mathematics, (Academic Press Inc., New York, 1982).

Nonassociative
Algebraic
Models

ISBN 1-56072-050-6

Jordan Two-Graded H*-Algebras

José Antonio Cuenca Mira and Cándido Martín González
Departamento de Algebra, Geometría y Topología,
Universidad de Málaga, Apartado 59, 29080 Malaga, Spain

The associative two-graded algebras were studied by C.T.C. Wall [12] and by the authors together with A.García [4], who found the link between the binary structure of the algebra and the ternary structure of its odd part. The structure of the associative real and complex two-graded H^*-algebras is also known [3]. After these works on associative algebras, the question of determining the structure of the Jordan ones arises. In this work we find the structure theory for real and complex Jordan two-graded H^*-algebras, a question which has turned out to be directly related with the determination of certain of its automorphisms.

1. Let K be a unitary commutative ring, a <u>two-graded</u> K<u>-algebra</u> A is a K-algebra which splits into the direct sum $A = A_0 \oplus A_1$ of K-submodules (called the even and the odd part respectively) satisfying $A_\alpha A_\beta \subset A_{\alpha+\beta}$ for all α,β in $\mathbb{Z}_2$. If A is a two-graded algebra, its underlying algebra (forgetting the grading) will be denoted by $\mathcal{U}n(A)$. A <u>homomorphism</u> f between graded algebras A and B is a homomorphism from $\mathcal{U}n(A)$ to $\mathcal{U}n(B)$ which preserves gradings i.e.: $f(A_\alpha) \subset B_\alpha$ for all $\alpha \in \mathbb{Z}_2$. The definitions of epimorphism, monomorphism and isomorphism are the obvious ones. Let A and B be two-graded algebras such that $B \subset A$ and the inclusion mapping $i : B \longrightarrow A$ is a two-graded algebra homomorphism; then B is called a <u>subalgebra</u> of A. If, besides, $B.A \subset B$ then B is called an <u>ideal</u> of A. Let

This work was supported by the "Plan andaluz para la consolidación de grupos de investigación y desarrollo tecnologico"

A be a $\mathbb{K}$-algebra with $\mathbb{K} = \mathbb{R}$ or $\mathbb{C}$ and $^* : A \longrightarrow A$ a map which is linear if $\mathbb{K}=\mathbb{R}$ and conjugate-linear if $\mathbb{K}=\mathbb{C}$, and for which $(x^*)^*=x$ and $(xy)^*=y^*x^*$ hold for any $x,y\in A$. Then * is called an <u>involution</u> of the algebra A. We recall that an <u>H^*-algebra</u> A over $\mathbb{K}$ ($= \mathbb{R}$ or $\mathbb{C}$) , is a nonassociative $\mathbb{K}$-algebra, which is also a Hilbert space over $\mathbb{K}$ with inner product $(\ |\ )$, endowed with an involution * such that $(xy|z)=(x|zy^*)=(y|x^*z)$ for all $x,y,z \in A$. A <u>two-graded H^*-algebra</u> , is an H^*-algebra which is a two-graded algebra whose even and odd part are selfadjoint closed orthogonal subspaces. We call the two-graded H^*-algebra A, <u>topologically simple</u> if $A^2\neq 0$ and it has no proper closed ideal. Throughout this work we shall denote by $\mathbb{H}$ and $\mathbb{O}$ the real composition division algebras of dimensions 4 and 8 respectively, and by $\mathbb{H}_{sp}$ and $\mathbb{O}_{sp}$ the split ones. We shall use the notation $\mathbb{H}_C$ and $\mathbb{O}_C$ for the complex composition algebras of dimensions 4 and 8. In the sequel an <u>J^*-algebra</u> will mean a Jordan H^*-algebra and the notation $M_\Lambda(\mathcal{D})$, with $\mathcal{D}$ any real or complex composition algebra, will denote the algebra of all $\Lambda\times\Lambda$ matrices $(a_{ij})_{i,j}$, $a_{ij} \in \mathcal{D}$ such that $\Sigma\|a_{ij}\|^2<\infty$. If $\square$ and $(|)$ denote the involution and inner product which endow $\mathcal{D}$ with an H^*-algebra structure (see [6]), then $M_\Lambda(\mathcal{D})$ is an H^*-algebra with the involution and inner product

$$(m_{ij}) \longrightarrow (m_{ji}^{\square})$$

$$((m_{ij})|(n_{ij}))=\textstyle\sum_{ij} (m_{ij}|n_{ij})$$

We shall denote by $M_\Lambda(\mathcal{D})^+$ the H^*-algebra with the same underlying Hilbert space as $M_\Lambda(\mathcal{D})$ and whose product is $(x,y)\longrightarrow(xy+yx)/2$, its involution and inner product being the same of $M_\Lambda(\mathcal{D})$. If t is an antiautomorphism in $\mathcal{D}$ and M an $\Lambda\times\Lambda$ matrix such that $M P M^{-1}\in M_\Lambda(\mathcal{D})$ for all $P \in M_\Lambda(\mathcal{D})$, then $H_\Lambda(\mathcal{D},t_M)$ denotes the subset of $M_\Lambda(\mathcal{D})$ of symmetric matrices relative to $t_M(a_{ij}) = M\ (t(a_{ji}))M^{-1}$ and $K_\Lambda(\mathcal{D},t_M)$ the subset of skew-symmetric matrices of $M_\Lambda(\mathcal{D})$ relative to t_M (see [6]). Whenever $M = Id$, the notations $H_\Lambda(\mathcal{D},t_M)$ and $K_\Lambda(\mathcal{D},t_M)$ will take the easier form $H_\Lambda(\mathcal{D},t)$ and $K_\Lambda(\mathcal{D},t)$. In case that $(MPM^{-1})^*=MP^*M^{-1}$ for all P in $M_\Lambda(\mathcal{D})$ then $H_\Lambda(D,t_M)$ is an H^*-subalgebra of $M_\Lambda(\mathcal{D})^+$.

2. The structure theory for complex and real Jordan H^*-algebras has been completely achieved by [7] and [6]. Following [7] (Proposition 1) it is easy to prove that every two-graded H^*-algebra A with continuous involution splits into the orthogonal direct sum $A = Ann(A)\perp \overline{L(A^2)}$, where

$Ann(A) := \{x \in A : xA = Ax = 0\}$, and $\overline{L(A^2)}$ is the closure of the vector span of A^2, which turns out to be a two-graded H^*-algebra with zero anhililator. Moreover, each two-graded H^*-algebra A with zero anhililator satisfies $A = \overline{\perp_\alpha I_\alpha}$ where $\{I_\alpha\}_\alpha$ denotes the family of (two-graded) minimal closed ideals of A, each of them being a topologically simple two-graded H^*-algebra. This reduces the study of these two-graded algebras to the study of the topologically simple ones. As in [2, proof of Theorem 1], we have two possibilities for one of these two-graded algebras A : (a) A is isomorphic to an orthogonal direct sum $B \perp B$ where B is a topologically simple H^*-algebra with $A_0 = B \perp \{0\}$, $A_1 = \{0\} \perp B$, and the product and involution are given by

$$(a_0, a_1)(b_0, b_1) = (a_0b_0 + a_1b_1, a_0b_1 + a_1b_0)$$
$$(a_0, a_1)^* = (a_0^*, a_1^*),$$

(b) A is topologically simple as ungraded H^*-algebra, that is to say $\mathcal{U}n(A)$ is topologically simple. If A is a nonassociative H^*-algebra, then $g(x_0+x_1) = x_0 - x_1$ defines an isometry $g \in Aut(A)$, commuting with the involution $*$, satisfying $g^2 = Id$ and such that $A_0 = Sym(A, g)$ and $A_1 = Skw(A, g)$. Reciprocally every $*$-preserving involutive automorphism induces a grading on any H^*-algebra. So the problem of the classification of topologically simple two-graded Jordan H^*-algebras reduces to the determination of these automorphism for topologically simple Jordan algebras. These automorphism may be obtained from [5] in the complex case. Following this as well as [7, Theorem 4], it is easy to give the following classification of couples (J, g) of complex topologically simple Jordan H^*-algebras and $*$-preserving involutive isometric automorphisms modulo the relation $(J_1, g_1) \approx (J_2, g_2)$ iff there is a $*$-preserving isomorphism $\varphi : J_1 \longrightarrow J_2$ such that $g_2 \varphi = \varphi g_1$:

(I) $J = M_\Lambda(\mathbb{C})^+$, card$(\Lambda) \neq 2$.

I.1 $g(M) = D.M.D$ where $D = diag(\varepsilon_i)_i$ and $\varepsilon_i \in \{1, -1\}$.

I.2 $g(M) = M^t$ (matrix transposition).

I.3 $g(M) = S\,M^t S^{-1}$ with $S = diag(S_i)_i$, each S_i being the 2x2 matrix

$$S_i = \begin{pmatrix} 0 & 1 \\ -1 & 0 \end{pmatrix}$$

and card(Λ) even if Λ is finite. S will be called in the sequel the $\Lambda \times \Lambda$ symplectic matrix, and g the symplectic involution.

(II) $J = \mathbb{C}.1 \oplus W$ the quadratic Jordan H^*-algebra associated with a complex Hilbert space W of dimension greater than 1 with isometric involutive conjugate-linear map $\square$ (see [7, Theorem 4]), and $g : J \longrightarrow J$ where $g|_{\mathbb{C}.1} = \mathrm{Id}$ and $g: W \longrightarrow W$ is an involutive isometry relative to the nondegenerate form on W, that is $W = W_0 \perp W_1$ and $g|_{W_0} = \mathrm{Id}$, $g|_{W_1} = -\mathrm{Id}$.

(III.a) $J = H_\Lambda(\mathbb{C}, \mathrm{Id})$ (symmetric matrices), $\mathrm{card}(\Lambda) \geq 3$.

III.a.1 $g(M) = S\, M\, S^{-1}$ the restriction of the symplectic involution to $H_\Lambda(\mathbb{C}, \mathrm{Id})$.

III.a.2 $g(M) = D.M.D$ with $D = \mathrm{diag}(\varepsilon_i)_i$ and $\varepsilon_i \in \{1, -1\}$.

(III.b) $J = H_\Lambda(\mathbb{C}, \mathrm{Id}_S)$ with S = symplectic and Id_S as defined in paragraph 1, $\mathrm{card}(\Lambda) \geq 6$ and if Λ is finite then $\mathrm{card}(\Lambda)$ is even.

III.b.1 $g(M) = M^t$ the restriction of transposition to $H_\Lambda(\mathbb{C}, \mathrm{Id}_S)$.

III.b.2 $g(M) = D.M.D$ with $D = \mathrm{diag}(\varepsilon_i)_i$ and $\varepsilon_i \in \{1, -1\}$.

(IV) $J = H_3(\mathbb{O}_C, -)$ with $-$ the Cayley antiautomorphism of $\mathbb{O}_C$.

IV.1 $g(M) = D.M.D$ with $D = \mathrm{diag}(1,-1,-1)$ or $D = \mathrm{Id}$.

IV.2 $g[(m_{ij})_{i,j}] = (\alpha(m_{ij}))_{i,j}$ where $\alpha \in Aut(\mathbb{O}_C)$ is described in a standard basis $\{e_1, \ldots, e_8\}$ of $\mathbb{O}_C$ as

$$\alpha\left(\Sigma_{i=1}^{8} \lambda_i e_i\right) = \Sigma_{i=1}^{4} \lambda_i e_i - \Sigma_{i=5}^{8} \lambda_i e_i .$$

As a consequence, using the notation of paragraph 1, we have the result :

Theorem 1. *The only complex topologically simple two-graded J^*-algebras are up to a positive factor of the inner product the following ones :*

1) $J \perp J$ with J a complex topologically simple J^*-algebra, the product and grading as in the paragraph 2 and involution $(a_0, a_1)^* = (a_0^*, a_1^*)$.

2) $A = M_\Lambda(\mathbb{C})^+$, $\mathrm{card}(\Lambda) \neq 2$, $A_0 = H_\Lambda(\mathbb{C}, Id)$, $A_1 = K_\Lambda(\mathbb{C}, Id)$.

3) $A = M_\Lambda(\mathbb{C})^+$, $\mathrm{card}(\Lambda) > 2$, $A_0 = \{M\colon SM^t = MS\}$, $A_1 = \{M\colon SM^t = -MS\}$, S symplectic and $\mathrm{card}(\Lambda)$ even if Λ is finite.

4) $A = M_{\Lambda \dot\cup B}(\mathbb{C})^+$, $\mathrm{card}(\Lambda \dot\cup B) > 2$, $A_0 = \begin{pmatrix} M_\Lambda(\mathbb{C}) & 0 \\ 0 & M_B(\mathbb{C}) \end{pmatrix}$, $A_1 = \begin{pmatrix} 0 & M_{\Lambda,B}(\mathbb{C}) \\ M_{B,\Lambda}(\mathbb{C}) & 0 \end{pmatrix}$.

5) $A = \mathbb{C}.1 \oplus W$ the quadratic Jordan H^*-algebra associated with a complex Hilbert space W of dimension greater than 1 with isometric involutive conjugate linear map $\Box$, $A_0 = \mathbb{C}.1 \oplus W_0$, $A_1 = W_1$ for any splitting of W into orthogonal subspaces $W = W_0 \perp W_1$, with $W_i^{\Box} \subset W_i$ $(i \in \{0,1\})$.

6) $A = H_{\Lambda}(\mathbb{C}, Id)$, card$(\Lambda) \geq 4$ and if Λ is finite then card(Λ) is even, $A_0 = H_{\Lambda}(\mathbb{C}, Id) \cap H_{\Lambda}(\mathbb{C}, Id_S)$, S = symplectic. $A_1 = H_{\Lambda}(\mathbb{C}, Id) \cap K_{\Lambda}(\mathbb{C}, Id_S)$.

7) $A = H_{\Lambda \cup B}(\mathbb{C}, Id)$, card$(\Lambda \cup B) \geq 3$ as graded subalgebra of $M_{\Lambda \cup B}(\mathbb{C})^+$.

8) $A = H_{\Lambda}(\mathbb{C}, Id_S)$, card$(\Lambda) \geq 6$, and if Λ is finite then card(Λ) is even, S = symplectic, $A_0 = H_{\Lambda}(\mathbb{C}, Id_S) \cap H_{\Lambda}(\mathbb{C}, Id)$, $A_1 = H_{\Lambda}(\mathbb{C}, Id_S) \cap K_{\Lambda}(\mathbb{C}, Id)$.

9) $A = H_{\Lambda \cup B}(\mathbb{C}, Id_S)$, S symplectic and card$(\Lambda \cup B) \geq 6$, regarded as graded subalgebra of $M_{\Lambda \cup B}(\mathbb{C})^+$. If $\Lambda \cup B$ is finite then card$(\Lambda \cup B)$ is even.

10) $A = H_3(\mathbb{O}_C, -)$, $A_0 = \{ (m_{ij}) \in A : m_{12}=m_{21}=m_{13}=m_{31}=0 \}$, and
$A_1 = \{ (m_{ij}) \in A : m_{ii}=0, m_{23}=0, i=1,2,3 \}$

11) $A = H_3(\mathbb{O}_C, -)$ with grading
$A_0 = \{(m_{ij}) : m_{ij} \in \mathbb{C}\langle e_1, .., e_4\rangle\}$, $A_1 = \{(m_{ij}) : m_{ij} \in \mathbb{C}\langle e_5, .., e_8\rangle\}$
being $\{e_i\}_{i=1}^{8}$ a standard basis of $\mathbb{O}_C$.

12) $A = H_3(\mathbb{O}_C, -)$ with trivial grading $A_0 = A$, $A_1 = 0$.

3. We now give the corresponding results for topologically simple real Jordan H^*-algebras. These have been determined in [6, Theorem 3] and fall into two classes : a) the ones which are complex topologically simple regarded as reals, and b) the purely real ones. The automorphism from which the grading comes are in the first case easy to determine because they are the given ones in 2 or these composed with the involution of the H^*-algebra (see [2, lemma 1]). So we have

Theorem 2. *Let J be a real two-graded topologically simple J^*-algebra which is a complex topologically simple J^*-algebra. Then up to positive factor of the inner product, J is *-isometrically isomorphic to one of the following*

1) *Any of the cases 2) - 12) of the clasification for two-graded topologically simple complex J^*-algebras regarded as real H^*-algebras.*

2) *A complex quadratic J^*-algebra $J = \mathbb{C}.1 \perp W$ associated with a complex Hilbert space W of dimension greater than 1 with isometric involutive conjugate linear map $\Box$, and grading*

$$J_0 = \mathbb{R} \perp Sym(W_0,\square)\perp Skw(W_1,\square)$$
$$J_1 = i\mathbb{R} \perp Skw(W_0,\square)\perp Sym(W_1,\square)$$

for any splitting of W *into orthogonal closed subspaces such that* $W_\alpha^\square \subset W_\alpha$ *for* $\alpha \in \{0,1\}$.

3) $J = M_{A\cup B}(\mathbb{C})^+$, $card(A)\geq 3$

$$J_0 = \left\{ \begin{pmatrix} a & b \\ -b^* & d \end{pmatrix} : a \in H_A(\mathbb{C},-),\ d \in H_B(\mathbb{C},-),\ b \in M_{A,B}(\mathbb{C}) \right\}$$

$$J_1 = \left\{ \begin{pmatrix} a & b \\ b^* & d \end{pmatrix} : a \in K_A(\mathbb{C},-),\ d \in K_B(\mathbb{C},-),\ b \in M_{A,B}(\mathbb{C}) \right\}$$

4) $J = M_A(\mathbb{C})^+$, $card(A)\neq 2$, $J_0 = M_A(\mathbb{R})^+$, $J_1 = i\, M_A(\mathbb{R})^+$.

5) $J = M_{A\cup A}(\mathbb{C})^+, card(A)\geq 2$

$$J_0 = \left\{ \begin{pmatrix} a & b \\ -\bar{b} & \bar{a} \end{pmatrix} : a,b \in M_A(\mathbb{C})\right\},\quad J_1 = \left\{ \begin{pmatrix} a & b \\ \bar{b} & -\bar{a} \end{pmatrix} : a,b \in M_A(\mathbb{C}) \right\}$$

6) $J = H_{A\cup A}(\mathbb{C},Id)$, $card(A)\geq 2$ as graded subalgebra of 5) or 3) with $A = B$.

7) $J = H_A(\mathbb{C},Id_S)$, $card(A)\geq 6$, S symplectic, as graded subalgebra of 4) and with $card(A)$ being even if A is finite.

8) $J = H_{A\cup B}(\mathbb{H}_C,s)$, $card(A\cup B)\geq 3$,

$$J_0 = J \cap \left\{ \begin{pmatrix} a & b \\ -b^* & d \end{pmatrix}: a\in M_A(\mathbb{H}_C),\ d\in M_B(\mathbb{H}_C),\ b\in M_{A,B}(\mathbb{H}_C),\ a^*=a,\ b^*=b \right\}$$

$$J_1 = J \cap \left\{ \begin{pmatrix} a & b \\ b^* & d \end{pmatrix}: a\in M_A(\mathbb{H}_C),\ d\in M_B(\mathbb{H}_C),\ b\in M_{A,B}(\mathbb{H}_C),\ a^*=-a,\ b^*=-b \right\}$$

9) $J = H_3(\mathbb{O}_C,-)$, with

$J_0 = \{(m_{pq}) \in J : m_{pp}, m_{23}, m_{32} \in \mathbb{R}e_1 + i\mathbb{R}\langle e_2,\dots,e_8\rangle,$
$m_{12}, m_{13} \in i\mathbb{R}e_1 + \mathbb{R}\langle e_2,\dots,e_8\rangle\}$

$J_1 = \{(m_{pq}) \in J : m_{pp}, m_{23}, m_{32} \in i\mathbb{R}e_1 + \mathbb{R}\langle e_2,\dots,e_8\rangle,$
$m_{12}, m_{13} \in \mathbb{R}e_1 + i\mathbb{R}\langle e_2,\dots,e_8\rangle\}$

with $\{e_p\}_{p=1}^8$ being a standard basis of $\mathbb{O}_C$.

10) $J = H_3(\mathbb{O}_C,-)$,

$$J_0 = \{(m_{pq}) \in J : m_{pq} \in \mathbb{R}\langle e_1,\dots,e_4\rangle + i\mathbb{R}\langle e_5,\dots,e_8\rangle\}$$
$$J_1 = \{(m_{pq}) \in J : m_{pq} \in i\mathbb{R}\langle e_1,\dots,e_4\rangle + \mathbb{R}\langle e_5,\dots,e_8\rangle\}$$

with $\{e_i\}_{i=1}^8$ being a standard basis of $\mathbb{O}_C$.

11) $J = H_3(\mathbb{O}_C,-)$, $J_0 = \{(m_{pq}) \in J : m_{pp} \in \mathbb{R}e_1,\ m_{pq} \in \mathbb{R}\langle e_1,\dots,e_8\rangle,\ p\neq q \}$
$J_1 = \{(m_{pq}) \in J : m_{pp} \in i\mathbb{R}e_1,\ m_{pq} \in i\mathbb{R}\langle e_1,\dots,e_8\rangle,\ p\neq q \}$.

4. Now we have to determine the automorphisms in the purely real case. If J is quadratic the grading automorphism g is similar to the one given for complex algebras. If J is excepcional then the grading automorphism may be obtained from [9, p.387 and Theorem 9, p.388]. If $J=A^+$ for any associative topologically simple real H^*-algebras, then by [11] A is essentially $M_A(\mathcal{D})$ with $\mathcal{D}=\mathbb{R},\mathbb{C}$ or $\mathbb{H}$, and the *-preserving involutive automorphism g of J has to be either an automorphism or an antiautomorphism of A (see [8]). The following proposition is a consequence of [10, Isomorphism Theorem] and gives an account of the couples (A,g) formed by a real associative topologically simple H^*-algebra A, and an involutive *-preserving automorphism g, modulo the relation $(A,g)\approx(A',g')$ iff there is a *-preserving isomorphism φ from A to A' such that $g'\circ\varphi=\varphi\circ g$.

Proposition 1. *Let $\mathcal{D}$ be one of the real associative division algebras $\mathbb{R},\mathbb{C}$ or $\mathbb{H}$ and $M_A(\mathcal{D})$ the real associative H^*-algebra given in 1 . Let $g\in Aut(M_A(\mathcal{D}))$ be an involutive *-preserving isometry. Then modulo the relation $\approx$, g has the following possibilities*

1) *$g(T)=D\,T\,D^{-1}$ for all $T\in M_A(\mathcal{D})$, with $D=diag(\varepsilon_i)_i$, $\varepsilon_i\in\{-1,1\}$.*
2) *$g[(a_{ij})_{i,j}]=(s(a_{ij}))_{i,j}$ for all $(a_{ij})_{i,j}$ with $s\in Aut(\mathcal{D})$, $s^2=Id$ $s\neq Id$ and commuting with the Cayley antiautomorphism of $\mathcal{D}$.*
3) *$g[(a_{ij})_{ij}]=S\,(\bar{a}_{ij})_{ij}\,S^{-1}$ for all $(a_{ij})_{i,j}$ being $\mathcal{D}\neq\mathbb{H}$, and where S is the symplectic matrix and - the Cayley antiautomorphism of $\mathcal{D}$.*

Corollary 1. *Under the same conditions as before but g being an anti-automorphism rather than an automorphism , then modulo a relation as $\approx$, g is :*

1) *$g\,[(a_{ij})_{i,j}]=D\,(\bar{a}_{ji})_{i,j}\,D^{-1}$ for all $(a_{ij})_{i,j}$ with $D=diag(\varepsilon_i)_i$ and $\varepsilon_i\in\{-1,1\}$.*
2) *$g[(a_{ij})_{i,j}]=(s(\bar{a}_{ji}))_{i,j}$ for all $(a_{ij})_{i,j}$ with $s\in Aut(\mathcal{D})$, $s^2=Id$, $s\neq Id$ and commuting with the Cayley antiautomorphism of $\mathcal{D}$.*
3) *$g[(a_{ij})_{ij}]=S\,(a_{ji})_{ij}\,S^{-1}$ for all $(a_{ij})_{i,j}$ being $\mathcal{D}\neq\mathbb{H}$ and where S is the symplectic matrix and - the Cayley antiautomorphism of $\mathcal{D}$.*

5. If $J=H_A(\mathcal{D},t_M)$ with $\mathcal{D}=\mathbb{R},\mathbb{C}$ or $\mathbb{H}$, then the automorphism g may be found by extending it to $M_A(D,t_M)$, applying the Isomorphism Theorem (see [10]) and doing a suitable change of orthonormal basis. Summarizing the results in paragraphs 4 and 5 we have

Theorem 3. *Let J be a real two-graded topologically simple J^*-algebra which is not a complex topologically simple J^*-algebra. Then up to positive factor of the inner product, J is $*$-isometrically isomorphic to one of the following*

1) $J = A \perp A$ being A a real topologically simple J^*-algebra, $J_0 = A \perp \{0\}$, $J_1 = \{0\} \perp A$, the product $(a_0,a_1)(b_0,b_1) = (a_0b_0+a_1b_1, a_0b_1+a_1b_0)$ and the involution $(a_0,a_1)^* = (a_0^*, a_1^*)$.

2) $J = \mathbb{R}.1 \perp W$ a quadratic Jordan H^*-algebra associated with a real Hilbert space W of dimension greater than 1, endowed with an isometric involutive linear mapping $\square$ and grading $J_0 = \mathbb{R}.1 \perp W_0$, $J_1 = W_1$ for any splitting of W into orthogonal subspaces W_i with $W_i^{\square} \subset W_i$ and $i \in \{0,1\}$.

3) $J = M_{A\cup B}(\mathcal{D})^+$, $\mathcal{D} = \mathbb{R}$ or $\mathbb{H}$, card$(A\cup B)\geq 2$ if $\mathcal{D}=\mathbb{H}$ and card$(A\cup B)\neq 2$ if $\mathcal{D} = \mathbb{R}$.

$$J_0 = \begin{pmatrix} M_A(\mathcal{D}) & 0 \\ 0 & M_B(\mathcal{D}) \end{pmatrix}^+, \quad J_1 = \begin{pmatrix} 0 & M_{A,B}(\mathcal{D}) \\ M_{B,A}(D) & 0 \end{pmatrix}^+$$

4) $J = M_{A\cup B}(\mathcal{D})^+$; $\mathcal{D} = \mathbb{R}$ or $\mathbb{H}$, card$(A\cup B)\geq 2$ if $\mathcal{D}=\mathbb{H}$ and card$(A\cup B)>2$ if $\mathcal{D}=\mathbb{R}$

$$J_0 = \left\{ \begin{pmatrix} a & b \\ -b^* & c \end{pmatrix} : a^* = a, c^* = c \right\}, \quad J_1 = \left\{ \begin{pmatrix} a & b \\ b^* & c \end{pmatrix} : a^* = -a, \ c^* = -c \right\}$$

5) $J = M_A(\mathbb{H})^+$, card$(A) \geq 2$.

$$J_\alpha = \{ (m_{ij}) \in J : \nu(m_{ij}) = (1-2\alpha)\, m_{ij} \}$$

with $\alpha \in \{0,1\}$ and $\nu \in Aut(\mathbb{H})$, $\nu^2 = Id$.

6) $J = M_A(\mathbb{H})^+$, with $J_\alpha = \{ (m_{ij}) \in J : \sigma(m_{ij}) = (1-2\alpha)\, m_{ji} \}$, $\alpha \in \{0,1\}$ and σ being an involutive antiautomorphism which is not the Cayley one.

7) $J = H_{A\cup B}(\mathbb{H},\sigma)$, σ as in 6), J as graded subalgebra of 3), 4) or 5) and card$(A\cup B)\geq 2$.

8) $J = H_{A\cup B}(\mathcal{D}, s_M)$ with $\mathcal{D} = \mathbb{R}, \mathbb{C}, \mathbb{H}, \mathbb{H}_{sp}$, $\mathbb{O}$ or $\mathbb{O}_{sp}$, s is the Cayley antiautomorphism in $\mathcal{D}$, $M = diag(\varepsilon_i)$, $\varepsilon_i = \pm 1$ (the identity if $\mathcal{D}$ is split), card$(A\cup B)\geq 3$, dim$(\mathcal{D}) \neq 8$ if card$(A\cup B) > 3$ and

$$J_0 = J \cap \begin{pmatrix} M_A(\mathcal{D}) & 0 \\ 0 & M_B(\mathcal{D}) \end{pmatrix}, \quad J_1 = J \cap \begin{pmatrix} 0 & M_{A,B}(\mathcal{D}) \\ M_{B,A}(D) & 0 \end{pmatrix}$$

9) $J = H_{A\cup A}(\mathcal{D}, s_M)$, $\mathcal{D} = \mathbb{R}, \mathbb{C}$ or $\mathbb{H}$, M and s as before and

$$J_0 = J \cap \left\{ \begin{pmatrix} a & b \\ \varepsilon b^{\vee} & a^{\vee} \end{pmatrix} : a, b \in M_A(\mathcal{D}) \right\}, \quad J_1 = J \cap \left\{ \begin{pmatrix} a & b \\ -\varepsilon b^{\vee} & -a^{\vee} \end{pmatrix} : a, b \in M_A(\mathcal{D}) \right\}$$

with ε and $\vee$ having the following possibilities :

$(\varepsilon=1, \nu=Id)$, $(\varepsilon=1, \nu=\blacksquare, \mathcal{D} \neq \mathbb{H})$, $(\varepsilon=-1, \nu=\blacksquare, \mathcal{D} \neq \mathbb{H})$ or $(\varepsilon=-1, \nu=Id, \mathcal{D}\neq\mathbb{C})$ where $\square$ is an involutive automorphism of $\mathcal{D}$, $\square \neq Id$ and $(m_{pq})^{\blacksquare}$ denoting the matrix $(m_{pq}{}^{\square})$.

10) $J = H_{\Lambda}(\mathcal{D}, s_M)$, $\mathcal{D}$, s and M as in 8), and

$$J_\alpha = \{ (m_{ij})\in J : \nu(m_{ij}) = (1-2\alpha)\, m_{ij}\}$$

with $\alpha = 0,1$ and $\nu \in Aut(\mathbb{H})$, $\nu^2=Id$.

11) $J = M_{\Lambda\cup\Lambda}(\mathbb{R})^{+}$, $card(\Lambda)\geq 2$ with grading

$$J_0 = \left\{ \begin{pmatrix} a & b \\ -b & a \end{pmatrix} : a,b\in M_{\Lambda}(\mathbb{R})\right\},\quad J_1 = \left\{ \begin{pmatrix} a & b \\ b & -a \end{pmatrix} : a,b \in M_{\Lambda}(\mathbb{R})\right\}$$

12) J as before but with grading

$$J_0 = \left\{ \begin{pmatrix} a & b \\ c & a^t \end{pmatrix} : b^t=-b,\ c^t=-c\right\},\quad J_1 = \left\{ \begin{pmatrix} a & b \\ c & -a^t \end{pmatrix} : b^t=b,\ c^t= c \right\}$$

13) $J = H_{\Lambda}(\mathbb{H}_{sp}, s)$, with s = Cayley antiautomorphism of $\mathbb{H}_{sp}$ and $card(\Lambda)\geq 3$, $D=diag(\varepsilon_i)$ with $\varepsilon_i=\pm 1$ and grading

$$J_0=J \cap H_{\Lambda}(\mathbb{H}_{sp}, \sigma_D) \text{ and } J_1=J \cap K_{\Lambda}(\mathbb{H}_{sp}, \sigma_D),$$

with σ an involutive antiautomorphism of $\mathbb{H}_{sp}$ which is not the Cayley one.

14) $J = H_{\Lambda}(\mathcal{D}, s)$, with $card(\Lambda)\geq 4$ and $card(\Lambda)$ even if Λ is finite, $\mathcal{D} = \mathbb{H}$ or $\mathbb{H}_{sp}$, s is the Cayley antiautomorphism of $\mathcal{D}$, σ an involutive antiautomorphism which is not the Cayley one and grading

$$J_0=J \cap H_{\Lambda}(\mathcal{D}, \sigma_S) \text{ and } J_1=J \cap K_{\Lambda}(\mathcal{D}, \sigma_S).$$

where S is the symplectic matrix.

15) $J = H_{\Lambda\cup\Lambda}(\mathbb{C}, s_D)$ with $card(\Lambda\cup\Lambda)\geq 4$, s the Cayley antiautomorphism of $\mathbb{C}$, $D = diag\ \{\varepsilon_p : p \in \Lambda\}$ with $\varepsilon_p = \pm \begin{pmatrix} 1 & 0 \\ 0 & 1 \end{pmatrix}$ and the grading

$$J_o = H_{\Lambda}(\mathbb{C}, s_D) \cap H_{\Lambda}(\mathbb{C}, Id_Q),\quad J_1 = H_{\Lambda}(\mathbb{C}, s_D) \cap K_{\Lambda}(\mathbb{C}, Id_Q),$$

where Q is the $\Lambda\times\Lambda$-matrix $Q = SD$ and S is the symplectic one.

References

1 W. Ambrose : *Structure theorem for a special class of Banach algebras.* Trans. Amer. Math. Soc. 57 (1945) 364-386.

2 M. Cabrera, J.Martinez y A.Rodriguez : *Nonassociative real H^*-algebras.* Publicacions Matemàtiques, 32, (1988), 267-274

3 A. Castellón, J.A. Cuenca and C. Martín: *H^*-algebras ternarias y H^*-algebras asociativas dos graduadas.* Primeras Jornadas Hispano - Marroquies de Matemáticas.

4 J.A. Cuenca, A.García and C. Martín : *Prime two-graded algebras with non zero socle.* Preprint.

5 J.A. Cuenca and A. Sanchez : *Conjugate linear automorphisms in complex topologically simple Jordan H^*-algebras.* Preprint.

6 J.A. Cuenca and A. Sanchez : *Structure theory for real noncommutative Jordan H^*-algebras.* Preprint.

7 J.A. Cuenca and A. Rodriguez : *Structure theory for noncommutative Jordan H^*-algebras.* Journal of Algebra 106 (1987) 1-14.

8 I.N. Herstein : *Jordan homomorphisms.* Trans. Amer. Math. Soc. 81 (1956) 331-341

9 N. Jacobson : *Structure and Representations of Jordan Algebras.* Amer. Math. Soc. Coll. Publ., vol 39. Amer. Math. Soc., Providence R.I 1968.

10 N. Jacobson : *Structure of rings.* Coll. Publ. vol 37. Amer. Math. Soc. second edition.

11 I.Kaplansky : *Dual rings*, Ann. of Math 49 no.3 (1948) 698-701.

12 C.T.C Wall : *Graded Brauer Groups.* J. reine angew. Math. Vol 213 1963, 187-199.

Nonassociative
Algebraic
Models

ISBN 1-56072-050-6

Zygotic Algebra for K-Linked Multialelic Loci With Sexually Different Recombination Rates

Ana Isabel Durand-Alegría, Jesús López-Sánchez, and Alberto Pérez de Vargas*
Departamento de Matemática Aplicada II,
Universidad Complutense de Madrid, 28040 Madrid, Spain

1. Introduction

If biological populations with one or several genetic traits are characterized by probabilistic vectors where the components are the frequencies of these traits, the convex combinations are formal populations where the addition and the multiplication with constants can be interpreted. The theory of genetic algebras is introduced by Etherington [3] in order to establish a mathematical formulation of the Mendel's laws. Subsequently the Etherington's basic algebraic model has been applied to more and more bilogical occurrences.

For an exposition on algebras in genetics as well as an extensive collection of papers see Wörz-Busekros [10].

---

*Partially supported by DGA

An algebra $\mathcal{A}$ is called *genetic* -Gonshor [5]- if it admits a basis, called a *canonical basis*, $\{c_i\}_{0\leq i\leq n}$ with constant coefficients σ_{ijk} defined by:

$$c_i c_j = \sum_{k=0}^{n} \sigma_{ijk} \, c_k$$

which have the following properties:

$$\sigma_{000} = 1$$
$$\sigma_{0jk} = 0 \text{ if } k < j$$
$$\sigma_{ijk} = 0 \text{ if } k \leq \max\{i,j\}, \text{ for } i,j > 0.$$

The multiplicative constants $\sigma_{000}, \sigma_{011}, \ldots, \sigma_{0nn}$ are called the *train roots* of $\mathcal{A}$. A *baric algebra* $\mathcal{A}$ is an algebra with a non-trivial algebra homomorphism ω over its underlying field, which is called the *weigth* of $\mathcal{A}$.

A baric algebra $\mathcal{A}$ with a weight ω is called a *special train algebra* iff ker ω is nilpotent and the *principal powers* of ker ω defined by:

$$(\ker \omega)^1 := \ker \omega$$
$$(\ker \omega)^r := < xy;\ x \in \ker \omega\ ,\ y \in (\ker \omega)^{r-1} >;\ r > 1,\ r \in \mathbb{N}$$

are ideals of $\mathcal{A}$.

If $\mathcal{U}$ is an ideal of $\mathcal{A}$ and $\mathcal{U}^{<n>}$ is the set of all linear combinations of products of the $\mathcal{A}$ elements with at least n factors belonging to $\mathcal{U}$, then $\mathcal{U}^{<n>}$ is an ideal of $\mathcal{A}$. Gonshor [6] showed that a baric algebra is genetic iff there is an $r \in \mathbb{N}$, $r \geq 2$, such that $(\mathrm{Ker}\ \omega)^{<r>} = <0>$.

Every special train algebra is genetic (Etherington [4]).

Algebras for k-linked loci have been studied by Reiersøl [10], Holgate [8] and Heuch [7]. In a recent paper [1] Campos, Machado and Holgate have studied the theory of sex dependent genetic recombination rates for an arbitrarily large number of loci and two alleles per locus. Also in [2] we have described a model for two linked loci when considering sex dependent recombinations and mutation rates.

In this paper, we handle populations of diploid organisms which differ in k linked loci with an arbitrary but finite number of alleles and different recombination rates in males and females. The model here constructed is a generalization of the López-Sánchez and Pérez de Vargas model [9].

2. The Algebra $\mathcal{Z}^k$

As above mentioned, populations of diploid individuals which differ in k autosomal linked loci each one with a finite arbitrary number of alleles are to be considered. Let :

$$L_x := \{\ell_x^1, \ell_x^2, \ldots, \ell_x^{r_x}\}\ ,\quad 1 \le x \le k$$

be the k linked loci under consideration, and $\ell(i)$ that gamete whose xth locus has the allele $\ell_x^{i_x}$, where $1 \le i_x \le r_x$ for $1 \le j \le k$ (x is the *localization index* of the allele in the locus) and $\ell(i) := \ell_1^{i_1}\ell_2^{i_2}\ldots\ell_x^{i_x}\ldots\ell_k^{i_k}$. We note $1(k) := 1..\overset{k}{.}..1$.

In the following we assume that in individuals composing populations the meiotic recombination phenomenon can be seen. Each recombination has been described by one out of 2^{k-1} non-ordered partitions of two elements $U_{\cdot} := (U'_{\cdot}, U''_{\cdot}) = (U''_{\cdot}, U'_{\cdot})$, obtained from the set of loci $\mathfrak{L} = \{L_1, \ldots, L_k\}$. This way, the linkage distribution is $\lambda(U)_{U \in W(L)}$ where W(L) is the set of non-ordered partitions and

$$\sum_{U \in W(l)} \lambda(U) = 1.$$

The partition $U_1 = (U'_1, U''_1) = (U''_1, U'_1)$ corresponds to the case in which recombination is not present.

Let us consider the following vector spaces (see [8] for k=2):

$$S := \langle m, f \rangle_{\mathbb{C}} \qquad (m : \text{males} \ , \ f : \text{females})$$

$$G^k := \langle \{\ell(i)\}_{1 \le i_x \le r_x,\ 1 \le x \le k} \rangle_{\mathbb{C}}$$

$$H^k := \langle \{m \otimes \ell(i) \otimes \ell(j) - m \otimes \ell(j) \otimes \ell(i)\}_{1 \le i_x, j_x \le r_x;\ 1 \le x \le k} ,$$

$$\{f \otimes \ell(i) \otimes \ell(j) - f \otimes \ell(j) \otimes \ell(i)\}_{1 \le i_x, j_x \le r_x;\ 1 \le x \le k} \rangle_{\mathbb{C}}$$

$$\mathcal{Z}^k := S \otimes G^k \otimes G^k / H^k.$$

The gametic segregation as originated from the zygote $[\ell(i), \ell(j)]$ when considering only the recombination event is represented by the expression:

$$[\ell(i), \ell(j)]_\lambda = \frac{1}{2} \sum_{\substack{t=1 \\ U_t \in W(L)}}^{2^{k-1}} \lambda(U_t)[\ell(i(U'_t), j(U''_t)) + \ell(i(U''_t), j(U'_t))]$$

where e.g. $\ell(i(U'_t), j(U''_t))$ is that gamete with the i type allele in loci belonging to U'_t and the j type allele in the remaining U''_t loci; on the other hand, λ is the recombination factor ($\lambda := \mu$ for males and $\lambda := \upsilon$ for females).

We shall employ the same notation to denote both elements of $\mathcal{Z}^k$ and elements of $S \otimes G^k \otimes G^k$; so an element of $\mathcal{Z}^k$ is

$$\sum_{\substack{(i_1,\dots,i_k)\le(j_1,\dots,j_k)\\ 1\le i_x,j_x\le r_x;\ 1\le x\le k}} \alpha_{ij}\, m_{ij} \quad + \sum_{\substack{(u_1,\dots,u_k)\le(v_1,\dots,v_k)\\ 1\le u_x,v_x\le r_x;\ 1\le x\le k}} \beta_{uv}\, f_{uv}$$

where the following notation has been used:

$$m_{ij} := m\otimes\ell(i)\otimes\ell(j),\ f_{uv} := f\otimes\ell(u)\otimes\ell(v)$$

and the $(i_1, \dots, i_k) \le (j_1, \dots, j_k)$ ordination means:

$$(i_1, \dots, i_k) = (j_1, \dots, j_k) \Leftrightarrow i_p = j_p,\ \forall\ p \in \{1, \dots,k\}$$

$$(i_1, \dots, i_k) < (j_1, \dots, j_k) \Leftrightarrow \exists\ q \in \{1, \dots,k\}\ /\ i_q < j_q,\ i_n = j_n,\ \forall\ n < q.$$

Theorem 1

$\mathcal{Z}^k$ is an $2\begin{pmatrix} \prod_{x=1}^{k} r_x + 1 \\ 2 \end{pmatrix}$-dimensional commutative algebra over $\mathbb{C}$ but not a genetic algebra.

Proof

Sexual reproduction leads to a multiplication table (m:males, f:females)

$$(*) \qquad m_{ij}\cdot m_{uv} = f_{ij}\cdot f_{uv} = 0,$$

$$m_{ij}\cdot f_{uv} = (m\cdot f)\otimes[\ell(i),\ell(j)]_\mu\otimes[\ell(u),\ell(v)]_\nu = \frac{1}{2}\,(m+f)\otimes$$

$$\otimes\,\frac{1}{2}\left\{\sum_{\substack{t=1\\ U_t\in W(L)}}^{2^{k-1}} \mu(U_t)[\ell(i(U_t'),j(U_t'')) + \ell(i(U_t''),j(U_t'))]\right\}\otimes$$

$$\otimes\,\frac{1}{2}\left\{\sum_{\substack{s=1\\ U_s\in W(L)}}^{2^{k-1}} \nu(U_s)[\ell(u(U_s'),v(U_s'')) + \ell(u(U_s''),v(U_s'))]\right\} =$$

$$= \frac{1}{8} \sum_{t,s=1}^{2^{k-1}} \mu(U_t)\nu(U_s) \Big(m_{i(U'_t)j(U''_t)u(U'_s)v(U''_s)} + m_{i(U'_t)j(U''_t)u(U''_s)v(U'_s)} +$$

$$+\, m_{i(U''_t)j(U'_t)u(U'_s)v(U''_s)} + m_{i(U''_t)j(U'_t)u(U''_s)v(U'_s)} +$$

$$+\, f_{i(U'_t)j(U''_t)u(U'_s)v(U''_s)} + f_{i(U'_t)j(U''_t)u(U''_s)v(U'_s)} +$$

$$+\, f_{i(U''_t)j(U'_t)u(U'_s)v(U''_s)} + f_{i(U''_t)j(U'_t)u(U''_s)v(U'_s)} \Big)$$

where $(i_1,\ldots,i_k) \leq (j_1,\ldots,j_k)$, $(u_1,\ldots,u_k) \leq (v_1,\ldots,v_k)$, $1 \leq i_x, j_x, u_x, v_x \leq r_x$, $1 \leq x \leq k$ and $U_t = (U'_t, U''_t)$, $U_s = (U'_s, U''_s)$.

Since by (*), $\omega(m_{ij}^2)=[\omega(m_{ij})]^2=0 \Rightarrow \omega(m_{ij})=0$ (and the same for f_{ij}) does not exist a non-trivial algebra homomorphism of $\mathcal{Z}^k$ over $\mathbb{C}$ then $\mathcal{Z}^k$ is not a baric algebra which, in turn, implies that it is not a genetic algebra.

3. The Algebra $\mathcal{A}^k$

Let us consider the subalgebra of $\mathcal{Z}^k$:

$$\mathcal{A}^k := \left\{ \sum_{\substack{(i_1,\ldots,i_k)\leq(j_1,\ldots,j_k) \\ 1\leq i_x, j_x \leq r_x;\ 1\leq x\leq k}} \alpha_{ij}\, m_{ij} + \sum_{\substack{(u_1,\ldots,u_k)\leq(v_1,\ldots,v_k) \\ 1\leq u_x, v_x \leq r_x;\ 1\leq x\leq k}} \beta_{uv}\, f_{uv} \,, \right.$$

$$\left. \sum_{(i_1,\ldots,i_k)\leq(j_1,\ldots,j_k)} \alpha_{ij} = \sum_{(u_1,\ldots,u_k)\leq(v_1,\ldots,v_k)} \beta_{uv} \right\}.$$

The biological populations are in $\mathcal{A}^k$ because for these,

$$\sum \alpha_{ij} = \sum \beta_{uv} = 1 \quad , \text{ where } \alpha_{ij}, \beta_{uv} \geq 0$$

Let us define the following elements:

$$d^m_{ij} := m_{ij} - m_{11}, \; d^f_{ij} := f_{11} - f_{ij}$$

with

$$(i_1,\dots,i_k) \leq (j_1,\dots,j_k) \neq (1,\dots,1)$$
$$1 \leq i_x, j_x \leq r_x \,, \qquad 1 \leq x \leq k,$$

where

$$m_{11} = m \otimes 1(k) \otimes 1(k) \quad , \quad f_{11} = f \otimes 1(k) \otimes 1(k)$$

A basis for $\mathcal{A}^k$ is

$$\left\{ \{d^m_{ij} \, , \, d^f_{ij}\}_{\substack{(i_1, \dots, i_k) \leq (j_1, \dots, j_k) \neq (1, \dots, 1) \\ 1 \leq i_x, j_x \leq r_x, \, 1 \leq x \leq k}} \, , \, m_{11} + f_{11} \right\}$$

hence the dimension of A^k is $2\begin{pmatrix} \prod_{x=1}^{k} r_x + 1 \\ 2 \end{pmatrix} - 1.$

The weight:

$$\omega : \mathcal{A}^k \longrightarrow \mathbb{C}$$
$$\sum \alpha_{ij} m_{ij} + \sum \beta_{uv} f_{uv} \longmapsto \sum \alpha_{ij} = \sum \beta_{uv}$$

is a non-trivial algebra homomorphism, therefore $\mathcal{A}^k$ is a baric algebra, and

$$\left\{ \{d^m_{ij}\} \; , \; \{d^f_{ij}\} \right\}$$

is a basis of Ker ω , hence

$$\dim \operatorname{Ker} \omega = 2 \left[\left(\frac{\prod_{x=1}^{k} r_x + 1}{2} \right) - 1 \right].$$

Lemma 1

$(\operatorname{Ker} \omega)^2$ is not an ideal of $\mathscr{A}^k$

Proof

If

$$D_{ij} := d^m_{ij} - d^f_{ij} \; , \; (i_1, \ldots, i_k) \le (j_1, \ldots, j_k) \ne (1, \ldots, 1), \quad 1 \le i_x, j_x \le r_x, \; 1 \le x \le k,$$

then

$$d^m_{ij} \cdot d^m_{uv} = d^f_{ij} \cdot d^f_{uv} = 0$$

$$d^m_{ij} \cdot d^f_{uv} = \frac{1}{8} \sum_{1 \le s,t \le 2^{k-1}} \mu(U_s) \upsilon(U_t) \Big(2D_{1(k)i(U'_s)j(U''_s)} + 2D_{1(k)i(U''_s)j(U'_s)} + 2D_{1(k)u(U'_t)v(U''_t)} + 2D_{1(k)u(U''_t)v(U'_t)} - D_{i(U'_s)j(U''_s)u(U'_t)v(U''_t)} - D_{i(U'_s)j(U''_s)u(U''_t)v(U'_t)} - D_{i(U''_s)j(U'_s)u(U'_t)v(U''_t)} - D_{i(U''_s)j(U'_s)u(U''_t)v(U'_t)} \Big)$$

If we note by

$$I_s := \left\{ i(U'_s)j(U''_s) \ , \ i(U''_s)j(U'_s) \right\}$$

$$J_t := \left\{ u(U'_t)v(U''_t) \ , \ u(U''_t)v(U'_t) \right\}$$

$$\Delta^1_{ij} := D_{1i} + D_{1j} - D_{ij}$$

then we have

$$d^m_{ij} \cdot d^f_{uv} = \frac{1}{8} \sum_{1 \leq s, t \leq 2^{k-1}} \mu(U_s)\upsilon(U_t) \sum_{(m,n) \in I_s \times I_t} \Delta^1_{ij}$$

and if we note

$$\mathcal{U}_1 := <\{\Delta^1_{ij}\}_{(1,\dots,1) \neq (i_1,\dots,i_k) \leq (j_1,\dots,j_k)}>$$

we have $(\text{Ker } \omega)^2 = \mathcal{U}_1$ and

$$\dim (\text{Ker } \omega)^2 = \begin{pmatrix} \prod_{x=1}^{k} r_x \\ 2 \end{pmatrix}.$$

Now, let $c_0 := m_{11} + f_{11}$ be an element of $\mathcal{A}^k - (\text{Ker } \omega)^2$; it can be deduced that the product $c_0 \cdot \Delta^1_{ij} \notin (\text{Ker } \omega)^2$ because

$$c_0 D_{ij} = \frac{1}{4} \sum_{1 \leq t \leq 2^{k-1}} [\mu(U_t) + \upsilon(U_t)] \cdot \left(D_{1(k)i(U'_t)j(U''_t)} + \right.$$

$$\left. + D_{1(k)i(U''_t)j(U'_t)} \right)$$

and then

$$c_0 \Delta^1_{ij} = \frac{1}{4} \sum_{1 \leq t \leq 2^{k-1}} [\mu(U_t) + \upsilon(U_t)] \cdot \left(D_{1(k)1(U'_t)i(U''_t)} + \right.$$

$$D_{1(k)1(U''_t)i(U'_t)} + D_{1(k)1(U'_t)j(U''_t)} + D_{1(k)1(U''_t)j(U'_t)} -$$

$$\left. D_{1(k)i(U'_t)j(U''_t)} - D_{1(k)i(U''_t)j(U'_t)} \right) \notin (\mathrm{Ker}\ \omega)^2$$

$$\text{where} \quad (1 \overset{k}{\ldots\ldots} 1) \neq (i_1 \ldots\ldots i_k) \leq (j_1 \ldots\ldots j_k)$$
$$\text{and} \quad 1 \leq i_x, j_x \leq r_x \quad , \quad 1 \leq x \leq k$$

Corollary 1

$\mathcal{A}^k$ is not a special train algebra.

Lemma 2

$(\mathrm{Ker}\ \omega)^{n+1}$ for $1 \leq n \leq k\text{-}1$ are not ideals of $\mathcal{A}^k$.

Proof

We have:

$$\ell(i) := \ell_1^{i_1} \ell_2^{i_2} \ldots \ell_x^{i_x} \ldots \ell_k^{i_k}$$
$$m_{ij} := m \otimes \ell(i) \otimes \ell(j), \; f_{uv} := f \otimes \ell(u) \otimes \ell(v)$$
$$d^m_{ij} := m_{ij} - m_{11}, \; d^f_{ij} := f_{11} - f_{ij}$$
$$D_{ij} := d^m_{ij} - d^f_{ij} \; , \; (i_1, \ldots, i_k) \leq (j_1, \ldots, j_k) \neq (1, \ldots, 1),$$
$$1 \leq i_x, j_x \leq r_x, \; 1 \leq x \leq k,$$
$$\Delta^1_{ij} := D_{1i} + D_{1j} - D_{ij}$$

and we define for an arbitrary index j

$$j[t_1 t_2 t_3 \ldots\ldots t_u] := \begin{cases} j & \text{if there is } v \text{ where } 1 \leq v \leq u \text{ so that} \\ & t_v = x \text{ being } x \text{ the } \textit{localization index} \\ 1 & \text{otherwise} \end{cases}$$

(where $1 \leq t_1 < t_2 < \ldots < t_u \leq k$, $1 \leq u \leq k$).

and

$$j[(t^1_1 t^1_2 \ldots t^1_{u_1}) \ldots\ldots (t^s_1 t^s_1 \ldots t^s_{u_s})] = \begin{cases} j \text{ if there is } v_1,\ldots,v_s \text{ where} \\ \quad 1 \leq v_1 \leq u_1, \ldots, 1 \leq v_s \leq u_s \\ \quad \text{so that } t^1_{v_1} = \ldots\ldots = t^s_{v_s} = x \\ 1 \text{ otherwise} \end{cases}$$

(where $1 \leq t^q_1 < t^q_2 < \ldots\ldots < t^q_{j_q} \leq k$, $1 \leq j_q \leq k$, $1 \leq q \leq s$)

We consider the following elements

$$\Delta^n_{ij} := \Delta^{n-1}_{ij} - \sum_{1 \leq t_1 < \ldots < t_{n-1} \leq k} \Delta^{n-1}_{ij[t_1 \ldots t_{n-1}]}$$

where the matrix $j:(j_1 \ldots j_k)$ has not $k-(n-1)$ elements equals to 1 , $i:(i_1 \ldots\ldots i_k) \neq (1 \ldots\ldots 1)$ $\forall i$ and $\Delta^{k+1}_{ij} = 0$.

We define the subalgebra $\mathcal{U}_n$ (where $1 \leq t_1 < \ldots < t_k \leq k$, $i := (i_1 \ldots i_k) \neq 1(k) := (1 \overset{k}{\ldots\ldots} 1)$, $j := (j_1 \ldots j_k) \neq 1(k)$)

$$\mathcal{U}_n = < \left\{ \Delta^n_{ij[t_1 \ldots t_k]} \right\}_{i \leq j[t_1 \ldots t_k]} ,$$

$$\left\{ \Delta^n_{i[1]j[t_1 \ldots t_k]} \right\}_{(i_1 1 \ldots 1) > j[t_1 \ldots t_k]} ,$$

$$\left\{ \Delta^n_{i[1 s_2]j[t_1 \ldots t_k]} \right\}_{i[1 s_2] > j[t_1 \ldots t_k]} , \ldots\ldots\ldots\ldots\ldots\ldots$$

$$\ldots\ldots\ldots \left\{ \Delta^n_{i[1 s_2 s_3 \ldots s_{n-1}]j[t_1 \ldots t_k]} \right\}_{i[1 s_2 \ldots s_{n-1}] > j[t_1 \ldots t_k]} .$$

$$\left\{ \Delta^n_{ij[t_1 \ldots t_{k-1}]} \right\}_{i \leq j} , \left\{ \Delta^n_{i[s_1]j[t_1 \ldots t_{k-1}]} \right\}_{i > j} \ldots\ldots\ldots,$$

$$\ldots\ldots, \left\{ \Delta^n_{i[s_1 s_2 .. s_{n-1}]j[t_1 .. t_{k-1}]} \right\}_{i>j}, \ldots\ldots,$$

$$\left\{\Delta^n_{ij[t_1..t_n]}\right\}_{i\leq j}, \left\{\Delta^n_{i[s_1]j[t_1..t_n]}\right\}_{i>j}, \ldots\ldots,$$

$$\ldots\ldots, \left\{\Delta^n_{i[s_1..s_{n-1}]j[t_1..t_n]}\right\}_{i>j} >_{\mathbb{C}}$$

We can prove that with the induction hypothesis (Ker $\omega)^{n+1} = \mathcal{U}_n$ it follows that (Ker $\omega)^{n+2} =$ (Ker $\omega)\cdot\mathcal{U}_n = \mathcal{U}_{n+1}$ but the calculations are heavy.

If we define the following elements

$$D^1_{ij} := D_{i(k)j}, \quad D^{n+1}_{ij} := D^n_{i(k)j} - \sum_{1\leq t_1 <..< t_n \leq k} D^n_{i(k)j[t_1..t_n]}$$

(where the matrix $j:(j_1,...j_k)$ has not k-n elements equals to 1), particularly $D^{k+1}_{1j}=0$, and if $c_0 := m_{1(k)1(k)} + f_{1(k)1(k)} \in \mathcal{A}^k$ - (Ker $\omega)^{n+1}$ we have

$$c_0\cdot\Delta^n_{ij} = \frac{1}{4}\sum_{1\leq\alpha\leq 2^{k-1}} [\mu(U_\alpha)+\upsilon(U_\alpha)]\left[\prod_{t=1}^{n-1}\left[1-\binom{k}{t}\right]D_{1(k)i(U_\alpha^{\cdot})1(U_\alpha^{\cdot\prime})} + \right.$$

$$+D^{n+1}_{1(k)1(U_\alpha^{\cdot})j(U_\alpha^{\cdot\prime})} - D^{n+1}_{1(k)i(U_\alpha^{\cdot})j(U_\alpha^{\cdot\prime})} +$$

$$+\sum_{r=1}^{n-1}\sum_{\substack{1\leq t_1^s <..< t_{p_s}^s \leq k \\ 1\leq s\leq r \\ 1\leq p_1 <..< p_r \leq n-1}} (-1)^{r+1} D^{n+1}_{1(k)i(U_\alpha^{\cdot})j[(t_1^1..t_{p_1}^1)..(t_1^r..t_{p_r}^r)](U_\alpha^{\cdot\prime})} +$$

$$+\prod_{t=1}^{n-1}\left[1-\binom{k}{t}\right]D_{1(k)i(U_\alpha^{\cdot\prime})1(U_\alpha^{\cdot})} +$$

$$+D^{n+1}_{1(k)1(U_\alpha^{\cdot\prime})j(U_\alpha^{\cdot})} - D^{n+1}_{1(k)i(U_\alpha^{\cdot})j(U_\alpha^{\cdot\prime})} +$$

$$+\sum_{r=1}^{n-1} \sum_{\substack{1 \le t_1^s < .. < t_{p_s}^s \le k \\ 1 \le s \le r \\ 1 \le p_1 < .. < p_r \le n-1}} (-1)^{r+1} D^{n+1}_{1(k)i(U_{\dot\alpha}')j[(t_1^1..t_{p_1}^1)..(t_1^r..t_{p_r}^r)](U_{\dot\alpha}')} \Bigg]$$

that is not an element of $(\mathrm{Ker}\omega)^{n+1} = \mathcal{U}_n$ $(n \ge 2)$.

The same conclusion follows for n+2 and $c_0 \cdot \Delta^{n+1}_{ij} \notin (\mathrm{Ker}\ \omega)^{n+2}$.

It is sufficient now to apply the Lemma 1.

LEMMA 3

$(\mathrm{Ker}\ \omega)^{k+1} = \mathcal{U}_k$ is an ideal of $\mathcal{A}^k$ and $(\mathrm{Ker}\ \omega)^{k+2} = \mathcal{U}_{k+1} := \langle 0 \rangle_{\mathbb{C}}$

PROOF

Let c_0 be the element $m_{1(k)1(k)} - f_{1(k)1(k)} \in \mathcal{A}^k - (\mathrm{Ker}\ w)^{k+1}$

Since $D^{k+1}_{1(k)j} = 0$ where $2 \le j_x \le r_x$, $1 \le x \le k$ we are not be able to derive from the Lemma 2 that $c_0 \cdot \Delta^k_{ij} = 0$ where $1 \le i_x \le r_x$, $2 \le j_x \le r_x$, $1 \le x \le k$. Moreover $m_{ij} \Delta^k_{uv} = f_{ij} \Delta^k_{uv} = 0$ where $1 \le i_x j_x u_x \le r_x$, $2 \le v_x \le r_x$, $1 \le x \le k$ because $\Delta^{k+1}_{uv} = 0$. Therefore $(\mathrm{Ker}\ \omega)^{k+1} = \mathcal{U}_k$ is an ideal of $\mathcal{A}^k$ and,

$$\dim\ (\mathrm{Ker}\ \omega)^{k+1} = \left(\prod_{x=1}^{k} r_x - 1\right) \left[\prod_{x=1}^{k} (r_x - 1)\right] - \left(\begin{matrix}\prod_{x=1}^{k} (r_x - 1) \\ 2\end{matrix}\right).$$

Is easy to see that $(\mathrm{Ker}\ \omega)^{k+2} = \langle 0 \rangle_{\mathbb{C}}$.

Unfortunaly the fact that Ker ω is a nilpotent element is not sufficient for affirm that $\mathcal{A}^k$ is a genetic algebra.

LEMMA 4

$(\mathrm{Ker}\ \omega)^{\langle k+2 \rangle} = \langle 0 \rangle_{\mathbb{C}}$.

PROOF

This proof is essentially similar to the proof of Lemma 3.5 in [2] but the calculations are more complicated.

We can derive from Lemma 2

$$\mathcal{A}^k\cdot(\mathrm{Ker}\ \omega)^{n+1} = (\mathrm{Ker}\ \omega)^{n+2} \oplus <\{D^{n+1}_{1v}\}>_{\mathbb{C}}\ ,\ 1 \le n \le k\text{-}1.$$

Particularly,

$$\mathcal{A}^k\cdot(\mathrm{Ker}\ \omega)^2 = (\mathrm{Ker}\ \omega)^3 \oplus <\{D^2_{1v}\}>_{\mathbb{C}}\ .$$

Moreover, it can be easily seen

$$(\mathrm{Ker}\ \omega)\cdot[\mathcal{A}^k\cdot(\mathrm{Ker}\ \omega)^2] = (\mathrm{Ker}\ \omega)^3$$

and

$$\mathcal{A}^k\cdot[\mathcal{A}^k\cdot(\mathrm{Ker}\ \omega)^2] = \mathcal{A}^k\cdot(\mathrm{Ker}\ \omega)^2.$$

Therefore ,

$$(\mathrm{Ker}\ \omega)^{<2>} = (\mathrm{Ker}\ \omega)^2 \oplus <\{D^2_{1v}\}>_{\mathbb{C}}$$

hence

$$\dim\ (\mathrm{Ker}\ \omega)^{<2>} = \dim\ (\mathrm{Ker}\ \omega)^2 + \Big[\prod_{x=1}^{k}(r_x-1) +$$

$$+ \sum_{1\le t_1<\dots<t_{k-1}\le k}\ \prod_{x=1}^{k-1}(r_{t_x}-1) + \dots + \sum_{1\le t_1<t_2\le k}\ \prod_{x=1}^{2}(r_{t_x}-1)\Big].$$

By induction,

$$(\mathrm{Ker}\ \omega)^{<n+1>} = (\mathrm{Ker}\ \omega)^{n+1} \oplus <\{D^{n+1}_{1v}\}_{\substack{1\le v_x \le r_x,\ 1\le x\le k,\\ \text{there are not } k-n \text{ elements}\\ \text{equal to one in } (v_1,\dots,v_k)}}>_{\mathbb{C}}$$

hence

$$\dim (\mathrm{Ker}\ \omega)^{<n+1>} = \dim (\mathrm{Ker}\ \omega)^{n+1} + \Big[\prod_{x=1}^{k}(r_x-1) +$$

$$+ \sum_{1\le t_1<\dots<t_{k-1}\le k} \prod_{x=1}^{k-1} (r_{t_x}-1) + \dots + \sum_{1\le t_1<\dots<t_{n+1}\le k} \prod_{x=1}^{n+1} (r_{t_x}-1) \Big].$$

Since $(\mathrm{Ker}\ \omega)^{k+1}$ is an ideal of $\mathcal{A}^k$ and $(\mathrm{Ker}\ \omega)^{k+2} = <0>_{\mathbb{C}}$ we can conclude,

$$(\mathrm{Ker}\ \omega)^{<2>} = (\mathrm{Ker}\ \omega)^{2} \oplus <\{D^{2}_{1(k)v}\}_{\substack{\text{there are not } k-1 \text{ elements}\\ \text{equal to one in } (v_1,\dots,v_k)}}>_{\mathbb{C}}$$

....

....

$$(\mathrm{Ker}\ \omega)^{<n+1>} = (\mathrm{Ker}\ \omega)^{n+1} \oplus <\{D^{n+1}_{1(k)v}\}_{\substack{\text{there are not } k-n \text{ elements}\\ \text{equal to one in } (v_1,\dots,v_k)}}>_{\mathbb{C}}$$

....

....

$$(\mathrm{Ker}\ \omega)^{<k+1>} = (\mathrm{Ker}\ \omega)^{k+1}$$
$$(\mathrm{Ker}\ \omega)^{<k+2>} = <0>_{\mathbb{C}}\ .$$

THEOREM 2

$\mathcal{A}^k$ is a genetic algebra

PROOF

It is an outcome of the Gonshor's Theorem and Lemma 4.

4. FINAL REMARKS

The nilpotence of Ker ω is not a sufficient condition for the algebra $\mathcal{A}^k$ be a genetic one so that the use of the Gonshor's Theorem is needed. $\mathcal{A}^k$ is a good example of an algebra which is genetic but not special train.

A canonical basis and the train roots of $\mathcal{A}^k$ can be obtained through a process of basis completion from $(\text{Ker }\omega)^{<k+1>}$ and by essentially the same ideas of [2], though the calculations are so long, heavy and complicated that we think they are not of sufficient interest to publish here.

The train roots of $\mathcal{A}$ are,

$$\Bigg\{1\ ;\ \tfrac{1}{2},\ \Big(\sum_{x=1}^{k}(r_x-1)\ \text{times}\Big)\ ;\ \Bigg\{\frac{1}{4}\Bigg[\sum_{\substack{1\leq\alpha\leq 2^{k-1}\\ t_1,\ldots,t_n\in U'_\alpha}}\Big(\mu(U_\alpha)+\nu(U_\alpha)\Big)+$$

$$+\sum_{\substack{1\leq\alpha\leq 2^{k-1}\\ t_1,\ldots,t_n\in U''_\alpha}}\Big(\mu(U_\alpha)+\nu(U_\alpha)\Big)\Bigg],\ \Big(\prod_{x=1}^{n}(r_{t_x}-1)\ \text{times}\Big)\Bigg\}_{\substack{1\leq t_1<\ldots<t_n\leq k\\ 2\leq n\leq k-1}}\ ;$$

$$\frac{1}{4}\Big(\mu(U_1)+\nu(U_1)\Big)\ ,\Big(\prod_{m=1}^{k}(r_x-1)\ \text{times}\Big)\ ;\ 0,\ \Big(\prod_{m=1}^{k}r_x^2-1\ \text{times}\Big)\Bigg\}\ .$$

REFERENCES

[1] T.M.M. Campos, S.D.A. Machado and P. Holgate , Sex Dependent Genetic Recombination Rates for Several Loci. Linear Alg. Appl., 136:165-172 (1990)

[2] A.I. Durand-Alegría, J. López-Sánchez and A. Pérez de Vargas, Zygotic Algebra for Two-Linked Loci with Sexually Different Recombination and Mutation Rates , Linear Alg. Appl. 121:385-399 (1989).

[3] I. M. H. Etherington , Genetic Algebras , Proc. Roy. Soc. Edinburgh 59 (1939) 242-258.

[4] I. M. H. Etherington , Special Train Algebras , Quart. J. Math. Oxford, Ser. (2) 12 (1941) 1-8.

[5] H. Gonshor , Contributions to Genetic Algebras , Proc. Edinburgh Math. Soc. (2) 17 (1971) 289-298.

[6] H. Gonshor , Contributions to Genetic Algebras II , Proc. Edinburgh Math. Soc. (2) 18 (1973) 273-279.

[7] I. Heuch , Genetic Algebras for Systems with Linked Loci , Math. Biosciences 34 (1977) 35-47. Errat. ibid. 37 (1977) 279.

[8] P. Holgate , The Genetic Algebra of k Linked Loci, Proc. London Math. Soc. (3) 18 (1968) 315-327.

[9] J. López-Sánchez and A. Pérez de Vargas , Zygotic Algebra for two Linked-Loci with Sexually Different Recombination Rates , Bull Math. Biol. 47 (1985) 771-782.

[10] O. Reiersøl , Genetic Algebras Studied Recursively and by Means of Differential Operators, Math. Scand. 10 (1962) 25-44.

[11] A. Worz-Busekros , Algebras in Genetics, Lect. Notes in Biomath. **36.** Springer V. 1980.

Nonassociative
Algebraic
Models

ISBN 1-56072-050-6

A Note on the Frattini Subalgebra of a Nonassociative Algebra

Alberto Elduque*
Departamento de Matemáticas, Universidad de Zaragoza, 50009 Zaragoza, Spain

Abstract: The behaviour of the Frattini subalgebra and ideal of a nonassociative algebra under extensions of the base field is studied. It turns out that, over perfect fields, the Frattini ideal of an alternative or Malcev algebra is preserved under an extension of the field.

1. Introduction

The Frattini subgroup of a group has been extensively studied and used since the end of the last century. An analogous theory in other algebraic structures has been developed more recently, mainly for Lie algebras (see [11], [14] and [15]), but also for associative and Malcev algebras ([2], [3], [9], [13] and [15]).

The purpose of this note is the investigation, for these classes of algebras (we will include also the alternative algebras), of the behaviour of the Frattini subalgebra and ideal under extensions of the base field. With some restrictions we will prove

* Partially supported by the DGICYT (PS 87-0054)

that the Frattini ideal of the extended algebra is the extension of the Frattini ideal of the original algebra. The Frattini subalgebra, in those cases in which it is not an ideal, does not behave so in general, although we will see some relationships.

We will study also what happens with the subalgebras $\sigma(A)$ and $T(A)$ defined in [16].

In what follows only finite dimensional nonassociative (that is, not necessarily associative) algebras over fields will be considered. In most cases, we shall restrict ourselves to either:

i) alternative algebras over arbitrary fields,

ii) Lie algebras over arbitrary fields or (*)

iii) Malcev algebras over fields of characteristic not 2.

The Frattini subalgebra of an algebra A is defined to be the intersection of the maximal subalgebras of A, and is denoted by $\Phi(A)$. The largest ideal contained in $\Phi(A)$ is called the Frattini ideal of A and is denoted by $\varphi(A)$. The algebra A is said to be φ-free if $\varphi(A) = 0$.

The following result, which will be used later on, was obtained essentially by Towers ([15]):

Theorem 1.1. *Let A be a nonassociative algebra, then:*

a) *If B is an ideal of A with $B^2 = 0$ and $B \cap \varphi(A) = 0$, then there is a subalgebra S of A such that $A = B \oplus S$.*

b) *If A is as in (*), then $\varphi(A)$ is nilpotent.* ∎

2. Extensions of the base field

If V is a vector space over the field F and K is a field extension of F, the extension $V \otimes_F K$ will be denoted by V_K.

Lemma 2.1. *Let A be a nonassociative algebra over a field F, K a field extension of F and B an ideal of A with $B^2 = 0$. Then there is a subalgebra S of A with $A = B \oplus S$ if and only if there is a subalgebra T of A_K such that $A_K = B_K \oplus T$.*

Proof. It can be checked that the existence of such a subalgebra S is equivalent to the existence of some solution in a system of linear equations with coefficients in F. ∎

Proposition 2.2. *Let A be an algebra as in (*) over the field F and K a field extension of F. If A_K is φ-free, then so is A.*

Proof. If A is not φ-free then there is, by the nilpotency of $\varphi(A)$, an ideal B of A with $B^2 = 0$ and $B \subseteq \varphi(A)$. This latter condition implies that there is no subalgebra S in A with $A = B \oplus S$. Now the last Lemma and Theorem 1.1 give a contradiction. ∎

The converse of this result is false in general if further restrictions are not imposed on F (Corollary 2.9) as shown by the next example:

Example 2.3.

Let F be a non–perfect field of characteristic not 2, $\alpha \in F - F^p$ and let K be the field extension obtained by adjoining a pth root of α, $K = F[\xi]$, with $\xi^p = \alpha$. Let L be the Lie algebra of trace zero 2×2 matrices over K, $L = sl(2, K) \cong sl(2, F) \otimes_F K$, but considered as an algebra over F. Then L is a simple Lie algebra over F with centroid K and it is easy to check that $\Phi(L) = \varphi(L) = 0$. Let us consider the extension $L_K = L \otimes_F K \cong sl(2, F) \otimes_F (K \otimes_F K)$.

If I denotes the ideal of $K \otimes_F K$ generated by $\xi \otimes 1 - 1 \otimes \xi$, then I is nilpotent since $(\xi \otimes 1 - 1 \otimes \xi)^p = \alpha \otimes 1 - 1 \otimes \alpha = 0$. Moreover $K = I \oplus (F \otimes_F K)$. Hence $L_K \cong (sl(2, F) \otimes_F I) \oplus (sl(2, F) \otimes_F F \otimes_F K)$. Thus L_K is the direct sum of a nilpotent ideal and a simple subalgebra isomorphic to $sl(2, K)$. By [15; Theorem 6.5] we know that $0 \neq I^2 \subseteq \varphi(L_K)$. Therefore L is φ-free but L_K is not.

Actually it can be checked that $\varphi(L_K) = I^2$. ∎

The next Lemma is well known:

Lemma 2.4. *Let V be a vector space over a field F and let K be a field extension of F so that F is the fixed field by the group $Aut_F(K)$ of F-automorphisms of K. If W is a subspace of V_K fixed by $Aut_F K$ then there is a subspace S of V such that $W = S_K$.* ∎

Proposition 2.5. *Let A be a nonassociative algebra over a field F, let K be a field extension of F such that F is the fixed field for $Aut_F K$ and let M be a maximal subalgebra of A. Let $\Delta = \{V : V$ is a maximal subalgebra of A_K with $M_K \subseteq V\}$. Then $\cap\Delta = M$. In consequence, $\Phi(A_K) \subseteq \Phi(A)_K$. Moreover $\Phi(A_K) = S_K$ for some subalgebra S of A.*

Proof. $Aut_F(K)$ permutes the elements of Δ, so $\cap\Delta = N_K$ for some subalgebra N of A by Lemma 2.4. But $M_K \subseteq N_K$, thus $M \subseteq N$ and $M = N$ since M is maximal. The rest is straightforward. ∎

Given a subalgebra S of A, the largest ideal of A contained in S is denoted by $Core(S)$. In order to investigate the behaviour of the Frattini ideal the following result is needed:

Lemma 2.6. *Let S be a subalgebra of the nonassociative algebra A over the field F and let K be a field extension of F. Then $Core(S_K) = Core(S)_K$.*

Proof. Let $M(A)$ be the multiplication algebra (with the identity) of A (see [12; Ch. II]), then $Core(S) = \{x \in A : xM(A) \subseteq S\}$. If π denotes a projection of A onto S (that is, π is an F-endomorphism of A such that $A\pi = S$ and $\pi^2 = \pi$) and $\{\sigma_1, \ldots, \sigma_n\}$ is a basis of $M(A)$ over F (so it is a basis of $M(A_K)$ over K too), then $Core(S) = \cap_{i=1}^{n} \ker \sigma_i(1-\pi)$, and the Lemma follows. ∎

Theorem 2.7. *Let A be an algebra as in (*) over the field F and K a field extension of F such that F is the fixed field for $Aut_F K$. Then $\varphi(A_K) = \varphi(A)_K$.*

Proof. We know (Proposition 2.5) that $\Phi(A_K) \subseteq \Phi(A)_K$ and $\Phi(A_K) = S_K$ for some subalgebra S of A. Then $\varphi(A_K) = Core(S_K) = Core(S)_K$, so $Core(S) \subseteq \Phi(A)$ and $\varphi(A_K) \subseteq \varphi(A)_K$. Therefore $(A/Core(S))_K$ is φ-free. By Proposition 2.2 we have that $A/Core(S)$ is φ-free, so $\varphi(A) \subseteq Core(S)$ and, as a consequence, $\varphi(A_K) = \varphi(A)_K$. ■

Proposition 2.8. *Let A be an algebra as in (*) over the field F, K a field extension of F and Ω a field extension of K such that F is the fixed field for $Aut_F\Omega$. Then $\varphi(A_K) = \varphi(A)_K$.*

Proof. The algebra $A_\Omega/\varphi(A_\Omega) = (A/\varphi(A))_\Omega = ((A/\varphi(A))_K)_\Omega$ is φ-free because of Theorem 2.7. Proposition 2.2 implies that $A_K/\varphi(A)_K$ is φ-free too, so $\varphi(A_K) \subseteq \varphi(A)_K$. If $\varphi(A)_K$ were an ideal strictly larger than $\varphi(A_K)$ we could take an ideal B of A_K, $\varphi(A_K) \subset B \subseteq \varphi(A)_K$, with $(B/\varphi(A_K))^2 = 0$ ($\varphi(A_K)$ is nilpotent). But $B_\Omega/\varphi(A_K)_\Omega \subseteq \varphi(A)_\Omega/\varphi(A_K)_\Omega = \varphi(A_\Omega/\varphi(A_K)_\Omega)$. Then there is no subalgebra $T/\varphi(A_K)_\Omega$ of $A_\Omega/\varphi(A_K)_\Omega$ such that $A_\Omega/\varphi(A_K)_\Omega = (B_\Omega/\varphi(A_K)_\Omega) \oplus (T/\varphi(A_K)_\Omega)$. By Lemma 2.1 there is no subalgebra $S/\varphi(A_K)$ of $A_K/\varphi(A_K)$ such that $A_K/\varphi(A_K) = (B/\varphi(A_K)) \oplus (S/\varphi(A_K))$ and this is a contradiction with Theorem 1.1. Therefore $\varphi(A_K) = \varphi(A)_K$. ■

Now, if F is a perfect field and Ω is an algebraically closed field containing F, then F is the fixed field for $Aut_F\Omega$. Hence:

Corollary 2.9. *Let A be an algebra as in (*) over the perfect field F and K a field extension of F. Then $\varphi(A_K) = \varphi(A)_K$.* ■

We have seen in Example 2.3 that Corollary 2.9 is not valid, in general, for non–perfect fields. In the next example we will check that we cannot substitute the Frattini ideal for the Fratttini subalgebra in the assertion of the last Corollary. This example is related to the first example of simple Lie algebras of dimension greater than 3 with all its proper subalgebras abelian ([6]).

Example 2.10.

Let us consider the algebra of quaternions B over $\mathbf{Q}(\sqrt{2})$, where $\mathbf{Q}$ denotes the field of rationals, with a basis $\{1, e_1, e_2, e_3\}$ and multiplication given by $1x = x1 = x$

for all x, $e_1e_2 = -e_2e_1 = e_3$, $e_2e_3 = -e_3e_2 = \sqrt{2}e_1$, $e_3e_1 = -e_1e_3 = e_2$ and $e_1^2 = -1$, $e_2^2 = e_3^2 = -\sqrt{2}$.

Let A be this associative algebra, considered as an algebra over $\mathbf{Q}$. Thus A is simple with center $\mathbf{Q}(\sqrt{2})1$. Let $V = \mathbf{Q}(\sqrt{2})e_1 + \mathbf{Q}(\sqrt{2})e_2 + \mathbf{Q}(\sqrt{2})e_3$, V is a simple Lie algebra with the product given by $[x, y] = xy - yx$.

Let M be a maximal subalgebra of A. There appear two possibilities: either $\mathbf{Q}(\sqrt{2})M = M$ or $\mathbf{Q}(\sqrt{2})M = A$. In the latter case M contains a $\mathbf{Q}(\sqrt{2})$-basis of A. If there exists an element $0 \neq v \in A$ with $\mathbf{Q}(\sqrt{2})v \subseteq M$, then $A = Av = (\mathbf{Q}(\sqrt{2})M)v = M(\mathbf{Q}(\sqrt{2})v) \subseteq M$. Therefore every $\mathbf{Q}(\sqrt{2})$-basis of A contained in M is a $\mathbf{Q}$-basis of M. Moreover, $1 \in M$ since, otherwise, M would be an ideal of A. Hence M contains a $\mathbf{Q}$-basis $\{1, \alpha_1 + v_1, \alpha_2 + v_2, \alpha_3 + v_3\}$, where $\alpha_i \in \mathbf{Q}(\sqrt{2})$, $v_i \in V$, $i = 1, 2, 3$. But $(\alpha_i + v_i)^2 = (\alpha_i^2 + v_i^2) + 2\alpha_i v_i \in M$, so $\alpha_i \in \mathbf{Q}$, $i = 1, 2, 3$. Thus M contains a $\mathbf{Q}$-basis $\{1, v_1, v_2, v_3\}$ with $v_i \in V$, $i = 1, 2, 3$. From this we get that $\mathbf{Q}v_1 + \mathbf{Q}v_2 + \mathbf{Q}v_3$ is a subalgebra of the Lie algebra V and this is a contradiction with [6; Example 6].

Therefore if M is a maximal subalgebra of A then $\mathbf{Q}(\sqrt{2})M = M$, so the sets of maximal subalgebras of A and B coincide. Now it follows that $\Phi(A) = \mathbf{Q}(\sqrt{2})1$. But $A \otimes_{\mathbf{Q}(\sqrt{2})} \mathbf{C} \cong M_2(\mathbf{C}) \oplus M_2(\mathbf{C})$, where $\mathbf{C}$ denotes the complex field and $M_2(\mathbf{C})$ the algebra of the 2×2 matrices over $\mathbf{C}$. Now $\Phi(M_2(\mathbf{C}) \oplus M_2(\mathbf{C})) = \mathbf{C}1$, so $\mathbf{C}1 = \Phi(A_{\mathbf{C}}) \subset \Phi(A)_{\mathbf{C}} \cong \mathbf{Q}(\sqrt{2}) \otimes_{\mathbf{Q}} \mathbf{C} \cong \mathbf{C} \oplus \mathbf{C}$. ∎

3. The subalgebras $\sigma(A)$ and $T(A)$

In [16], Towers studied two ideals related to the Frattini ideal of an algebra. If A is an algebra over F, $\sigma(A)$ will denote the intersection of all maximal subalgebras of A which are also ideals of A ($\sigma(A) = A$ if no such subalgebras exist), and $T(A)$ will be the intersection of the maximal subalgebras of A which are not ideals of A. It is denoted by $\tau(A)$ the largest ideal of A contained in $T(A)$.

If A is an algebra as in (*) over the field F and M is a maximal subalgebra of A and an ideal of A, then A/M does not contain any proper subalgebra, so A/M has dimension 1. Now, if K is a field extension of F such that F is the fixed field for $Aut_F K$, M_K is a maximal subalgebra and an ideal of A_K. On the other hand, if M is a maximal subalgebra of A which is not an ideal and if V is a maximal subalgebra of A_K containing M_K, then for all $\sigma \in Aut_F K$, V^σ is also a maximal subalgebra of A_K containing M_K, and it is an ideal if V is. Thus $\bigcap\{V^\sigma \;:\; \sigma \in Aut_F K\} = M_K$ and this is not an ideal, hence V is not an ideal of A_K. Therefore we have proved the following:

Proposition 3.1. *Let A be an algebra as in (*) over a field F and let K be a field extension of F such that F is the fixed field for $Aut_F K$. Then $\sigma(A_K) \subseteq \sigma(A)_K$, $T(A_K) \subseteq T(A)_K$ and $\tau(A_K) \subseteq \tau(A)_K$.* ■

Example 3.2.

Let us consider the associative algebra $A = \mathbf{Q} \oplus \mathbf{Q}(\sqrt{2})$ over $\mathbf{Q}$. It is straightforward to check that $\sigma(A) = \mathbf{Q}(\sqrt{2})$, $T(A) = \mathbf{Q} \oplus \mathbf{Q}$ and $\tau(A) = \mathbf{Q} \oplus 0$. But $A_{\mathbf{C}} \cong \mathbf{C} \oplus \mathbf{C} \oplus \mathbf{C}$, so $\sigma(A_{\mathbf{C}}) = 0$, $T(A_{\mathbf{C}}) = \mathbf{C}1$ and $\tau(A_{\mathbf{C}}) = 0$. Hence all the relationships in Proposition 3.1 may be strict. ■

In [4] it is proved that if A is an anticommutative algebra then $\sigma(A) = A^2$ and $\tau(A) = \{x \in A \;:\; xA \subseteq \varphi(A)\}$. This and Corollary 2.9 give:

Proposition 3.3. *Let A be either a Lie algebra over an arbitrary perfect field F or a Malcev algebra over a perfect field F of characteristic not 2, and let K be a field extension of F. Then, $\sigma(A_K) = \sigma(A)_K$ and $\tau(A_K) = \tau(A)_K$.* ■

4. The Frattini subalgebra of a Malcev algebra

Barnes gave an example of a Lie algebra over a field of characteristic 2 in which the Frattini subalgebra is not an ideal, but in characteristic not 2 no such examples are known. Moreover, we have (see [1], [2], [5] and [15]):

Proposition 4.1.

i) If A is a Malcev algebra over a field of characteristic 0, then $\Phi(A)$ is an ideal of A.

ii) If A is a solvable Malcev algebra over a field of characteristic not 2, then $\Phi(A)$ is an ideal of A. ■

When dealing with Lie algebras, we can omit the restriction on the characteristic in assertion ii) of Proposition 4.1.

The last Proposition will indicate that in order to check if the Frattini subalgebra of any Malcev algebra over a field of characteristic not 2 is an ideal it is enough to deal with the semisimple algebras:

Proposition 4.2. *Let us suppose that $\Phi(A) = 0$ for all semisimple Malcev algebras over a field F of characteristic not 2. Then the Frattini subalgebra of any Malcev algebra over F is an ideal.*

Proof. Let A be a counterexample of minimal dimension. Then $\varphi(A) = 0 \neq \Phi(A)$ and A is not semisimple. Thus A contains ideals which square to 0. Let us denote by $Zsoc(A)$ the sum of all the minimal ideals of A which square to 0, then $Zsoc(A)$ is called the zero socle of A. By Theorem 1.1 there is a subalgebra S of A such that $A = Zsoc(A) \oplus S$.

Let $Zsoc(A) = Z_1 \oplus \cdots \oplus Z_n$ with the Z_i's minimal ideals of A. Since $M_i = (Z_1 \oplus \overset{i}{\ldots} \oplus Z_n) \oplus S$ is a maximal subalgebra of A and, since if T is a maximal subalgebra of S, then $Zsoc(A) \oplus T$ is a maximal subalgebra too, we get that $\Phi(A) \subseteq \Phi(S)$. By the minimality of A we have $\Phi(S) = \varphi(S)$ which is a nilpotent ideal of S. Now, for all i, $Z_i\Phi(S) = Z_i(Zsoc(A) + \Phi(S))$ is an ideal of A (see [2]), so it is equal either to 0 or to Z_i. Let us suppose that $Z_i\Phi(S) = Z_i$ for $i = 1, \ldots, r$ and $Z_i\Phi(S) = 0$ for $i = r+1, \ldots, n$.

Let $A = A_0 \oplus A_1$ be the Fitting decomposition of A relative to the nilpotent subalgebra $\Phi(S)$ (see [10]). For $i = 1, \ldots, r$, $Z_i\Phi(S) = Z_i$, so $Z_i \subseteq A_1$ and, since $\Phi(S)$ is a nilpotent ideal of S, $A_0 = Z_{r+1} \oplus \cdots \oplus Z_n \oplus S$ and $A_1 = Z_1 \oplus \cdots \oplus Z_r$. But $J(A_0, A_0, A_1) = 0$, so that $J(Z_i, A, A) = 0$ for $i = 1, \ldots, r$. That is, Z_i is contained in the J–nucleus of A for $i = 1, \ldots, r$. Therefore, if $z \in Z_i$ $(i = 1, \ldots, r)$, then the endomorphism $I + R_z$, where I is the identity and R_z is the right multiplication by the element z, is an automorphism of A. Since $\Phi(A)(I + R_z) = \Phi(A)$, we get that $\Phi(A)z \subseteq \Phi(S) \cap Zsoc(A) = 0$. In this way we get that $\Phi(A)Zsoc(A) = 0$. Moreover, if $i = 1, \ldots, r$, Z_i is contained in the J–nucleus of A, so $C_A(Z_i) = \{x \in A : xZ_i = 0\}$ is an ideal of A and $\Phi(A) \subseteq C_A(Zsoc(A)) \cap \Phi(S)$, which is a nilpotent ideal of A with trivial intersection with $Zsoc(A)$. And this is a contradiction since every nilpotent ideal contains a nonzero ideal which squares to 0. ∎

In connection with the last result, let us notice that in [7] it has been proved that any form of a classical Lie algebra over a field of characteristic not 2 or 3 has trivial Frattini subalgebra. Incidentally this provides an alternative proof of the fact that the Frattini subalgebra of any Lie algebra over a field of characteristic 0 is an ideal. Also in [5] it has been shown that the Frattini subalgebra of any simple non–Lie Malcev algebra over a field of characteristic not 2 is trivial. On the other hand, if F is a perfect field of characteristic not 2 and A is a semisimple Malcev algebra then A is a direct sum (as algebra) of a semisimple Lie algebra and of a direct sum of simple non Lie Malcev algebras (see [8]). Hence, in order to prove that the Frattini subalgebra of any Malcev algebra over these fields is an ideal, it is enough to check that the Frattini subalgebra of any semisimple Lie algebra is trivial.

An analogous result to Proposition 4.2 is valid, with a slight modification of the proof, for $T(A)$.

REFERENCES

[1] D.W. Barnes and H. Gastineau-Hills: "On the theory of soluble Lie algebras". Math. Z. **106** (1968), 343-354.

[2] A.A. El Malek: "The Frattini subalgebra of a Malcev algebra". Arch. Math. **37** (1981), 306-315.

[3] A. Elduque: "A Frattini theory for Malcev algebras". Algebras Groups Geom. **1** (1984), 247-266.

[4] A. Elduque: "Ideals related to the Frattini ideal of a Malcev algebra". Proceedings of the "X Jornadas Hispano–Lusas de Matemáticas" (1985), Section I, 208-217.

[5] A. Elduque: "On maximal subalgebras of central simple Malcev algebras". J. Algebra **103** (1986), 216-227.

[6] A. Elduque: "A note on noncentral simple minimal nonabelian Lie algebras". Commun. Algebra **15** (1987), 1313-1318.

[7] A. Elduque: "Sobre los generadores de un álgebra de Lie semisimple" (Spanish). Proceedings of the "XII Jornadas Luso–Españolas de Matemáticas (1987), Vol. II, 56-61.

[8] A. Elduque: "On semisimple Malcev algebras". Proc. Amer. Math. Soc. **107** (1989), 73-82.

[9] S. González: "Subálgebras de Frattini y de Cartan, y formaciones en álgebras asociativas" (Spanish. Doctoral Dissertation). University of Zaragoza 1977.

[10] E.N. Kuzmin: "Malcev algebras and their representations". Algebra and Logic **7** (1968), 233-244.

[11] E.I. Marshall: "The Frattini subalgebra of a Lie algebra". J. London Math. Soc. **42** (1967), 416-422.

[12] R.D. Schafer: *An introduction to nonassociative algebras* (Academic Press, New York, 1966).

[13] E.L. Stitzinger: "A non-imbedding theorem of associative algebras". Pacific J. Math. **30** (1969), 529-531.

[14] E.L. Stitzinger: "The Frattini subalgebra of a Lie algebra". J. London Math. Soc. (2) **2** (1970), 429-438.

[15] D.A. Towers: "A Frattini theory for algebras". Proc. London Math. Soc. (3) **27** (1973), 440-462.

[16] D.A. Towers: "Two ideals of an algebra closely related to its Frattini ideal". Arch. Math. **35** (1980), 112-120.

Nonassociative
Algebraic
Models

ISBN 1-56072-050-6

The Compact Malcev Algebra and Torsion and Curvature Tensors of Connections on S^7

Alberto Elduque*
Departamento de Matemáticas, Universidad de Zaragoza, 50009 Zaragoza, Spain

Hyo Chul Myung
*Department of Mathematics, University of Northern Iowa,
Cedar Falls, Iowa 50614, USA*

1. Introduction

Let $\boldsymbol{g}$ be a Lie algebra with multiplication $[\, ,\,]$ over a field F and $\boldsymbol{h}$ be a subalgebra of $\boldsymbol{g}$. The pair $(\boldsymbol{g}, \boldsymbol{h})$ is called a *reductive pair* if there is a subspace $\boldsymbol{m}$ of $\boldsymbol{g}$ such that $\boldsymbol{g} = \boldsymbol{m} \oplus \boldsymbol{h}$ (vector space direct sum) and $[\boldsymbol{h}, \boldsymbol{m}] \subseteq \boldsymbol{m}$. In this case, $\boldsymbol{h}$ is called a *reductive subalgebra* of $\boldsymbol{g}$. Note that any semisimple subalgebra of a finite-dimensional Lie algebra of characteristic 0 is reductive. Reductive pairs naturally arise from differential geometry of affine connections on homogeneous spaces. let G be a connected Lie group with Lie algebra $\boldsymbol{g}$ and H be a closed Lie subgroup of G with Lie algebra $\boldsymbol{h}$. The homogeneous space G/H is called reductive if there is a decomposition as above $\boldsymbol{g} = \boldsymbol{m} \oplus \boldsymbol{h}$ with $Ad\,H(\boldsymbol{m}) \subseteq \boldsymbol{m}$. This implies that the pair

* Partially supported by the DGICYT (PS 87-0054)

$(\boldsymbol{g}, \boldsymbol{h})$ is reductive (the converse is true if H is connected). In this case, if $\boldsymbol{g} = \boldsymbol{m} \oplus \boldsymbol{h}$ is a fixed decomposition with $[\boldsymbol{h}, \boldsymbol{m}] \equiv (ad\,\boldsymbol{h})\boldsymbol{m} \subseteq \boldsymbol{m}$, then there is an analytic structure on G/H having $\boldsymbol{m}$ as the tangent space $T(G/H, \bar{e})$ at $\bar{e} = eH$. Then, the relation between G–invariant affine connections on G/H and algebra multiplications $*$ on $\boldsymbol{m}$ is given by the Fundamental Existence Theorem for affine connections (see [2,8] and references therein):

Theorem 1. *Let G/H be a reductive homogeneous space with a fixed decomposition $\boldsymbol{g} = \boldsymbol{m} \oplus \boldsymbol{h}$ such that $(Ad\,H)\boldsymbol{m} \subseteq \boldsymbol{m}$. Then, there exists a bijective correspondence between the set of all G–invariant affine connections ∇ on G/H and the set of all algebra multiplications $*$ on $\boldsymbol{m}$ such that $Ad\,H \subseteq Aut(\boldsymbol{m}, *)$ (which implies $ad\,\boldsymbol{h} \subseteq Der(\boldsymbol{m}, *)$). The correspondence is given by $X * Y = (\nabla_{\tilde{X}}\tilde{Y})(\bar{e})$ for $X, Y \in \boldsymbol{m}$, where $\tilde{X}, \tilde{Y}$ are vector fields uniquely determined by X and Y on a neighborhood $\tilde{N}$ of $\bar{e}$.*
■

In this case, $(\boldsymbol{m}, *)$ is called a *connection algebra* of ∇. Geometric properties of ∇, such as geodesics, torsions, curvatures, holonomy, etc., are given in terms of the connection algebra $(\boldsymbol{m}, *)$ [2]. For $X, Y \in \boldsymbol{m}$, let

$$[X, Y] = [X, Y]_{\boldsymbol{m}} + [X, Y]_{\boldsymbol{h}},$$

where $[X, Y]_{\boldsymbol{m}}$ and $[X, Y]_{\boldsymbol{h}}$ are the projections of $[X, Y]$ on $\boldsymbol{m}$ and $\boldsymbol{h}$ respectively. From the Lie algebra identities of $\boldsymbol{g}$, it is readily seen that $(\boldsymbol{m}, \frac{1}{2}[\,,\,]_{\boldsymbol{m}})$ is an anti-commutative algebra with $Ad\,H \subseteq Aut(\boldsymbol{m}, \frac{1}{2}[\,,\,]_{\boldsymbol{m}})$. Thus, by Theorem 1 there is a unique G–invariant connection ∇ on G/H whose associated algebra is given by $(\boldsymbol{m}, \frac{1}{2}[\,,\,]_{\boldsymbol{m}})$. This connection is called the canonical connection on G/H of the first kind which necessarily has zero torsion and whose geodesics through $\bar{e}$ have the form $(exp\,tX)H$ $(t \in \mathbb{R})$ for some $X \in \boldsymbol{m}$ (see [2] and references therein). The algebra $(\boldsymbol{m}, [\,,\,]_{\boldsymbol{m}})$ is also called the *reductive algebra* associated with the decomposition $\boldsymbol{g} = \boldsymbol{m} \oplus \boldsymbol{h}$.

Let $\mathbb{O}$ denote the Cayley algebra (over $\mathbb{R}$). Thus, $Aut\,\mathbb{O}$ is the compact Lie group G_2 with Lie algebra $\boldsymbol{g}_2 = Der\,\mathbb{O}$. It has been shown in [2] that the sphere S^6 is realized as the reductive homogeneous space $G_2/SU(3)$ such that the reductive subalgebra $su(3)$ has the unique complementary subspace $\boldsymbol{m}$ with $[su(3), \boldsymbol{m}] \subseteq \boldsymbol{m}$, where $\boldsymbol{m} = su(3)^{\perp}$, the orthogonal complement relative to the Killing form, and that the reductive algebra $(\boldsymbol{m}, [\,,\,]_{\boldsymbol{m}})$ is isomorphic to the compact vector color algebra of dimension 6. Furthermore, all G_2- invariant affine connections on S^6 have been determined by the connection algebras given by complex multiples of the compact vector color algebra. In [3], we have used these results to give explicit formulas for the torsion and curvature tensors of G_2–invariant connections on $S^6 = G_2/SU(3)$ in terms of its connection algebras.

Following a similar approach employed for S^6, in [4] the sphere S^7 is realized as the reductive homogeneous space $Spin(7)/G_2$ such that the reductive subalgebra $\mathfrak{g}_2$ has the unique complementary subspace $\boldsymbol{m} = \mathfrak{g}_2^{\perp}$ with $[\mathfrak{g}_2, \boldsymbol{m}] \subseteq \boldsymbol{m}$. The reductive algebra $(\boldsymbol{m}, [\ ,\ ]_{\boldsymbol{m}})$ is shown to be isomorphic to the compact Malcev algebra of dimension 7 (the Malcev algebra of trace zero elements in $\mathbf{O}$). It turns out that all connection algebras on $S^7 = Spin(7)/G_2$ are parameterized by real numbers in terms of the compact Malcev algebra [4].

The purpose of this note is to give explicit formulas for the torsion and curvature tensors of all $Spin(7)$–invariant connections on S^7 by means of its connection algebras. Also, some relevant results are briefly reviewed from [4]. Our approach is purely algebraic.

2. Connection Algebras on $S^7 = Spin(7)/G_2$

Let $\mathbf{O}$ denote the division Cayley algebra (over $\mathbf{R}$) with trace t and norm n, and let $(\ ,\ )$ be the bilinear form associated with n. Then $\mathbf{O}$ decomposes as

$$\mathbf{O} = \mathbf{R}1 \oplus V$$

where $V = \{x \in \mathbf{O} : t(x) = 0\}$, and V has an orthonormal basis $e_1, \ldots, e_7$ relative to $(\ ,\ )$ having multiplication given by

$$e_i^2 = -1, \qquad e_i e_{i+1} = -e_{i+1} e_i = e_{i+3}$$

with indices modulo 7 for $i = 1, \ldots, 7$. For $u, v \in V$, let $u \times v$ denote the projection of uv onto V where uv denotes the product in $\mathbf{O}$. Linearizing the quadratic equation $x^2 - t(x)x + n(x)1 = 0$ with $x = u + v$ and using $t(u) = t(v) = 0$, we obtain

$$uv = -(u, v)1 + u \times v,$$

$$\frac{1}{2}[u, v] \equiv \frac{1}{2}(uv - vu) = u \times v$$

for $u, v \in V$. Hence, $(V, \times)$ becomes one of the two simple real Malcev algebras of dimension 7, and is called the *compact Malcev algebra* of dimension 7 (since the form $(\,,\,)$ is positive definite).

If we let

$$Q : V \longrightarrow \mathbf{R} \quad : \quad u \mapsto -n(u)$$

then Q is a nondegenerate negative quadratic form on V with $Q(e_i) = -1$ for $i = 1, \ldots, 7$. Let $C(V, Q)$ be the Clifford algebra associated with V and Q, and let $C^+(V, Q)$ be the even Clifford algebra. It is well known [1, p.49; 6, p.234] that

$$C(V, Q) \cong M_8(\mathbf{R}) \oplus M_8(\mathbf{R}),$$

$$C^+(V, Q) \cong M_8(\mathbf{R})$$

where $M_8(\mathbf{R})$ denotes the algebra of 8×8 matrices over $\mathbf{R}$.

For $x \in \mathbf{O}$, let L_x and R_x denote the left and right multiplications in $\mathbf{O}$ by x, i.e., $L_x(y) = xy$ and $R_x(y) = yx$, $y \in \mathbf{O}$. If we let

$$\rho_L : V \longrightarrow End_{\mathbf{R}}(\mathbf{O}) \quad : \quad u \mapsto L_u,$$

then by the alternative identities in $\mathbf{O}$, we have $(\rho_L(u))^2 = L_u^2 = L_{u^2} = Q(u)I$ since $u^2 = Q(u)1$. Hence, there is a unique extension to a homomorphism (actually an isomorphism) $\rho_L : C^+(V, Q) \longrightarrow End_{\mathbf{R}}(\mathbf{O})$ which induces a representation of the spin group

$$Spin(Q) = Spin(7) = \{x_1 \cdot \cdots \cdot x_{2r} \in C^+(V, Q) : x_i^2 = -1\},$$

acting on $\mathbf{O}$ [1,6]. We denote by $x \cdot y$ the multiplication in $C(V, Q)$. Since $Spin(Q)$ generates $C^+(V, Q)$, ρ_L is irreducible on $Spin(Q)$. It is shown [1, p.89] that the Lie algebra $\mathcal{L}$ of $Spin(Q)$ is given by

$$\begin{aligned} \mathcal{L} &= \{\sum_i u_i \cdot v_i : u_i, v_i \in V \text{ and } \sum_i (u_i, v_i) = 0\} \\ &= \mathbf{R}\langle e_i \cdot e_j : i < j \text{ and } i, j = 1, \ldots, 7\rangle \subseteq C^+(V, Q) \end{aligned}$$

where $\mathbf{R}\langle\ :\ \rangle$ means the linear span over $\mathbf{R}$. Note that $\dim \mathcal{L} = 21$.

Therefore, $\mathbf{g} \equiv \rho_L(\mathcal{L})\,(\cong \mathcal{L})$ is the Lie algebra of the Lie group $\rho_L(Spin(Q))$ which is shown to be the compact Lie group of type B_3 [1, Proposition 6.1.7]. The compact Lie algebra $Der\,\mathbf{O}$ of type G_2 occurs as the subalgebra of $\mathbf{g}$ given by [4]:

Proposition 2. *$Der\,\mathbb{O}$ is the subspace of $\mathbf{g} = \rho_L(\mathcal{L})$ spanned by the elements $\rho_L(e_i \cdot e_j - e_h \cdot e_k)$ such that $e_i e_j = e_h e_k$ for $i, j, h, k = 1, \ldots, 7$ with $i \neq j$ and $h \neq k$.*
■

Using this and the algebraic properties of $\mathbb{O}$ and $(V, \times)$, as for $S^6 = G_2/SU(3)$ [2], the following result is proved in [4]:

Theorem 3. (i) *$Der\,\mathbb{O} = \{\theta \in \rho_L(\mathcal{L}) = \mathbf{g} : \theta(1) = 0\}$.*

(ii) *$Aut\,\mathbb{O} = \{\phi \in \rho_L(Spin(Q)) : \phi(1) = 1\}$.*

(iii) *$Spin(Q)$ acts via ρ_L analytically and transitively on $S^7 = \{x \in \mathbb{O} : n(x) = 1\}$, and*

$$S^7 = Spin(Q)/G_2$$

as a reductive homogeneous space. ■

To describe the reductive structure of S^7, for $x, y \in \mathbb{O}$ and $u, v, w \in V$, we define the operators

$$D_{x,y} = L_{[x,y]} - R_{[x,y]} - 3[L_x, R_y], \tag{1}$$

$$d(u,v)w = u \times (v \times w) - v \times (u \times w) + (u \times v) \times w. \tag{2}$$

It is shown that $Der\,\mathbb{O} = D_{\mathbb{O},\mathbb{O}}$ and that $Der(V, \times) = Der\,\mathbb{O}|_V = d(V, V)$ for any octonion algebra $\mathbb{O}$ over a field of characteristic $\neq 2, 3$ [7,9]. Using these, it can be proven [4] that $(\mathbf{g}, Der\,\mathbb{O})$ is a reductive pair with decomposition

$$\mathbf{g} = (L + 2R)_V \oplus Der\,\mathbb{O} = \mathbf{m} \oplus \mathbf{h} \tag{3}$$

where $\mathbf{m} = (L + 2R)_V = \mathbb{R}\langle L_u + 2R_u : u \in V\rangle$, $\mathbf{h} = Der\,\mathbb{O}$ and $[\mathbf{h}, \mathbf{m}] \subseteq \mathbf{m}$. Hence, $\dim \mathbf{m} = 7$, and since $\dim \mathbf{h} > \frac{1}{2} \dim \mathbf{g}$, it follows that $\mathbf{m} = \mathbf{h}^{\perp}$ is the unique complementary subspace of $\mathbf{h}$ with $[\mathbf{h}, \mathbf{m}] \subseteq \mathbf{m}$ [2, Proposition 2.3]. The reductive algebra $(\mathbf{m}, [\,,\,])$ is determined by the identity [4]

$$[L_u + 2R_u, L_v + 2R_v] = 2D_{u,v} - 2(L_{u\times v} + 2R_{u\times v}) \tag{4}$$

for $u, v \in V$, and hence by the multiplication

$$[L_u + 2R_u, L_v + 2R_v]_{\mathbf{m}} = -2(L_{u\times v} + 2R_{u\times v}). \tag{5}$$

Therefore, we have [4]

Theorem 4. *The map φ defined by*

$$\varphi : (\mathbf{m}, [\ ,\ ]_{\mathbf{m}}) \longrightarrow (V, \times) : L_u + 2R_u \mapsto -2u \tag{6}$$

is an algebra isomorphism as well as an $\mathbf{h}$–module isomorphism. ∎

As for the case of S^6, we have the following correspondence theorem [4]:

Theorem 5. *There is a one-one correspondence between the set of $Spin(7)$–invariant affine connections on S^7 and the set of algebra multiplications $*$ on V with $\mathbf{h} \subseteq Der(V, *)$ which are given by*

$$u * v = \eta(u \times v) \tag{7}$$

for $u, v \in V$ and $\eta \in \mathbf{R}$. ∎

Thus, the set of algebras $(V, *)$ with $\mathbf{h} \subseteq Der(V, *)$ is parameterized by $\mathbf{R}$ in terms of $(V, \times)$. For the proof of Theorem 5, we have used Theorem 1, the isomorphism φ given by (6) and the relation $\dim Hom_{\mathbf{h}}(V \otimes_{\mathbf{R}} V, V) = 1$ [7, Lemma 3.13]. The set of algebras $(V, *)$ given by (7) is in one-one correspondence with the set of algebras $(\mathbf{m}, \circ)$ with $ad\,\mathbf{h} \subseteq Der(\mathbf{m}, \circ)$ under the relation

$$\varphi(L_u + 2R_u) * \varphi(L_v + 2R_v) = \varphi((L_u + 2R_u) \circ (L_v + 2R_v)). \tag{8}$$

3. Torsion and Curvature Tensors of Connections on S^7

Consider the sphere S^7 as the reductive homogeneous space $S^7 = Spin(7)/G_2$ with decomposition (3), and as usual, identify $\boldsymbol{m} = T(S^7, \bar{e})$, the tangent space at $\bar{e} = eG_2$. For any $\eta \in \mathbf{R}$, we denote by η_V the multiplication $*$ on V given by (7) and by $\eta_{\boldsymbol{m}}$ the unique multiplication $\circ$ on $\boldsymbol{m}$ determined by (8). Thus, $\varphi : (\boldsymbol{m}, \eta_{\boldsymbol{m}}) \longrightarrow (V, \eta_V)$ is an isomorphism and (8) is expressed by

$$\eta_{\boldsymbol{m}}(X,Y) = \varphi^{-1}(\eta_V(\varphi X, \varphi Y)) \tag{9}$$

for $X, Y \in \boldsymbol{m}$. Note that $1_{\boldsymbol{m}}(X,Y) = [X,Y]_{\boldsymbol{m}}$, $(\frac{1}{2})_{\boldsymbol{m}}(X,Y) = \frac{1}{2}[X,Y]_{\boldsymbol{m}}$ for $X, Y \in \boldsymbol{m}$ and $1_V(u,v) = u \times v$ for $u, v \in V$.

In view of Theorem 5, the set of $Spin(7)$–invariant connections on S^7 is in one-one correspondence with $\mathbf{R}$. We shall denote by ∇^η the invariant connection on S^7 associated with $(\boldsymbol{m}, \eta_{\boldsymbol{m}})$. Thus $\nabla^{\frac{1}{2}}$ and ∇^0 are the canonical connections on S^7 of the first and second kinds.

Let G/H be a reductive homogeneous space with decomposition $\boldsymbol{g} = \boldsymbol{m} \oplus \boldsymbol{h}$. For any G–invariant affine connection ∇ on G/H, the torsion and curvature tensors of ∇ are determined by its values at $\bar{e}$ which are given by

$$T(X,Y) = X \circ Y - Y \circ X - [X,Y]_{\boldsymbol{m}}, \tag{10}$$

$$R(X,Y)Z = X \circ (Y \circ Z) - Y \circ (X \circ Z) - [X,Y]_{\boldsymbol{m}} \circ Z - [[X,Y]_{\boldsymbol{h}}, Z] \tag{11}$$

for $X, Y, Z \in \boldsymbol{m}$, where $(\boldsymbol{m}, \circ)$ is the connection algebra of ∇.

For a $Spin(7)$–invariant affine connection ∇^η on S^7, let T^η and R^η denote the torsion and curvature tensors of ∇^η. Using the isomorphism φ of (6), we only need to compute T^η_V and R^η_V defined by

$$T^\eta_V(u,v) = \varphi(T^\eta(\varphi^{-1}u, \varphi^{-1}v)), \tag{12}$$

$$R^\eta_V(u,v)w = \varphi(R^\eta(\varphi^{-1}u, \varphi^{-1}v)\varphi^{-1}w) \tag{13}$$

for $u, v, w \in V$, which are expressed in terms of $(V, \times)$ and $\eta \in \mathbf{R}$.

To compute the last term on the right side of (11), we need

Lemma 6. *For $u, v, w \in V$, we have*

$$D_{u,v}(w) = 2d(u,v)w = -2(u \times v) \times w + 6\big((u,w)v - (v,w)u\big). \tag{14}$$

Proof. We use (1) and (2) to compute

$$\begin{aligned} D_{u,v}(w) &= [u,v]w - w[u,v] - 3u(wv) + 3(uw)v \\ &= [[u,v],w] - 3u(-(w,v)1 + w \times v) + 3(-(u,w)1 + u \times w)v \\ &= 4(u \times v) \times w - 3u \times (w \times v) + 3(u, w \times v)1 + 3(w,v)u \\ &\qquad - 3(u,w)v - 3(u \times w, v)1 + 3(u \times w) \times v \\ &= 4(u \times v) \times w - 3u \times (w \times v) + 3(u \times w) \times v + 3(v,w)u - 3(u,w)v, \end{aligned}$$

since $uv = -(u,v)1 + u \times v$, $[u,v] = 2(u \times v)$ and $(\ ,\ )$ is an invariant form on $(V, \times)$. We linearize the identity $(u \times v) \times v = (u,v)v - (v,v)u$ [2] to obtain

$$\begin{aligned} (u \times v) \times w + (u \times w) \times v &= (u,w)v + (u,v)w - 2(v,w)u, \\ (v \times u) \times w + (v \times w) \times u &= (v,w)u + (u,v)w - 2(u,w)v. \end{aligned}$$

Subtracting the second from the first gives

$$2(u \times v) \times w + (u \times w) \times v - u \times (w \times v) = 3(u,w)v - 3(v,w)u. \tag{15}$$

Therefore,

$$D_{u,v}(w) = 2(u \times v) \times w + 2u \times (v \times w) - 2v \times (u \times w) = 2d(u,v)w$$

which by (15) implies the second equality of (14). ∎

For $u \in V$, let $P_u = -(L_u + 2R_u)$. From (4) we obtain $[P_u, P_v] = 2D_{u,v} + P_{u \times v}$ and hence

$$[P_u, P_v]_{\boldsymbol{m}} = 2P_{u \times v}, \quad [P_u, P_v]_{\boldsymbol{h}} = 2D_{u,v}. \tag{16}$$

Since $D_{u,v}$ is a derivation of $\mathbb{O}$, it follows from (14) and (16) that

$$\begin{aligned} [[P_u, P_v]_{\boldsymbol{h}}, P_w] &= 2[D_{u,v}, P_w] = 2P_{D_{u,v}(w)} = 4P_{d(u,v)w} \\ &= -4P_{(u \times v) \times w} + 12(u,w)P_v - 12(v,w)P_u \end{aligned} \tag{17}$$

for $u, v, w \in V$. The last term on the right side of (11) is now computed as

$$\begin{aligned} [[\varphi^{-1}u, \varphi^{-1}v]_{\boldsymbol{h}}, \varphi^{-1}w] &= \frac{1}{8}[[P_u, P_v]_{\boldsymbol{h}}, P_w] \\ &= -\varphi^{-1}((u \times v) \times w) + 3(u,w)\varphi^{-1}v - 3(v,w)\varphi^{-1}u \\ &= \varphi^{-1}\big(-(u \times v) \times w + 3(u,w)v - 3(v,w)u\big) \end{aligned} \tag{18}$$

for $u, v, w \in V$, using (16) and (17). We use these results to obtain the formulas:

Theorem 7. *The torsion and curvature tensors of the connection* ∇^η *on* S^7 *are given by*

$$T_V^\eta(u,v) = (2\eta - 1)\, u \times v,$$

$$R_V^\eta(u,v)w = (1-\eta-2\eta^2)(u\times v)\times w + 3(\eta^2-1)\big((u,w)v - (v,w)u\big)$$

for $u, v, w \in V$.

Proof. From (6), (7), (9), (10) and (12), we have

$$\begin{aligned} T_V^\eta(u,v) &= \varphi\big(\eta\mathbf{m}(\varphi^{-1}u,\varphi^{-1}v) - \eta\mathbf{m}(\varphi^{-1}v,\varphi^{-1}u) - [\varphi^{-1}u,\varphi^{-1}v]_{\mathbf{m}}\big) \\ &= \eta_V(u,v) - \eta_V(v,u) - u\times v \\ &= \eta(u\times v) - \eta(v\times u) - u\times v \\ &= (2\eta-1)(u\times v). \end{aligned}$$

For $R_V^\eta(u,v)w$, we use (7), (9), (11), (13) and (18) to compute

$$\begin{aligned} R_V^\eta(u,v)w &= \varphi\big(\eta\mathbf{m}(\varphi^{-1}u,\eta\mathbf{m}(\varphi^{-1}v,\varphi^{-1}w)) - \eta\mathbf{m}(\varphi^{-1}v,\eta\mathbf{m}(\varphi^{-1}u,\varphi^{-1}w)) \\ &\quad \eta\mathbf{m}([\varphi^{-1}u,\varphi^{-1}v]_{\mathbf{m}},\varphi^{-1}w) - [[\varphi^{-1}u,\varphi^{-1}v]_{\mathbf{h}},\varphi^{-1}w]\big) \\ &= \eta_V(u,\eta_V(v,w)) - \eta_V(v,\eta_V(u,w)) - \eta_V(u\times v,w) \\ &\quad - \big(-(u\times v)\times w + 3(u,w)v - 3(v,w)u\big) \\ &= \eta^2\big(u\times(v\times w) - v\times(u\times w)\big) - \eta((u\times v)\times w) \\ &\quad + (u\times v)\times w - 3(u,w)v + 3(v,w)u, \end{aligned}$$

which combines with (15) to give the desired formula for $R_V^\eta(u,v)w$. ∎

We note that the connection ∇^η has zero torsion if and only if $\eta = \frac{1}{2}$, that is, ∇^η is the canonical connection $\nabla^{\frac{1}{2}}$ of the first kind. Thus, $\nabla^{\frac{1}{2}}$ is the unique Riemannian connection on S^7, and its curvature tensor reduces to

$$R_V^{\frac{1}{2}}(u,v)w = \frac{9}{4}\big((v,w)u - (u,w)v\big). \tag{19}$$

For simplicity, denote $R = R_V^{\frac{1}{2}}$. It follows easily from (19) that $R(u,v)$ is an orthogonal linear transformation on V relative to $(\ ,\ )$ for all $u,v \in V$. Hence $R(V,V) = \mathbf{R}\langle R(u,v) : u,v \in V\rangle$ is the compact Lie algebra of type B_3 with basis

$$\{R(e_i,e_j) : i<j,\ i,j = 1,\ldots,7\}.$$

Using (19), we compute the multiplication in $B_3 = R(V,V)$ as

$$[R(s,t),R(u,v)] = \frac{9}{4}((s,v)R(t,u) + (t,u)R(s,v) - (s,u)R(t,v) - (t,v)R(s,u))$$

for $s,t,u,v \in V$.

If we denote by $[x,y]^{\cdot} = x \cdot y - y \cdot x$ the commutator in $C(V,Q)$, then from a formula given in [5, p.231] we obtain

$$R(u,v)w = \frac{9}{16}[[v,u]^{\cdot},w]^{\cdot}$$

for $u,v,w \in V$. Using this and (19), the curvature tensor R_V^{η} of ∇^{η} is expressed as

$$\begin{aligned} R_V^{\eta}(u,v)w &= (1-\eta-2\eta^2)(u \times v) \times w - \frac{4}{3}(\eta^2-1)R(u,v)w \\ &= (1-\eta-2\eta^2)(u \times v) \times w + \frac{3}{4}(\eta^2-1)[[u,v]^{\cdot},w]^{\cdot}. \end{aligned}$$

REFERENCES

1. A. Crumeyrolle, *Orthogonal and Symplectic Clifford Algebras. Spinor Structures*, Kluwer Academic Publishers, 1990.

2. A. Elduque and H.C. Myung, Color algebras and affine connections on S^6, J. Algebra, to appear.

3. A. Elduque and H.C. Myung, Note on the Cayley algebra and connections on S^6, Proc. of the 5th International Conference on Hadronic Mechanics and Nonpotential Interactions, H.C. Myung, ed., Nova Science Publ., to appear.

4. A. Elduque and H.C. Myung, The reductive pair (B_3, G_2) and affine connections on S^7, to appear.

5. N. Jacobson, *Lie Algebras*, Interscience Publishers, New York, 1962.

6. N. Jacobson, *Basic Algebra* II, W.H. Freeman, San Francisco, 1980.

7. H.C. Myung, *Malcev-Admissible Algebras*, Birkhäuser, Boston, 1986.

8. K. Nomizu, Invariant affine connections on homogeneous spaces, Amer. J. Math. **76** (1954), 33-65.

9. R.D. Schafer, *Introduction to Nonassociative Algebras*, Academic Press, New York, 1966.

Nonassociative
Algebraic
Models

ISBN 1-56072-050-6

Prime Nondegenerate Jordan Triple Systems with Minimal Inner Ideals

Antonio Fernández López, Eulalia García Rus and Esperanza Sánchez Campos
Departamento de Algebra, Geometría y Topología, Universidad de Málaga, 29071, Málaga, Spain

1. Introduction

The aim of this paper is to classify prime nondegenerate Jordan triple systems (over a field of characteristic not 2 or 3) with nonzero socle. This is done by using the classification of prime nondegenerate Jordan triple systems given by Zelmanov in [18], and the later version of this theorem due to D' Amour [1] where prime nondegenerate Jordan triple systems of hermitian type are trapped between the symmetric elements of a *-prime associative triple system (of the first kind) T and those of a certain quotient triple Q(T), and by determining the quotient triples of prime associative triple systems (of the first kind) with minimal inner ideals and their involutions. As a particular case we get the structure of all the simple Jordan triple systems with minimal inner ideals, which extends a previous result of the authors [6] where reduced simple Jordan triple systems over an algebraically closed field were determined. Since a nondegenerate Jordan triple system J has dcc on principal inner ideals if, and only if, J coincides with its socle [5,7,11], equivalently, J is a direct sum of simple ideals each of which contains a minimal inner ideal, nondegenerate Jordan triple systems satisfying descending chain condition on principal inner ideals are also determined. Module some

---

Supported by the DGICYT(Ps 89-0143).

extra definitions, the corresponding structure theorems for Jordan pairs follow almost immediately from the Jordan triple results. In particular, under our restriction on the characteristic, the structure theorem due to Loos **[10]** for nondegenerate Jordan pairs having descending chain condition and ascending chain condition on principal inner ideals can be derived from our results.

Finally we note that the structure theorem for simple Jordan triple systems with minimal inner ideals is closely related to the structure due to Neher **[16]** for simple Jordan triple systems which are covered by a grid. However these two classes are distinct.

2. Preliminaries and notation

A *Jordan triple system* (JTS) over an arbitrary ring of scalars Φ is a Φ-module J with a product $P(x)y$ quadratic in x and linear in y, such that the identities

$$V(P(x)y, y) = V(x, P(y)x) \quad (1)$$
$$P(x)V(y, x) = V(x, y)P(x) \quad (2)$$
$$P(P(x)y) = P(x)P(y)P(x) \quad (3)$$

hold in all scalar extensions, where $V(x, y)z = \{x\ y\ z\} = P(x, z)y$ in terms of the polarization $P(x, z) = P(x+z) - P(x) - P(z)$. Definitions not explicitly given are taken from Loos' book **[10]**. The linear operator $B(x,y) = Id - V(x,y) + P(x)P(y)$ satisfies the following identity of Macdonald type

$$P(B(x,y)z) = B(x,y)P(z)B(y,x). \quad (4)$$

A unital Jordan algebra is a JTS with unit element 1 ($P(1) = Id$). Every Jordan algebra J defined by a quadratic map $U : J \to End_{\Phi}(J)$ and a squaring operation $x \to x^2$ gives rise to a JTS by defining $P(a) = U(a)$. Other examples of JTS are provided by associative triple systems.

A unital Φ-module B with a trilinear composition $<a\,b\,c>$ is called an *associative triple system* (ATS) *of the first* (respectively *second*) *kind* if satisfies:

$$<a\,b<c\,d\,e>> = <a<b\,c\,d>e> = <<a\,b\,c>d\,e> \quad (5)$$

respectively

$$<a\,b<c\,d\,e>> = <a<d\,c\,b>e> = <<a\,b\,c>d\,e>. \quad (6)$$

Every ATS B gives rise to a JTS B^+ by defining $P(a)b = < a\, b\, a >$. A JTS J is called *special* if it is a subtriple of A^+, where A is an associative algebra. Otherwise J is called *exceptional.* We can carry over notions on JTS to ATS. There is an interesting relationship between JTS and Jordan algebras. Fixed an element v in a Jordan triple system J, a squaring and a quadratic operator are defined by

$$x^2 = x^{(2,v)} = P(x)v, \qquad U(x) = U^{(v)}(x) = P(x)P(v) \tag{7}$$

with these operators J becomes a Jordan algebra $J^{(v)}$ called the v-*homotope* of J

The multiplication algebra of a JTS J is defined to be the subalgebra $\mathcal{M}(J)$ of $End_\Phi(J)$ generated by $P(J) \cup V(J, J)$ and the identity. A submodule N of J is called an outer ideal if $\mathcal{M}(J)N \subset N$. By (4) N is an outer ideal as soon as N is invariant under all $P(x)$, $B(x,y)$, $x, y \in J$. A submodule I of J is called an *inner ideal* if $P(I)J \subset I$. Inner ideals of the form $P(x)J$ are called *principal inner ideals.* Outer ideals that are inner ideals too are called *ideals*. If $1/2 \in \Phi$ then there is no difference between outer ideals and ideals. A JTS J is called *simple* if $P(J) \neq 0$ and J does not contain proper ideals. Unlike the case of Jordan algebras, the mere product $P(N)M$ of two ideals N, M of J is not an ideal. Accordingly, the *product* of N and M is defined as the ideal

$$N * M = P(N)M + P(J)P(N)M. \tag{8}$$

A JTS J is *semiprime* (*prime*) if $N * N = 0$ implies $N = 0$ ($N * M = 0$ implies $N = 0$ or $M = 0$) N, M ideals of J.

The *centroid* $\Gamma(J)$ of a JTS J is defined to be the set of all $\gamma \in End_\Phi(J)$ such that $T\gamma = \gamma T$ for all $T \in \mathcal{M}(J)$ and $P(\gamma x) = \gamma^2 P(x)$. If J is prime then $\Gamma(J)$ is a commutative domain that acts on J without torsion and we can embed J in its *central localization* $\Gamma(J)^{-1}J$, which is a prime JTS over the field of fractions $\Gamma(J)^{-1}\,\Gamma(J)$. If J is simple then $\Gamma(J)$ is a field.

An element a in a JTS J is said to be *trivial* if $P(a)J = 0$. A JTS without nonzero trivial elements is called *nondegenerate* . An ATS B is semiprime if and only if it is nondegenerate (see [**4**, (1.9.3)]) but this is not true for general Jordan triple systems.

We recall (see [**13**]) that the *annihilator* of a subset X of J is defined to be

(10) $$Ann_J(X) = \{ z \in J : \{z\, X\, J\} = \{ X\, z\, J \} = \{ z\, J\, X \} = P(z)X = P(X)z$$
$$= P(z)P(X)J = P(X)P(z)J = P(z)P(J)X = P(X)P(J)z = 0 \}. \tag{9}$$

If N is an ideal then $Ann_J(N)$ is an ideal. Moreover, for a nondegenerate ideal N of J its annihilator has a simple expresion [**13**, (1.7i)]

$$Ann_J(N)= \{ z \in J : P(z)N = 0 \}. \qquad (10)$$

Let J be a JTS over Φ. An element $x \in J$ is called (von Neumann) *regular* if there exists $y \in J$ such that $P(x)y = x$. Since $P(x)y = x^{(2,y)}$ in the Jordan algebra $J^{(y)}$ this means that x is an idempotent of $J^{(y)}$. It is well-known that for every regular element $x \in J$ there exists $z \in J$ such that (x, z) is an *idempotent* in the sense of [**10**, p.42], that is, $P(x)z = x$ and $P(z)x = z$. Every idempotent (x,z) of J gives rise to two Peirce decompositions $J = J_2^+ \oplus J_1^+ \oplus J_0^+$ and $J = J_2^- \oplus J_1^- \oplus J_0^-$ where the first one is the associated to the orthogonal projections $E_2 = P(x)P(z)$, $E_1=V(x, z) - 2E_2$ and $E_0 = B(x, z)$ and the second one is obtained by interchanging the roles of x and z. If $1/2 \in \Phi$ then the first Peirce decomposition coincides with the usual of the Jordan algebra $J^{(z)}$ relative to the idempotent x, where $J_2=J_1^{(z)}(x)$, $J_1 = J_{1/2}^{(z)}(x)$ and $J_0 = J_0^{(z)}(x)$.

Following [**6**], the sum of all minimal inner ideals of a nondegenerate JTS J is an ideal, Soc(J), called the *socle* of J. If J has minimal inner ideals then Soc(J) is a direct sum of simple ideals each of which contains a minimal inner ideal [**6**, Theorem 2.3]. If T is a semiprime (equivalently nondegenerate) ATS (2º kind) then $Soc(T) = Soc(T)^+$ is an ideal of T equals the sum of all minimal left (right) ideals of T [**4**, (4.3)]. The same is true for associative triple systems of the first kind.

3. Associative triple systems with minimal inner ideals and with involutions

In the recent interpretation given by D'Amour [**1**] of Zelmanov's theory, special prime nondegenerate JTS with a nonzero hermitian part are described in terms of Martindale quotient triples of prime ATS (1º kind) and of their involutions. In this section we determine the involutions of a prime ATS (1º kind) with nonzero socle, and compute its Martindale quotient triple.

Associative triple systems of the first kind were introduced by Lister [**9**] as ternary rings. He showed that every ternary ring can be realized as the anti-symmetric elements $\mathcal{A}(A, \delta)$ of an associative algebra A with respect to an automorphism δ, and used this embedding to classify finite-dimensional ternary algebras over an algebraically closed field. By using other methods, Cuenca, García and Martín obtained in [**2**] the structure of prime (in particular simple) ATS of the first kind with nonzero socle. We recall that an associative triple system A is called *prime* (respectively, *strongly prime*) if $< B\ A\ C >=0$ implies $B = 0$ or $C = 0$, B, C ideals of A (respectively, $< a\ A\ b >=0$ implies $a = 0$ or $b = 0$, $a, b \in A$). Unlike the associative algebras primeness and strong primeness are not equivalent notions in ATS. An ATS A with minimal inner ideals is strongly prime iff it is prime and $Soc(A)$ is *strongly simple* (it does not contain nontrivial *outer ideals* B, $< B A A > + < A A B > \subset B$).

Let (X,X',f) be a pair of dual vector spaces over a division associative algebra Δ (see [8, p.12]) and let $\sigma : \Delta \to \Delta$ be an automorphism. Consider the set $\mathcal{L}_\sigma(X,X',f)$ of all σ-linear operators $a : X \to X$ having a (unique) adjoint $a^\# : X' \to X'$,

$$f(xa, x') = \sigma(f(x, a^\# x')). \tag{11}$$

If $\alpha \in \Delta$ satisfies $\sigma(\alpha) = \alpha$ and $\sigma^2(\beta) = \alpha^{-1}\beta\alpha$ ($\beta \in \Delta$), then $\mathcal{L}_\sigma(X, X', f)$ is a strongly prime ATS (1º kind) with respect to the triple product

$$< a\, b\, c > = abch_\alpha = h_\alpha\, abc \tag{12}$$

where $xh_\alpha = \alpha x$ ($x \in X$). We denote this ATS by $\mathcal{L}_\sigma(X, X', f, \alpha)$. Moreover, the socle of $\mathcal{L}_\sigma(X, X', f, \alpha)$ coincides with the ideal $\mathcal{F}_\sigma(X, X', f, \alpha)$ of all finite-rank operators a in $\mathcal{L}_\sigma(X, X', f, \alpha)$.

Suppose now that (X, X', f), (Y, Y', g) are two pairs of dual vector spaces over Δ and let $\mathcal{L}(X, Y)$ be the set of all linear operators $a : X \to Y$ having a (unique) adjoint $a^\#: Y' \to X'$ ($g(xa, y') = f(x, a^\# y')$, $x \in X$, $y' \in Y'$). Then $\mathcal{L}(X, Y) \oplus \mathcal{L}(Y, X)$ under the triple product

$$< (a_1, a_2)\, (b_1, b_2)\, (c_1, c_2) > = (a_1 b_2 c_1, a_2 b_1 c_2) \tag{13}$$

is a prime (not strongly prime) ATS of first kind whose socle is $\mathcal{F}(X, Y) \oplus \mathcal{F}(Y, X)$. Conversely

Theorem 1 [2, Corollary 3]. Every prime associative triple system of first kind with minimal inner ideals is one of the following:

(i) A subtriple B of some $\mathcal{L}_\sigma(X, X', f, \alpha)$ containing $\mathcal{F}_\sigma(X, X', f, \alpha)$.

(ii) A subtriple B of some $\mathcal{L}(X, Y) \oplus \mathcal{L}(Y, X)$ containing $\mathcal{F}(X, Y) \oplus \mathcal{F}(Y, X)$.

We note that every nondegenerate self-dual vector space (X, g) over a division associative algebra Δ with involution $\tau : \Delta \to \Delta$ gives rise to a dual pair (X, X, g)

where the second X is regarded as a right vector space under $x.\beta = \tau(\beta)x$, $x \in X$, $\beta \in \Delta$.

Proposition 2. Let (Y, g) be a nondegenerate (hermitian or alternate) self-dual vector space over a division associative algebra with involution (Δ, τ), $\sigma : \Delta \to \Delta$ an automorphism and $\alpha \in \Delta$ such that $\sigma(\alpha) = \alpha$, $\sigma^2(\beta) = \alpha^{-1}\beta\alpha$ ($\beta \in \Delta$). If $\tau(\alpha) = \alpha$ and $\sigma\tau = \tau\sigma^{-1}$ then the mapping $a \to a^{\#}$ is an involution of $\mathcal{L}_\sigma(Y, g, \alpha)$. We note that if g is alternate then τ is the identity and hence $\sigma = \sigma^{-1}$.

Proof. Since both cases are similar we may assume g is hermitian.

$$\begin{aligned} &g(y_1< a\ b\ c >, y_2) = \alpha g(y_1abc, y_2) = \sigma(\alpha g(y_1ab, y_2c^{\#})) = \\ &\alpha\sigma^2(g(y_1a, y_2c^{\#}b^{\#})) = g(y_1a, y_2c^{\#}b^{\#})\alpha = \sigma(g(y_1, y_2c^{\#}b^{\#}a^{\#}))\alpha = \\ &\sigma(g(y_1, \tau(\alpha)(y_2c^{\#}b^{\#}a^{\#}))) = \sigma(g(y_1, \alpha(y_2c^{\#}b^{\#}a^{\#}))) = \\ &\sigma(g(y_1, y_2< c^{\#}b^{\#}a^{\#}>)) \end{aligned}$$

implies

$$< a\ b\ c >^{\#} = < c^{\#}\ b^{\#}\ a^{\#} >.$$

$$\begin{aligned} &g(y_1a^{\#}, y_2) = \tau(g(y_2, y_1a^{\#})) = \tau\sigma^{-1}(\sigma(g(y_2, y_1a^{\#}))) = \tau\sigma^{-1}(g(y_2a, y_1)) = \\ &\sigma\tau(g(y_2a, y_1)) = \sigma(g(y_1, y_2a)) \end{aligned}$$

implies $(a^{\#})^{\#} = a$, so the mapping $a \to a^{\#}$ is an involution.

Conversely, strongly prime ATS (1º kind) with minimal inner ideals and with involution are those described above. We stress that the proof of this result given below is intrinsic in the sense that it avoids imbedding into associative algebras.

Theorem 3. Every strongly prime associative triple system of the first kind A (over Φ, $1/2 \in \Phi$) with minimal inner ideals and with involution $* : A \to A$ is a subtriple of some $\mathcal{L}_\sigma(Y, g, \alpha)$ containing $\mathcal{F}_\sigma(Y, g, \alpha)$, $*$ being either # or -#.

Paragraph 1 of the proof. Let (a, b) be a *regular pair* ($a = < a\ b\ a >$, $b = < b\ a\ b >$) with a in a minimal inner ideal. For every $c \in A$ such that $< b\ c\ a > \neq 0$ the element $u = < a\ b\ c >$ generates a *nondegenerate* minimal inner ideal I ($< y\ I\ y > \neq 0$ for all $0 \neq y \in I$, and hence $< y\ I\ y > = I$ by minimality of I).

By [4, Lemma 4.2] $u = R(b, c)a$ lies in a minimal inner ideal, then we only need show that $< u\ u\ u > \neq 0$. The mapping $L(b, u) : < a\ b\ A > \to < b\ a\ A >$ is one-to-one since $L(b, u)a \neq 0$ and $< a\, b\, A >$, $< b\, a\, A >$ are minimal right ideals. Hence

$$0 \neq L(b, u)u = < b\, u\, u > = < b\, u < a\, b\, u >>$$

implies $< u\, a\, b > \neq 0$. Again $L(u, a) : < b\, a\, A > \rightarrow < a\, b\, A >$ is one-to-one, and hence

$$0 \neq L(u, a)< b\, u\, u > = < u\, u\, u >,$$

as required. We stress the following relations for a regular pair (u, v) in a nondegenerate minimal inner ideal I:

$$u = < u\, u\, v > = < v\, u\, u >, \quad v = < v\, v\, u > = < u\, v\, v > \text{ and } < u\, A\, u > = I = < v\, A\, v > . \qquad (14)$$

Paragraph 2. There exists a regular pair (u, v) in a nondegenerate minimal inner ideal and a nonzero hermitian or anti-hermitian element $z \in A$ satisfying one of the following conditions:

(i) $\quad < u\, z\, v^* > = z^* = < v\, z\, u^* >$

or

(ii) $\quad < u\, z\, v^* > = -z^* = < v\, z\, u^* >$

Suppose first that A has a nondegenerate minimal inner ideal I such that $I \cap \mathcal{H}(A,*) \neq 0$ ($I \cap \mathcal{A}(A,*) \neq 0$, respectively). We note that $0 \neq I \cap I^* \Rightarrow I = I^*$ by minimality of I. Let u be a nonzero hermitian (anti-hermitian, respectively) element of I. Then $u = < u\, y\, u >$ with $y \in I$ and we have $u = < u\, w\, u >$, where $w = 1/2(y+y^*)$ ($w=1/2(y-y^*)$, respectively) is a hermitian (anti-hermitian, respectively) element of I. If $v = < w\, u\, w >$ then (u, v) is a regular pair, with v a hermitian (anti-hermitian, respectively) element of I. By (14) $z = u$ satisfies (i).

Suppose to the contrary that every nondegenerate minimal inner ideal does not contain any nonzero hermitian or anti-hermitian element. Every element u in a minimal inner ideal I of A satisfies

$$< u^*\, u\, A > = 0. \qquad (15)$$

Suppose otherwise $< u^*\, u\, A > \neq 0$ and let (u, v) be a regular pair. By strongly primeness, $0 \neq < u\, A < u^*\, u\, A >> = << u\, v\, u > A < u^*\, u\, A >> = < u < v < u\, A\, u^* > u > A > \Rightarrow$ $< v < u\, A\, u^* > u > \neq 0$. Since $1/2 \in \Phi$, there exists a hermitian or anti-hermitian element b such that $< v < u\, b\, u^* > u > \neq 0$. Then $x = < u\, b\, u^* > = < u\, v\, x >$ is a hermitian or anti-hermitian element generating a nondegenerate minimal inner ideal by Paragraph 1, which yields a contradiction.

Let u be in a minimal inner ideal I and (u, v) a regular pair. If $y \in < u\, v\, A >$ then

$$< y\, y^*\, A > = 0. \tag{16}$$

This follows from (15) since y^* lies in the minimal inner ideal $L(y^*, v^*)I^*$.

Let (u, v) be a regular pair in a nondegenerate minimal inner ideal I (that exists by Paragraph 1 since A is strongly prime). Since $1/2 \in \Phi$ we have by strongly primeness of A that there exists a hermitian or anti-hermitian element $c \in A$ such that

$$0 \neq < u\, c\, v^* > = << u\, u\, v > c < v^*\, u^*\, v^* >> = < u < u < v\, c\, v^* > u^* > v^* > = < u\, z\, v^* >$$

where $z = < u < v\, c\, v^* > u^* >$ is a hermitian or anti-hermitian element in $< u\, v\, A > \cap < A\, v^*\, u^* >$. Linearizing (16) we get

$$< u\, z\, v^* > = -< u\, v\, z^* > = -z^* = -< v\, u\, z^* > = < v\, z\, u^* >$$

since $z, v, u \in < u\, v\, A >$.

Paragraph 3. Let u, v and z as in Paragraph 2. Without loss in generality we may assume that z is HERMITIAN since otherwise we can take $-*$ which is also an involution of triple systems. Write Δ to denote the division associative algebra defined in $< u\, A\, u >$ by

$$a.b = < a\, v\, b > \qquad (a, b \in < u\, A\, u >) \tag{17}$$

and let $Y = < u\, v\, A >$ be the left vector space over Δ defined by

$$a.y = < a\, v\, y > \qquad (a \in \Delta,\ y \in Y). \tag{18}$$

Since $< u\, z\, v^* > = \pm z \neq 0$, we have that the mapping $\eta = R(z, v^*) : < A\, v\, u > \longrightarrow < A\, v^*\, u^* >$ is an *isomorphism of* (minimal) *left ideals*, i.e, η is a linear isomorphism satisfying:

$$\eta < a\, b\, x > = < a\, b\, \eta(x) >$$

for all $a, b \in A$, $x \in < A\, v\, u >$. Now we define $\tau : \Delta \rightarrow \Delta$ by

$$\tau(a) = \pm < z\, a^* \eta^{-1}(v^*) > \qquad (a \in \Delta) \tag{19}$$

We have

$$\tau(a.b) = \pm< z < a\, v\, b >^* \eta^{-1}(v^*) > = \pm< z < b^*\, v^*\, a^* > \eta^{-1}(v^*) > =$$
$$\pm< z < b^* < \eta^{-1}(v^*)\, z\, v^* > a^* > \eta^{-1}(v^*) > =$$
$$\pm<< z\, b^*\, \eta^{-1}(v^*) >< v\, u\, z >< v^*\, a^*\, \eta^{-1}(v^*) >> =$$
$$< \tau(b)\, v << u\, z\, v^* > a^*\, \eta^{-1}(v^*) >>=< \tau(b)\, v < (\pm z)\, a^*\, \eta^{-1}(v^*) >>=\tau(b).\tau(a)$$

and

$$\tau(\tau(a)) = < z < z\, a^*\, \eta^{-1}(v^*) >^* \eta^{-1}(v^*) > =$$
$$< z < (\eta^{-1}(v^*))^*\, a\, z > \eta^{-1}(v^*) > = << u\, v\, z > < (\eta^{-1}(v^*))^*\, a\, z > \eta^{-1}(v^*) >=$$
$$<< u < v\, z\, (\eta^{-1}(v^*))^* > a > z\, \eta^{-1}(v^*) > =$$
$$<< u < \eta^{-1}(v^*)\, z\, v^* >^* a > z\, \eta^{-1}(v^*) > = << u\, (v^*)^*\, a > z\, \eta^{-1}(v^*) > =$$
$$<< u\, v\, a > z\, \eta^{-1}(v^*) > = < a\, z\, \eta^{-1}(v^*) > = << a\, v\, u > z\, \eta^{-1}(v^*) > =$$
$$< a\, v < u\, z\, \eta^{-1}(v^*) >> = < a\, v\ \eta^{-1} < u\, z\, v^* >> = < a\, v\, u > = a.$$

Thus $\tau : \Delta \to \Delta$ is an involution. Now we define $\sigma : \Delta \to \Delta$ by

$$\sigma(a) = < u\, a\, v > \qquad (a \in \Delta). \tag{20}$$

It is not difficult to verify that σ is an automorphism with $\sigma^{-1}(b) = < v\, b\, u >$ $(b \in \Delta)$. Moreover,

$$\tau\sigma^{-1} = \sigma\tau. \tag{21}$$

Indeed,

$$\tau\sigma^{-1}(a) = \pm < z < v\, a\, u >^* \eta^{-1}(v^*) > = \pm < z < u^*\, a^*\, v^* > \eta^{-1}(v^*) > =$$
$$\pm << u\, v\, z > < u^*\, a^*\, v^* > \eta^{-1}(v^*) > = \pm < u < v\, z\, u^* > < a^*\, v^*\, \eta^{-1}(v^*) >> =$$
$$<< u\, z\, a^* > v^*\, \eta^{-1}(v^*) > = << u\, z\, a^* > < \eta^{-1}(v^*)\, z\, v^* > \eta^{-1}(v^*) > =$$
$$<< u\, z\, a^* > < \eta^{-1}(v^*) < v\, u\, z > v^* > \eta^{-1}(v^*) > =$$
$$<< u\, z\, a^* > \eta^{-1}(v^*) < v < u\, z\, v^* > \eta^{-1}(v^*) >> =$$
$$\pm<< u\, z\, a^* > \eta^{-1}(v^*) < v\, z\, \eta^{-1}(v^*) >> = \pm<< u\, z\, a^* > \eta^{-1}(v^*)\, v > =$$
$$\pm< u < z\, a^*\, \eta^{-1}(v^*) > v > = \sigma\tau(a).$$

Now we define $g : Y \times Y \to \Delta$ by

$$g(y_1, y_2) = < y_1\, y_2{}^*\, \eta^{-1}(v^*) > \tag{22}$$

Then g is a nondegenerate hermitian or skew-hermitian inner product over Δ relative to the involution τ, i.e.,

(i) $g(y_1+y_2, y) = g(y_1, y) + g(y_2, y)$

(ii) $g(a.y_1, y_2) = a.g(y_1, y_2)$

(iii) $g(y, y_1+y_2) = g(y, y_1) + g(y, y_2)$

(iv) $g(y_1, a.y_2) = g(y_1, y_2).\tau(a)$

(v) $\tau(g(y_1, y_2)) = \pm\ g(y_2, y_1)$

(vi) $g(y, Y) = 0 \Rightarrow y = 0.$

We only show the conditions (iv), (v) and (vi):

(iv) $g(y_1, a.y_2) = < y_1 < a\ v\ y_2 >^* \eta^{-1}(v^*) > = < y_1 < y_2^*\ v^*\ a^* > \eta^{-1}(v^*) > =$
$< y_1 < y_2^* < \eta^{-1}(v^*)\ z\ v^* > a^* > \eta^{-1}(v^*) > =$
$<< y_1\ y_2^*\ \eta^{-1}(v^*) > < v\ u\ z > < v^*\ a^*\ \eta^{-1}(v^*) >>=$
$<< y_1\ y_2^*\ \eta^{-1}(v^*) > v << u\ z\ v^* > a^*\ \eta^{-1}(v^*) >> =$
$<< y_1\ y_2^*\ \eta^{-1}(v^*) > v < (\pm z)\ a^*\ \eta^{-1}(v^*) >> = \ g(y_1, y_2).\tau(a)$

(v) $\tau(g(y_1, y_2)) = \pm < z < y_1\ y_2^*\ \eta^{-1}(v^*) >^* \eta^{-1}(v^*) > =$
$\pm< z < (\eta^{-1}(v^*))^*\ y_2\ y_1^* > \ \eta^{-1}(v^*) > =$
$\pm<<< u\ v\ z > (\eta^{-1}(v^*))^*\ y_2 > y_1^*\ \eta^{-1}(v^*) > =$
$\pm<< u < \eta^{-1}(v^*)\ z\ v^* >^*\ y_2 > y_1^*\ \eta^{-1}(v^*) > =$
$\pm<< u\ (v^*)^*\ y_2 > y_1^*\ \eta^{-1}(v^*) > = \pm < y_2\ y_1^*\ \eta^{-1}(v^*) > = \pm\ g(y_2, y_1)$

and

(vi) $g(y, Y) = 0 \ \Rightarrow < y < u\ v\ A >^* \eta^{-1}(v^*) > = < y < A\ v^*\ u^* > \eta^{-1}(v^*) > =$
$< y\ A < v^*\ u^*\ \eta^{-1}(v^*) >> = < y\ A\ \eta^{-1}(v^*) > = 0 \quad \Rightarrow \quad y = 0,$

by strongly primeness of A. The hermitian case occurs when there exists a nondegenerate minimal inner ideal with nonzero hermitian or anti-hermitian elements. Otherwise we have that condition (16) holds and hence the skew-hermitian inner product is actually alternate. We have

$$\sigma(v) = v = \tau(v) \quad \text{and} \quad v.\sigma^2(b) = b.v \qquad (b \in \Delta). \tag{23}$$

Since u, v are in a nondegenerate minimal inner ideal, $\sigma(v) = v$ by (14). Moreover

$$\tau(v) = \pm< z\ v^*\ \eta^{-1}(v^*) > = \pm<< v\ u\ z > v^*\ \eta^{-1}(v^*) > =$$
$$\pm< v < u\ z\ v^* > \eta^{-1}(v^*) > = \pm< v\ (\pm z)\ \eta^{-1}(v^*) > = v$$

and

$v.\sigma^2(b) = <v\,v<u<u\,b\,v>v>> = <<v<v\,u\,u>b>v\,v> = <<v\,u\,b>v\,v> =$
$<b\,v\,v> = b.v.$

Now the Proposition applies, in virtue of (21) and (23), to get that $\mathcal{L}_\sigma(Y, g, v)$ is a strongly prime ATS (1º kind) with involution $a \to a^{\#}$. Next we show that A can be regarded as a subtriple of $\mathcal{L}_\sigma(Y, g, v)$ containing $\mathcal{F}_\sigma(Y, g, v)$, with $a^* = \pm a^{\#}$, $a \in A$. For $a \in A$ we define $\varphi_a : Y \to Y$ by

$$y\varphi(a) = y\varphi_a = <u\,y\,a> \quad (y \in Y) \tag{24}$$

$$g(y_1\varphi_a, y_2) = <<u\,y_1\,a>y_2{}^*\,\eta^{-1}(v^*)> = <<u\,y_1\,a>y_2{}^*<u^*\,v^*\,\eta^{-1}(v^*)>> =$$
$$<u<y_1<u\,y_2\,a^*>^*\,v^*>\eta^{-1}(v^*)> =$$
$$<u<y_1<u\,y_2\,a^*>^*<\eta^{-1}(v^*)\,z\,v^*>>\eta^{-1}(v^*)> =$$
$$<u<y_1<u\,y_2\,a^*>^*\,\eta^{-1}(v^*)><<v\,u\,z>v^*\,\eta^{-1}(v^*)>> =$$
$$<u<y_1<u\,y_2\,a^*>^*\,\eta^{-1}(v^*)><v\,(\pm z)\,\eta^{-1}(v^*)>> =$$
$$<u<y_1<u\,y_2\,(\pm a^*)>^*\,\eta^{-1}(v^*)><v\,z\,\eta^{-1}(v^*)>> = <u\,g(y_1, y_2\,\varphi_{\pm a^*})\,v> =$$
$$\sigma(g(y_1, y_2\varphi_{\pm a^*})).$$

Thus $\varphi_a \in \mathcal{L}_\sigma(Y, g, v)$ with $(\varphi_a)^{\#} = \varphi_{\pm a^*}$. For $a, b, c \in \mathcal{L}_\sigma(Y, g, v)$, $y \in Y$ we have

$$y<\varphi_a\,\varphi_b\,\varphi_c> = (<v\,v\,y>)\varphi_a\varphi_b\varphi_c = <u<u<u<v\,v\,y>a>b>c> =$$
$$<u<u<u\,v\,v>y><a\,b\,c>> = <u<u\,v\,y><a\,b\,c>> = <u\,y<a\,b\,c>> =$$
$$y\varphi_{<a\,b\,c>}$$

so $a \to \varphi_a$ is an homomorphism of A into $\mathcal{L}_\sigma(Y, g, v)$. If $\varphi_a = 0$ then $a = 0$ since $<u<u\,v\,A>a> = <u\,A\,a> = 0 \Rightarrow a = 0$ by strongly primeness of A. Finally we see that φ_A contains $\mathcal{F}_\sigma(Y, g, v)$. For $y_1, y_2 \in Y$ we define $y_1 \otimes y_2 : Y \to Y$ by $y(y_1 \otimes y_2) = \sigma(g(y, y_1))y_2$ $(y \in Y)$. It is easy to see that $y_1 \otimes y_2 \in \mathcal{L}_\sigma(Y, g)$ with $(y_1 \otimes y_2)^{\#} = y_2 \otimes y_1$. Moreover, we have the following relations for $a, b \in \mathcal{L}_\sigma(Y, g, v)$

$$<a\,b\,y_1 \otimes y_2> = (y_1 b^{\#} a^{\#}) \otimes (v.y_2) \tag{25}$$
$$<y_1 \otimes y_2\,a\,b> = y_1 \otimes ((v.y_2)ab) \tag{26}$$
$$<a\,y_1 \otimes y_2\,b> = (y_1 a^{\#}) \otimes ((v.y_2)b). \tag{27}$$

Hence the linear span of all $y_1 \otimes y_2$ is an ideal of $\mathcal{L}_\sigma(Y, g, v)$ equals $\mathcal{F}_\sigma(Y, g, v)$. Now we have

$$y(y_1 \otimes y_2) = \sigma(g(y, y_1)).y_2 = << u < y\, y_1{}^* \eta^{-1}(v^*) > v > v\, y_2 > =$$
$$< u\, y < y_1{}^* \eta^{-1}(v^*) < v\, v\, y_2 >>> = y\varphi\ (< y_1{}^* \eta^{-1}(v^*) < v\, v\, y_2 >>)$$

so $y_1 \otimes y_2 = \varphi\ (< y_1{}^* \eta^{-1}(v^*) < v\, v\, y_2 >>)$, which completes the proof.

Now we are going to describe the involutions of a prime (but not strongly prime) ATS of the first kind with nonzero socle.

Let (X, g), (Y, h) be self-dual (both hermitian or alternate) vector spaces over a division algebra with involution (Δ, τ). Consider the ATS (1º kind) $\mathcal{L}(X, Y) \oplus \mathcal{L}(Y, X)$. Then the mapping

$$(a_1, a_2) \rightarrow (a_2{}^{\#}, a_1{}^{\#}) \tag{28}$$

is an involution of $\mathcal{L}(X, Y) \oplus \mathcal{L}(Y, X)$.

Suppose now that (X, Y, f) is a pair of dual vector spaces over a division associative algebra Δ and suppose that $\tau : \Delta \rightarrow \Delta$ is an involution. We can consider the *opposite pair,* with respect to τ, (Y, X, f^{op}) defined by $\beta.y = y\tau(\beta)$, $x.\beta = \tau(\beta)x$ and $f^{op}(y, x) = \tau(f(x, y))$, $x \in X$, $y \in Y$, $\beta \in \Delta$. Then it is not difficult to see that the mapping

$$(a_1, a_2) \rightarrow (a_1{}^{\#}, a_2{}^{\#}) \tag{29}$$

is an involution of the ATS of first kind $\mathcal{L}(X, Y) \oplus \mathcal{L}(Y, X)$. Next theorem shows that these are essentially the only involutions that there exist in a prime (but not strongly prime) ATS of the first kind with nonzero socle.

Theorem 4. Let T be a prime (but not strongly prime) associative triple system of the first kind (over a field F) with nonzero socle and with an involution $^* : T \rightarrow T$. Then we have one of the following possibilities:

(i) There exist (X, g), (Y, h) self-dual (both hermitian or alternate) vector spaces over a division algebra with involution (Δ, τ) such that T is a subsystem of $\mathcal{L}(X, Y) \oplus \mathcal{L}(Y, X)$ under $< (a_1, a_2)\ (b_1, b_2)\ (c_1, c_2) > = (a_1 b_2 c_1, a_2 b_1 c_2)$ containing $\mathcal{F}(X, Y) \oplus \mathcal{F}(Y, X)$, and $^* : T \rightarrow T$ is the restriction of the involution $(a_1, a_2) \rightarrow (a_2{}^{\#}, a_1{}^{\#})$ defined in (28).

(ii) T is a subsystem of $\mathcal{L}(X, Y) \oplus \mathcal{L}(Y, X)$ containing $\mathcal{F}(X, Y) \oplus \mathcal{F}(Y, X)$, where (X,Y, f) is a pair of dual vector spaces over a division associative algebra Δ with involution $\tau : \Delta \rightarrow \Delta$, (Y, X, f^{op}) is the opposite pair of (X, Y, f), and $^* : T \rightarrow T$

is either the restriction of the involution $(a_1, a_2)^* = (a_1^{\#}, a_2^{\#})$ defined in (29) or minus the restriction of this involution (Δ being a field and $\tau : \Delta \to \Delta$ the identity in the last case).

Proof. By Theorem 1, T is a subsystem of $\mathcal{L}(X,Y)\oplus\mathcal{L}(Y,X)$ containing $\mathcal{F}(X,Y)\oplus\mathcal{F}(Y,X)$, (X, X', f), (Y, Y', g) being pairs of dual vector spaces over a division algebra Δ. Then $Soc(T) = \mathcal{F}(X, Y)\oplus\mathcal{F}(Y, X) = N_1 \oplus N_2$ is a direct sum of two nonzero outer ideals. Since Soc(T) is *-invariant and has no other proper nonzero outer ideals, either $N_1^* = N_2$ or $N_i^* = N_i$, $i = 1,2$. Now $\mathcal{F}(X, Y) = N_1$ (respectively, $\mathcal{F}(Y, X) = N_2$) become the Peirce space B_{12} (respectively, B_{21}) of the unital associative algebra over F.

$$B = (\mathcal{F}(X, X)+F1_X) \oplus \mathcal{F}(X, Y) \oplus \mathcal{F}(Y, X) \oplus (\mathcal{F}(Y, Y)+F1_Y)$$

with respect to the idempotents $e_1 = 1_X$, $e_2 = 1_Y$.

Suppose first that $N_i^* = N_i$, $i = 1, 2$. Then the involution $^* : Soc(T) \to Soc(T)$ extends to a (unique) involution of the algebra B by

$$(a_{11}+a_{12}+a_{21}+a_{22})^* = a_{22}^*+a_{12}^*+a_{21}^*+a_{11}^*$$

where

$$a_{11}^* = (\beta e_1+\Sigma b_{12}c_{21})^* = \beta e_2+\Sigma c_{21}^* b_{12}^* \in B_{22}$$
and
$$a_{22}^* = (\beta e_2+\Sigma b_{21}c_{12})^* = \beta e_1+\Sigma c_{12}^* b_{21}^* \in B_{11}.$$

By applying the structure theorem of primitive associative algebras with involution containing minimal right ideals ([8, p.17]), we get that $\star : B \to B$ is the adjoint involution of a nondegenerate (hermitian or alternate) inner product t on the vector space $X \oplus Y$ over a division algebra with involution (Δ, τ), that is

$$t(zb, z') = t(z, z'b^*) \quad (z, z' \in X \oplus Y,\ b \in B). \tag{31}$$

We note that $t(x, x') = t(y, y') = 0$. Indeed, $t(x, x') = t(xe_1, x') = t(x, x'e_2) = 0$. Similarly $t(y, y') = 0$. Define $h : X \times Y \to \Delta$ by $h(x, y) = t(x, y)$. Then (X, Y, h) is a pair of dual vector spaces over Δ, where Y is the right vector space defined by $y.\beta = \tau(\beta)y$. Now $Soc(T) = B_{12}\oplus B_{21} = \mathcal{F}(X, Y) \oplus \mathcal{F}(Y, X)$, relative to the dual pairs (X, Y, h), (Y, X, h^{op}). If t is hermitian we have by (31) that

$$\tau(t(x', xa_1)) = t(x, x'a_1^*) \Leftrightarrow h^{op}(xa_1, x') = h(x, x'a_1^*)$$

so $a_1^* = a_1^\#$, $a_1 \in \mathcal{F}(X, Y)$. Similarly $a_2^* = a_2^\#$ for all $a_2 \in \mathcal{F}(Y, X)$. This implies, by uniquenes of extensions of involutions on Soc(T), that $(a_1, a_2)^* = (a_1^\#, a_2^\#)$ for all $(a_1, a_2) \in T$. If t is alternate then Δ is a field and $\tau = 1_\Delta$, so by (31) we have

$$-t(x', xa_1) = t(x, x'a_1^*) \Leftrightarrow -h^{op}(xa_1, x') = h(x, x'a_1^*)$$

and hence $a_1^* = -a_1^\#$, $a_1 \in \mathcal{F}(X, Y)$. Similarly $a_2^* = -a_2^\#$ for all $a_2 \in \mathcal{F}(Y, X)$. Then $(a_1, a_2)^* = -(a_1^\#, a_2^\#)$ for all $(a_1, a_2) \in T$.

Suppose finally that $N_1^* = N_2$. Then, as above, the involution $* : \mathrm{Soc}(T) \to \mathrm{Soc}(T)$ extends to a (unique) involution of the algebra B by

$$(a_{11}+a_{12}+a_{21}+a_{22})^* = a_{11}^*+a_{21}^*+a_{12}^*+a_{22}^*$$

where now

$$a_{11}^* = (\beta e_1+\Sigma b_{12}c_{21})^* = \beta e_1+\Sigma c_{21}^* b_{12}^* \in B_{11}$$

and

$$a_{22}^* = (\beta e_2+\Sigma b_{21}c_{12})^* = \beta e_2+\Sigma c_{12}^* b_{21}^* \in B_{22}.$$

Again, by applying the structure theorem of primitive associative algebras with involution containing minimal right ideals, we get that $* : B \to B$ is the adjoint involution of a nondegenerate (hermitian or alternate) inner product t on the vector space $X \oplus Y$ over a division algebra with involution (Δ, τ) as in (31). Since $t(x, y) = t(xe_1, y) = t(x, e_1 y) = 0$, (X, t), (Y, t) remain self-dual (both hermitian or alternate) vector spaces. Then T is a subsystem of the associative triple system of the first kind $\mathcal{L}(X,Y) \oplus \mathcal{L}(Y,X)$ containing $\mathcal{F}(X,Y) \oplus \mathcal{F}(Y,X)$, and $* : T \to T$ is now the restriction of the involution $(a_1, a_2) \to (a_2^\#, a_1^\#)$ defined in (28).

For a prime ATS (1º kind) T, let Q(T) be the Martindale quotient triple of T with respect to the denominator filter of all nonzero ideals of T (see [**14**]). By [**14**, (3.20)] Q(T) is a prime ATS (1º kind) containing T, and since prime ATS are nondegenerate, we can consider the socles of T and Q(T).

Lemma 5. Let T be a prime associative triple system of the first kind with nonzero socle. Then its Martindale quotient triple Q(T) is the largest prime associative triple system of first kind with the same socle as T.

Proof. Write $M = Soc(T)$. By **[6**,(2.3)] M is a simple ideal of T. Now we are going to verify that M is actually an ideal in Q(T). By [**14**, (3.21)]

$$<q T M> + <M T q> \subset T$$

for all $q \in Q(T)$ because M is simple; but $M = <M M <M M M>>$ by simplicity of M again. Hence for all $q, v \in Q(T)$ we have

$$<q v M> = <q v <M M <M M M>>> = <q <v M M> <M M M>> \subset$$
$$<<q T M> M M> \subset <T M M> \subset M.$$

Similarly we get that both $<M v q>$ and $<q M v>$ are contained in M, which proves that M is an ideal of Q(T). Finally, by [**6**, (2.4)], $M \lhd Q(T)$ implies $M = Soc(M) = Soc(Q(T)) \cap M$, so $M \subset Soc(Q(T))$; but Soc(Q(T)) is a simple ideal because Q(T) is prime. Then $Soc(Q(T)) = M$. Suppose finally that B is a prime ATS (1º kind) such that $Soc(B) = M = Soc(T)$. Then $<b T M> + <M T b> \subset M \subset T$ for all $b \in B$, and hence, by [**14**, (3.21)], $B \subset Q(T)$, because $M = <M M M>$ is *doubly-T-sturdy* in B (in the sense of McCrimmon [**14**]) by primeness of B.

We end this section computing the Martindale quotient triple of a prime ATS (1º kind) with nonzero socle, and of a prime ATS (1º kind) with nonzero socle having an involution. A similar result was obtained in [**3**] for the case of an associative algebra, in order to derive Osborn-Racine theorem for prime nondegenerate Jordan algebras with nonzero socle from the more general Zelmanov theorem for prime nondegenerate Jordan algebras.

Theorem 6. Let T be a prime associative triple system of the first kind with nonzero socle. If T is strongly simple then

(i) $Q(T)= \mathcal{L}_\sigma(X, X', f, \alpha)$ under $P(a)b = abah_\alpha$, where (X, X', f) is a pair of dual vector spaces over a division associative algebra Δ, $\sigma : \Delta \to \Delta$ an automorphism, and $\alpha \in \Delta$ satisfies $\sigma(\alpha) = \alpha$, $\sigma^2(\beta) = \alpha^{-1}\beta\alpha$, for all $\beta \in \Delta$.

If T is not strongly simple then

(ii) $Q(T)= \mathcal{L}(X,Y) \oplus \mathcal{L}(Y,X)$ under $P(a_1, a_2)(b_1, b_2) = (a_1 b_2 a_1, a_2 b_1 a_2)$ with respect to (X,X', f), (Y,Y', g) pairs of dual vector spaces over a division associative algebra Δ.

Suppose now that T has an involution $^* : T \to T$. Then we have one of the following:

(iii) $Q(T)= \mathcal{L}_\sigma(Y, g, \alpha)$ under $P(a)b = abah_\alpha$ with $a^* = \pm a^\#$, where (Y, g) is a nondegenerate hermitian or alternate self-dual vector space over a division associative algebra Δ with involution $\tau : \Delta \to \Delta$, $\sigma : \Delta \to \Delta$ is an automorphism such that $\sigma\tau = \tau\sigma^{-1}$, and $\alpha \in \Delta$ satisfies $\sigma(\alpha) = \alpha = \tau(\alpha)$, $\sigma^2(\beta) = \alpha^{-1}\beta\alpha$, for all $\beta \in \Delta$.

(iv) $Q(T) = \mathcal{L}(X, Y) \oplus \mathcal{L}(Y, X)$ under $P(a_1, a_2)(b_1, b_2) = (a_1b_2a_1, a_2b_1a_2)$ with $(a_1, a_2)^* = (a_2^{\#}, a_1^{\#})$, where (X, g), (Y, h) are nondegenerate (both hermitian or alternate) self-dual vector spaces over a division associative algebra with involution (Δ, τ).

(v) $Q(T) = \mathcal{L}(X, Y) \oplus \mathcal{L}(Y, X)$ under $P(a_1, a_2)(b_1, b_2) = (a_1b_2a_1, a_2b_1a_2)$ with $(a_1, a_2)^* = (a_1^{\#}, a_2^{\#})$, where (X, Y, f) is a pair of dual vector spaces over a division associative algebra Δ, and (Y, X, f^{op}) is its opposite pair relative to an involution $\tau : \Delta \to \Delta$.

(vi) $Q(T) = \mathcal{L}(X, Y) \oplus \mathcal{L}(Y, X)$ under $P(a_1, a_2)(b_1, b_2) = (a_1b_2a_1, a_2b_1a_2)$ with $(a_1, a_2)^* = -(a_1^{\#}, a_2^{\#})$, with respect to a pair of dual vector spaces (X, Y, f) and its opposite (Y, X, f^{op}) over a field K.

Proof. By Lemma 5, Q(T) is a prime ATS (1º kind) with the same socle as T. If T is strongly prime then Q(T) is also strongly prime because Soc(T) is strongly simple. Then, by Theorem 1(i), there exists a pair of dual vector spaces (X, X', f) over a division associative algebra Δ, an automorphism $\sigma : \Delta \to \Delta$, and $\alpha \in \Delta$ satisfying $\sigma(\alpha) = \alpha$, $\sigma^2(\beta) = \alpha^{-1}\beta\alpha$, for all $\beta \in \Delta$, such that

$$\mathcal{F}_\sigma(X, X', f, \alpha) = \mathrm{Soc}(T) \lhd T \subset Q(T) \subset \mathcal{L}_\sigma(X, X', f, \alpha),$$

but Q(T) is actually equal to $\mathcal{L}_\sigma(X, X', f, \alpha)$ by maximality of Q(T) (Lemma 5). Similarly, if T is not strongly prime we have by Theorem 1(ii) that

$$\mathcal{F}(X,Y) \oplus \mathcal{F}(Y,X) = \mathrm{Soc}(T) \lhd T \subset Q(T) \subset \mathcal{L}(X,Y) \oplus \mathcal{L}(Y,X),$$

where $(X, X', f,)$, $(Y, Y', g,)$ are pairs of dual vector spaces. Again, by maximality of Q(T), we have $Q(T) = \mathcal{L}(X,Y) \oplus \mathcal{L}(Y,X)$.

Suppose now that T has an involution *, which extends to a unique involution on Q(T). As above we can distinguish two possibilities.

If T is strongly prime then we have by Theorem 3 that

$$\mathcal{F}_\sigma(Y, g, \alpha) = \mathrm{Soc}(T) \lhd T \subset Q(T) \subset \mathcal{L}_\sigma(Y, g, \alpha),$$

where $* : T \to T$ is now either the restriction of the involution $a \to a^{\#}$ or minus the restriction of this involution. Again $Q(T) = \mathcal{L}_\sigma(Y, g, \alpha)$ by maximality of $Q(T)$. If T is not strongly simple we apply Theorem 4 to get that either $Q(T) = \mathcal{L}(X,Y) \oplus \mathcal{L}(Y,X)$ where (X, g), (Y, h) are self-dual (both hermitian or alternate) vector spaces and $(a_1, a_2)^* = (a_2^{\#}, a_1^{\#})$, $Q(T) = \mathcal{L}(X,Y) \oplus \mathcal{L}(Y,X)$. with $(a_1, a_2)^* = (a_1^{\#}, a_2^{\#})$, with respect to a pair of dual vector spaces (X, Y, f) and its opposite (Y, X, f^{op}) over a division associative algebra with involution (Δ, τ), or $Q(T) = \mathcal{L}(X,Y) \oplus \mathcal{L}(Y,X)$ with $(a_1, a_2)^* = -(a_1^{\#}, a_2^{\#})$, with respect to a pair of dual vector spaces (X, Y, f) and its opposite (Y, X, f^{op}) over a field K.

4. Prime nondegenerate Jordan triple systems with nonzero socle.

In this section we prove the main result of this paper which determines prime nondegenerate JTS containing minimal inner ideals. The key tool to prove this result is Zelmanov's structure theorem for prime nondegenerate JTS. Since this theorem has been settled for JTS over a field of characteristic not 2 or 3, we must assume this restriction. We will adopt the recent interpretation of Zelmanov's theory given by D'Amour [1] to describe special prime nondegenerate JTS with a nonzero hermitian part in terms of quotient triples of prime ATS of the first kind, and of symmetric elements of these quotient triples with respect to an involution. This will allow us to apply the results proved in Section 3.

Theorem 7. Every prime nondegenerate Jordan triple system (over a field of characteristic not 2 or 3) with nonzero socle is isomorphic to one of the following:

(i) A simple exceptional finite-dimensional Jordan triple system over its centroid.

(ii) A Jordan triple system coming from the simple Jordan algebra $J = J(f)$ of a nondegenerate symmetric bilinear form, either by polarizing the Jordan triple system (J, U) or by changing the operator U by $P(a) = U(a)\eta$, where η is an involution of the Jordan pair obtained by duplicating the Jordan álgebra (J, U).

(iii) A subsystem of $\mathcal{L}_\sigma(X, X', f, \alpha)^+$ containing $\mathcal{F}_\sigma(X, X', f, \alpha)$ under $P(a)b = abah_\alpha$ with respect to a pair of dual vector spaces (X, X', f) over a division associative algebra Δ, an automorphism $\sigma : \Delta \to \Delta$, and $\alpha \in \Delta$ satisfying $\sigma(\alpha) = \alpha$, $\sigma^2(\beta) = \alpha^{-1}\beta\alpha$, for all $\beta \in \Delta$.

(iv) A subsystem of $(\mathcal{L}(X, Y) \oplus \mathcal{L}(Y, X))^+$ containing $\mathcal{F}(X, Y) \oplus \mathcal{F}(Y, X)$ under $P(a_1, a_2)(b_1, b_2) = (a_1b_2a_1, a_2b_1a_2)$, where (X, X', f), (Y, Y', g) are pairs of dual vector spaces over a division associative algebra Δ.

(v) A subsystem of $\mathcal{L}((X, g), (Y, h))^+$ containing $\mathcal{F}((X, g), (Y, h))$ under $P(a)b =$

$ab^{\#}a$, where (X, g) and (Y, h) are nondegenerate (hermitian or alternate) self-dual vector spaces over a division associative algebra with involution (Δ, τ).

(vi) A subsystem of $\mathcal{H}(\mathcal{L}_{\sigma}(Y, g, \alpha),\#)$ (respectively, $\mathcal{A}(\mathcal{L}_{\sigma}(Y, g, \alpha),\#))$ containing $\mathcal{H}(\mathcal{F}_{\sigma}(Y, g, \alpha),\#)$ (respectively, $\mathcal{A}(\mathcal{F}_{\sigma}(Y, g, \alpha),\#))$ under $P(a)b = abah_{\alpha}$, where (Y, g) is a nondegenerate (hermitian or alternate) self-dual vector space over a division associative algebra with involution (Δ, τ), $\sigma : \Delta \to \Delta$ is an automorphism such that $\sigma\tau = \tau\sigma^{-1}$, and $\alpha \in \Delta$ satisfies $\sigma(\alpha) = \alpha = \tau(\alpha)$, $\sigma^2(\beta) = \alpha^{-1}\beta\alpha$, for all $\beta \in \Delta$.

(vii) $\mathcal{H}(\mathcal{F}(X, Y),\#) \oplus \mathcal{H}(\mathcal{F}(Y, X),\#) \lhd J \subset \mathcal{H}(\mathcal{L}(X, Y),\#) \oplus \mathcal{H}(\mathcal{L}(Y, X),\#)$ under $P(a_1, a_2)(b_1, b_2) = (a_1b_2a_1, a_2b_1a_2)$, where (X, Y, f) is a pair of dual vector spaces over a division associative algebra Δ, and (Y, X, f^{op}) is its opposite pair relative to an involution $\tau : \Delta \to \Delta$.

(viii) $\mathcal{A}(\mathcal{F}(X, Y),\#) \oplus \mathcal{A}(\mathcal{F}(Y, X),\#) \lhd J \subset \mathcal{A}(\mathcal{L}(X, Y),\#) \oplus \mathcal{A}(\mathcal{L}(Y, X),\#)$ under $P(a_1, a_2)(b_1, b_2) = (a_1b_2a_1, a_2b_1a_2)$, with respect to a pair of dual vector spaces (X, Y, f) and its opposite (Y, X, f^{op}) over a field K.

Paragraph 1 of the proof. Zelmanov's structure theorem for prime nondegenerate JTS J [**18**, Theorem 4] asserts that J is either special or its central localization $\Gamma(J)^{-1}J$ is a simple exceptional finite-dimensional JTS over the field $\Gamma(J)^{-1}\Gamma(J)$. This result can be refined in the following way

Lemma 8. Let J be a prime nondegenerate Jordan triple system. If J contains a minimal ideal, then J is either special or a simple exceptional finite-dimensional Jordan triple system over its centroid.

Proof of Lemma 8 . By Zelmanov's theorem [**18**, Theorem 4] we may assume that J is exceptional. Let M be a minimal ideal of J. If $0 \neq \gamma \in \Gamma(J)$ then $0 \neq \gamma(M)$ is an ideal of J , but $\gamma(M) = M$ since $M = M \bullet M$ by minimality of M. Hence $\gamma^{-1} \otimes M = \gamma^{-1} \otimes \gamma(M) = 1 \otimes M$, which proves that M remains ideal in the central localization $\Gamma(J)^{-1}J$. Then $\Gamma(J)^{-1}J = M = J$ is a simple exceptional JTS which is finite-dimensional over its centroid.

Paragraph 2. Since J is prime, $M = \mathrm{Soc}(J)$ is a simple ideal. So we may assume by Lemma 8 that J is special (otherwise $J = M$ is a simple exceptional finite-dimensional JTS over its centroid, and we are in case (i) of Theorem 7). Then M is a simple special JTS and we have by Zelmanov's structure theorem for simple JTS [**18**, Theorem 1] (see also D'Amour's version [**1**, (5.5)]) that M is either of quadratic type (case (ii) of

Theorem 7) or it is of hermitian type: either $M = B^+$ or $M = \mathcal{H}(B, *)$, where B is a simple ATS (1º kind) and $* : B \to B$ is an involution of triple systems.

Paragraph 3. Suppose first that M is of quadratic type. Then it is easy to check that M contains an invertible element in the sense of [**10**, p. 5], i.e., there exists $x \in M$ such that $P(x) : M \to M$ is invertible; but this implies that $J = M$, as it is shown below.

Lemma 9. Let J be a prime nondegenerate Jordan triple system with a nonzero ideal M. If M contains an invertible element then $J = M$.

Proof of Lemma 9. Let x be invertible in M. Then there exists a unique element y in M such that $P(x)y = x$. Hence (x, y) is an idempotent and we can consider the Peirce decomposition $J = J_2^+ \oplus J_1^+ \oplus J_0^+$ associated to the orthogonal projections $E_2 = P(x)P(y)$, $E_1 = V(x, y) - 2E_2$ and $E_0 = B(x, y)$. Then $J_2^+ \oplus J_1^+ \subset M \subset J_2^+$ implies $M = J_2^+$ and $J_1^+ = 0$. Now for $x_0 \in J_0^+$ we have, by Peirce relations [**10**, p.44] that $P(x_0)M = 0$, and hence $x_0 \in Ann(M)$ by (10). Then $Ann(M) = 0$ because J is prime. Therefore $J = M$ as required.

Paragraph 4. Suppose finally that M is of hermitian type. Then J has a nonzero hermitian part and by [**1**, (4.1)] we have one of the following possibilities:

(4.1) J contains a nonzero ideal of the form T^+ where T is a prime ATS (1º kind) such that $T^+ \lhd J \subset Q(T)^+$, or

(4.2) J contains a nonzero ideal of the form $\mathcal{H}(T, *)$ where T is as above and $*$ is an involution of T such that $\mathcal{H}(T, *) \lhd J \subset \mathcal{H}(Q(T), *)$.

If J is as in (4.1) then $Soc(T) = Soc(T)^+ = Soc(J)$ by [**6**, (2.4)], so T is a prime associative triple system of the first kind with nonzero socle. Hence, by Theorem 6, either $Q(T) = \mathcal{L}_\sigma(X, X', f, \alpha)$, and we are in case (iii) of Theorem 7, or $Q(T) = \mathcal{L}(X,Y) \oplus \mathcal{L}(Y,X)$, case (iv) of Theorem 7.

Suppose now that J is as in (4.2). By [**6**, (5.2i)] T has nonzero socle as well. Hence, by Theorem 6, we have the following possibilities for $(Q(T), *)$.

If T is strongly prime (Theorem 6iii) then $Q(T) = \mathcal{L}_\sigma(Y, g, \alpha)$ with $a^* = \pm a^\#$, where (Y, g) is a nondegenerate (hermitian or alternate) self-dual vector space over a division associative algebra Δ with involution $\tau : \Delta \to \Delta$, $\sigma : \Delta \to \Delta$ is an automorphism such that $\sigma\tau = \tau\sigma^{-1}$, and $\alpha \in \Delta$ satisfies $\sigma(\alpha) = \alpha = \tau(\alpha)$, $\sigma^2(\beta) =$

$\alpha^{-1}\beta\alpha$, for all $\beta \in \Delta$. Then J is as in case (vi) of Theorem 7.

If T is not strongly prime we can still distinguish three possibilities (cases (iv), (v) and (vi) of Theorem 6).

(I) $Q(T) = \mathcal{L}(X,Y) \oplus \mathcal{L}(Y,X)$ with $(a_1, a_2)^* = (a_2^{\#}, a_1^{\#})$,

where (X, g) and (Y, h) are nondegenerate (hermitian or alternate) self-dual vector spaces over a division algebra Δ with involution $\tau : \Delta \to \Delta$; but now it is easy to see that the mapping $a \to (a, a^{\#})$ is an isomorphism from the JTS $\mathcal{L}((X, g),(Y, h))^+$, under $P(a)b = ab^{\#}a$, onto $H(Q(T), {}^*)$. So J is as in case (v) of Theorem 7.

(II) $Q(T) = \mathcal{L}(X,Y) \oplus \mathcal{L}(Y,X)$ with $(a_1, a_2)^* = (a_1^{\#}, a_2^{\#})$,

where (X, Y, f) is a pair of dual vector spaces over a division associative algebra Δ, and (Y, X, f^{op}) is the opposite pair relative to an involution $\tau : \Delta \to \Delta$. Then J is as in case (vii) of Theorem 7.

(III) $Q(T)= \mathcal{L}(X,Y) \oplus \mathcal{L}(Y,X)$ with $(a_1, a_2)^* = -(a_1^{\#}, a_2^{\#})$

with respect to a pair of dual vector spaces (X, Y, f) and its opposite (Y, X, f^{op}) over a field K. Then we are in case (viii) of Theorem 7, which completes the proof.

In a recent work the first two authors gave the following characterization of the socle of a nondegenerate Jordan triple system.

Theorem 10 ([7]). Let J be a prime nondegenerate Jordan triple system (over a field F of characteristic not 2 or 3). Then an element $x \in J$ is in the socle iff J satisfies dcc on principal inner ideals $P(y)J$, $y \in Fx+P(x)J$. Hence J coincides with its socle iff it has dcc on principal inner ideals.

While the proof of sufficiency of Theorem 10 is elementary (see [5]), the converse required Zelmanov's structure theory for simple JTS. However, Loos has given in [11] a new proof that works for an arbitrary ring of scalars.

Theorem 10 and socle theory reduce the study of nondegenerate JTS with dcc on principal inner ideals to studying simple JTS with minimal inner ideals. Since, by the last corollary of [18], simple Jordan pairs, and hence simple Jordan triple systems, over a field of characteristic not 2 or 3 are nondegenerate, we get as a particular case of

Theorem 7 the structure of all the simple JTS containing minimal inner ideals, thus extending a result of [6] where reduced simple JTS over an algebraically closed field were classified.

Theorem 11. Every simple Jordan triple system (over a field of characteristic not 2 or 3) with minimal inner ideal is isomorphic to one of the following:

(i) A simple exceptional finite-dimensional Jordan triple system over its centroid.

(ii) A Jordan triple system of quadratic type (cf. Theorem 7 (ii)).

(iii) A Jordan triple system $J = (\mathcal{F}_\sigma(X,X',f,\alpha))^+$ under $P(a)b = abah_\alpha$ with (X,X',f) a pair of dual vector spaces over a division associative algebra Δ, $\sigma : \Delta \to \Delta$ an automorphism, and $\alpha \in \Delta$ satisfying $\sigma(\alpha) = \alpha$, $\sigma^2(\beta) = \alpha^{-1}\beta\alpha$, for all $\beta \in \Delta$.

(iv) A Jordan triple system $J = (\mathcal{F}(X,Y) \oplus \mathcal{F}(Y,X))^+$ under $P(a_1, a_2)(b_1, b_2) = (a_1b_2a_1, a_2b_1a_2)$, where (X, X', f), (Y, Y', g) are pairs of dual vector spaces over a division associative algebra Δ.

(v) A Jordan triple system $J = \mathcal{F}((X, g), (Y, h))^+$ under $P(a)b = ab^\#a$, where (X,g), (Y,h) are nondegenerate (hermitian or alternate) self-dual vector spaces over a division associative algebra Δ with involution $\tau : \Delta \to \Delta$.

(vi) A Jordan triple system $J = \mathcal{H}(\mathcal{F}_\sigma(Y, g, \alpha),\#)$ or $J = \mathcal{A}(\mathcal{F}_\sigma(Y, g, \alpha),\#)$ under $P(a)b = abah_\alpha$, where (Y,g) is a nondegenerate (hermitian or alternate) self-dual vector space over a division associative algebra Δ with involution $\tau : \Delta \to \Delta$, $\sigma : \Delta \to \Delta$ is an automorphism such that $\sigma\tau = \tau\sigma^{-1}$, and $\alpha \in \Delta$ satisfies $\sigma(\alpha) = \alpha = \tau(\alpha)$, $\sigma^2(\beta) = \alpha^{-1}\beta\alpha$, for all $\beta \in \Delta$.

(vii) A Jordan triple system $J = \mathcal{H}(\mathcal{F}(X, Y),\#) \oplus \mathcal{H}(\mathcal{F}(Y, X),\#)$ under $P(a_1, a_2)(b_1, b_2) = (a_1b_2a_1, a_2b_1a_2)$, where (X, Y, f) is a pair of dual vector spaces over a division associative algebra Δ, and (Y, X, f^{op}) is its opposite pair with respect to an involution $\tau : \Delta \to \Delta$.

(viii) A Jordan triple system $J = \mathcal{A}(\mathcal{F}(X, Y), \#) \oplus \mathcal{A}(\mathcal{F}(Y, X),\#)$ under $P(a_1, a_2)(b_1, b_2) = (a_1b_2a_1, a_2b_1a_2)$ with respect to a pair of dual vector spaces (X,Y, f) and its opposite (Y, X, f^{op}) over a field K.

In our earlier paper [6] we proved that von Neumann regular Jordan Banach triple systems have dcc and acc on principal inner ideals, and classified such Jordan Banach triple systems in the complex case. Suppose now that J is a nondegenerate Jordan triple system having dcc and acc on principal inner ideals. By Theorem 10, J coincides with its socle; but acc on principal inner ideals forces that there can be at most finitely many simple ideals in $J = \mathrm{Soc}(J)$. Moreover, the different types of simple JTS listed in Theorem 11 must be modified according to this new restriction.

Theorem 12. Every nondegenerate Jordan triple system with dcc and acc on principal is a direct sum of finitely many simple ideals satisfying the same conditions. The simple Jordan triple systems (over a field of characteristic not 2 or 3) with dcc and acc on principal inner ideals are the following:

(i) A simple exceptional finite-dimensional Jordan triple system over its centroid.

(ii) A Jordan triple system of quadratic type (cf. Theorem 7 (ii)).

(iii) A Jordan triple system of square matrices $J = M_n(\Delta)$ under $P(a)b = a\alpha b^{\sigma}a$, where Δ is a division associative algebra, $\sigma : \Delta \to \Delta$ an automorphism, and $\alpha \in \Delta$ satisfies $\sigma(\alpha) = \alpha$, $\sigma^2(\beta) = \alpha^{-1}\beta\alpha$, for all $\beta \in \Delta$.

(iv) A Jordan triple system $J = (\mathfrak{F}(X,Y) \oplus \mathfrak{F}(Y,X))^+$ under $P(a_1, a_2)(b_1, b_2) = (a_1b_2a_1, a_2b_1a_2)$, where (X, X', f), (Y, Y', g) are two pairs of dual vector spaces over a division associative algebra Δ, and where X is finite-dimensional.

(v) A Jordan triple system $J = \mathfrak{F}((X,g), (Y,h))^+$ under $P(a)b = ab^{\#}a$, where (X,g), (Y,h) are nondegenerate (both hermitian or alternate) self-dual vector spaces over a division associative algebra with involution (Δ,τ), X being finite-dimensional

(vi) A Jordan triple system $J = \mathcal{H}(\mathfrak{F}_{\sigma}(Y, g, \alpha),\#)$ or $J = \mathcal{A}(\mathfrak{F}_{\sigma}(Y, g, \alpha),\#)$ under $P(a)b = abah_{\alpha}$, where (Y, g) is a finite-dimensional nondegenerate (hermitian or alternate) self-dual vector space over a division associative algebra with involution (Δ, τ), $\sigma : \Delta \to \Delta$ is an automorphism such that $\sigma\tau = \tau\sigma^{-1}$, $\alpha \in \Delta$ satisfies $\sigma(\alpha) = \alpha = \tau(\alpha)$, $\sigma^2(\beta) = \alpha^{-1}\beta\alpha$, for all $\beta \in \Delta$.

(vii) A polarized Jordan triple system of hermitian matrices $H_n(\Delta, \tau) \oplus H_n(\Delta, \tau)$ under $P(a_1,a_2)(b_1,b_2) = (a_1b_2a_1, a_2b_1a_2)$, with respect to a division associative algebra with involution (Δ, τ).

(viii) A polarized Jordan triple system of alternate matrices $A_n(K) \oplus A_n(K)$ under $P(a_1, a_2)(b_1, b_2) = (a_1b_2a_1, a_2b_1a_2)$ over a field K.

Remarks: (a) Following [**10**, (12.13)], the Jordan triple systems of type (i), (iii), (vi), (vii), and (viii) of the above theorem satisfy dcc and acc on all inner ideals. The Jordan triple systems of type (iv) and (v) have dcc and acc on all inner ideals if and only if the vector spaces occurring there are finite-dimensional. By McCrimmon [**12**], Jordan algebras (and therefore Jordan triple systems) defined by quadratic forms (type ii) have dcc and acc on all inner ideals if there exist no totally isotropic subspaces of infinite dimension.

(b) The structure theorem for simple Jordan triple systems with minimal inner ideals is closely related to the structure theorem due to Neher [**16**] for simple Jordan triple systems which are covered by a grid. However these two classes are distinct. On the one hand, a simple Jordan triple system covered by a grid need not have minimal inner ideals (see structure theorems IV.1.14 and IV.1.16 of **[16]**). On the other hand, a simple Jordan triple system with minimal inner ideals need not be covered by a grid, as example I.6.5 of **[16]** shows.

(c) Since a polarized Jordan triple system is simple if, and only if, it is the Jordan triple system of a simple Jordan pair, Theorems 11 and 12 give the classification of all the simple Jordan pairs having dcc on principal inner ideals, and dcc and acc on principal inner ideals respectively; results proved by Loos for Jordan pairs over an arbitrary ring of scalars **[10]**.

REFERENCES

[1] A. D'AMOUR, Quadratic Jordan systems of hermitian type. To appear

[2] J.A. CUENCA MIRA, A. GARCIA MARTIN and C.MARTIN GONZALEZ, Jacobson density for associative pairs and its aplications, *Commun. in Algebra* **17** (10) (1989), 2595-2610.

[3] A. FERNANDEZ LOPEZ, Prime nondegenerate Jordan algebras with nonzero socle and the symmetric Martindale algebra of quotients, *Collect. Math.* **39** (3) (1988), 249-256.

[4] A. FERNANDEZ LOPEZ and E. GARCIA RUS, Prime associative triple systems with nonzero socle, *Commun. in Algebra* **18** (1) (1990), 1-13.

[5] A. FERNANDEZ LOPEZ and E. GARCIA RUS, A characterization of the elements of the socle of a Jordan algebra, *Proc. Amer. Math. Soc.* **108** (1) (1990), 69-71.

[6] A. FERNANDEZ LOPEZ, E. GARCIA RUS and E. SANCHEZ CAMPOS, Von Neumann regular Jordan Banach triple systems, *J. London Math. Soc.* **42** (2) (1990), 32-48.

[7] A. FERNANDEZ LOPEZ and E. GARCIA RUS, An elementary characterization of the socle of a Jordan triple system. Preprint.

[8] I. N. HERSTEIN, *Rings with involutions,* The University of Chicago Press 1976.

[9] W. G. LISTER, Ternary rings, *Trans. Amer. Math. Soc.* **154** (1971), 37-55.

[10] O. LOOS, *Jordan Pairs,* Lecture Notes in Mathematics **460**, Springer Verlag 1975.

[11] O. LOOS, On the socle of a Jordan pair, *Collect. Math.* **40** (2) (1989), 109-125.

[12] K. McCRIMMON, Inner ideals in quadratic Jordan algebras, *Trans. Amer. Math. Soc.* **159** (1971), 445-468.

[13] K. McCRIMMON, Strong prime inheritance in Jordan systems, *Algebras, Groups and Geometries* **1** (1984), 217-234.

[14] K. McCRIMMON, Martindale systems of symmetric quotients, *Algebras, Groups and Geometries* **6** (1989), 153-237.

[15] K. MEYBERG, *Lectures on algebras and triple systems,* Lecture Notes, The University of Virginia, Charlottesville, 1972.

[16] E. NEHER, *Jordan triple systems by the grid approach,* Lecture Notes in Mathematics **1280**, Springer Verlag 1987.

[17] E.I. ZELMANOV, Absolute zero divisors in Jordam pairs and Lie algebras, *Math. Sbornik* **40** (4) (1981), 549-565.

[18] E.I. ZELMANOV, Primary Jordan triple systems III, *Sibirskii Math. Zh.* **26** (1985), 71-82.

Nonassociative
Algebraic
Models

ISBN 1-56072-050-6

Weakly Compact Homomorphisms in Nonassociative Algebras

José E. Galé*
Depatamento de Matemáticas, Universidad de Zaragoza, 50009 Zaragoza, Spain

It was proved in §3 of [6] that every weakly compact homomorphism from a C^*-algebra has a finite-dimensional and semisimple range. The purpose of this note is to show that the proof of the above statement also works in a nonassociative setting, so we can establish an analogous result for JB^*- algebras. Such a result, which is Theorem 2 below, will be obtained as a consequence of the following Theorem 1.

Let A be a unital, complete, normed, power-associative complex algebra. An element $h \in A$ is said to be *hermitian* if $||e^{ith}|| = 1$ for all $t \in R$. The set H of hermitian elements of A forms a closed real subspace of A. Let B be a complete, normed, nonassociative complex algebra. A homomorphism $\theta : A \longrightarrow B$ is called *weakly compact* if it takes bounded subsets of A to weakly relatively compact subsets of B.

Theorem 1. *Let A and B be as above. If $\theta : A \longrightarrow B$ is a weakly compact homomorphism, then there exists an integer $N \geq 1$ such that every element of $\overline{\theta(H + iH)}$ is algebraic over $\mathbb{C}$ of degree at most N.*

The first step of the proof of Theorem 1 consists of showing that

$$\theta(h) \text{ is algebraic for every } h \in H. \tag{1}$$

* The research of this paper has been partially supported by the Spanish DGICYT Project PS87-0059.

By considering the restriction of θ to the Banach subalgebra of A generated by h, it is clear that the proof of (1) is of an associative nature. The reader is referred to [6], where (1) is proved using that every weakly compact homomorphism from the Wiener algebra is a finite rank operator. We are going to see here that property (1) may also be obtained from a theorem on analytic semigroups in Banach algebras which appears in [5].

An analytic semigroup $(a^z)_{Re\ z>0}$ in a Banach algebra E is an analytic function $z \to a^z, \{Re\ z > 0\} \to E$ such that $a^{z+w} = a^z a^w$ for all $z, w \in \mathbb{C}$ with $Re\ z, Re\ w > 0$.

Lemma 1. *Let $(a^z)_{Re\ z>0}$ be an analytic semigroup in a Banach algebra E such that E is polynomially generated by a^1, and $\{a^{1+iy} : y \in R\}$ is relatively weakly compact. Then*

(i) The character space Φ_E of E is discrete, and

(ii) E is semisimple if and only if $E^\perp = (0)$, where $E^\perp = \{x \in E : xy = 0\ for\ all\ y \in E\}$.

Proof. It is just Theorem 3.1 and Proposition 3.2 of [5] ∎

Proposition 1. *Let A, B and θ be as in Theorem 1. If h is an hermitian element of A, then $\theta(h)$ is algebraic.*

Proof. Define $b^z = e^{z\theta(h)}$ $(z \in \mathbb{C})$, and let E be the Banach subalgebra of B which is polynomially generated by $b = b^1$. For each continuous functional l on B the entire function $l(b^z)$ is of exponential type and bounded on $i\mathbb{R}$. Hence, by a Phragmen-Lindelof theorem, it is also bounded on $\{Re\ z \geq 0\}$. If, moreover, we suppose that $l(b^n) = 0$ for every integer $n \geq 1$, all these conditions imply that $l(b^z) = 0$ $(z \in \mathbb{C})$. This, via the Hahn-Banach theorem, means that $b^z \in E$ for all $z \in \mathbb{C}$. In particular E is unital, and so Φ_E is compact. Also, $\theta(h) = lim_{z\to 0} z^{-1}(b^z - b^0) \in E$.

Now, $\{b^{1+iy} : y \in \mathbb{R}\} = \{b\theta(e^{iyh}) : y \in \mathbb{R}\}$ is clearly relatively weakly compact, whence, by (i) of Lemma 1, we get that Φ_E is finite. This means that the spectrum $\sigma_E(b)$ is finite, whence $\sigma_E(\theta(h))$ is also finite. Moreover, $E^\perp = (0)$ since E is unital. By (ii) of Lemma 1, E is semisimple and therefore $\theta(h)$ is algebraic. ∎

We are now ready to give the proof of Theorem 1, which follows word by word the proof of Theorem 3.2 in [6]. We only give an outline of the argument.

Proof of Theorem 1.- For $n \geq 1$, define $H_n = \{h \in H :$ there exist $\alpha_0, \ldots, \alpha_n \in \mathbb{C}, \sum_0^n |\alpha_j| = 1, \sum_0^n \alpha_j \theta(h)^j = 0\}$. Each H_n is closed in H, and $\cup_n H_n = H$ by Proposition 1. Then we can apply Baire's theorem to deduce that H_N has interior points, for some N. Now, we use this fact, and arguments which involve comples polynomials in one variable and the uniqueness principle to get that $\theta(h + ik)$ is algebraic of degree at most N, for every $h, k \in H$. ∎

As we said at the beginning, our goal is to describe the range of a weakly compact homomorphism θ whose domain is a JB^*-algebra A. If A is associative (in this case A is a C^*-algebra), $\theta(A)$ is of finite codimension ([6]). For general JB^*-algebras we can get exactly the same conclusion on $\theta(A)$ if we assume that θ is compact. This is an obvious consequence of the fact that every homomorphism from a JB^*-algebra has closed range. A proof of the last interesting property, which is a nonassociative extension of earlier results by Kaplansky and Cleveland, can be seen in [10, Lemma IV 1.3]. Nevertheless, we cannot expect the finite dimensionality of $\theta(A)$ if θ is an arbitrary weakly compact homomorphism: most of the so-called ***quadratic*** JB^*-algebras (see e.g. [8, p.439] for the definition and a complete description of these algebras), whose underlying Banach space is reflexive, have infinite dimension. If A is such an algebra the identity operator on A is of course weakly compact. We are going to see in the next theorem that the above example is canonical in some sense.

Theorem 2. *Let A be a JB^*-algebra, let B be a Jordan-Banach algebra, and let $\theta : A \longrightarrow B$ be a weakly compact homomorphism. Then $\theta(A)$ is a finite direct sum of finite-dimensional or quadratic JB^*-algebras.*

Proof. Adjoining identities if necessary, we can suposse that A, B and θ are unital, and that $\overline{\theta(A)} = B$. Since A is a JB^*-algebra, it follows from [11, Theorem 12] that $Ker\theta$ is a closed $*$-ideal of A and $A/Ker\theta$ is a JB^*-algebra and so it is semisimple. Also, $\theta(A) = B$ because of [10, Theorem IV 1.4]. Furthermore, again because A is a JB^*-algebra, $A = H + iH$ and then there exists $N \geq 1$ such that every element of $\theta(A)$ is algebraic of degree at most N. Thus $\theta(A)$ is a semisimple JB^*-algebra in which each element has finite spectrum. Then, from [3], we deduce that $\theta(A)$ is a direct sum of finite-dimensional or quadratic JB^*-algebras. The fact that $\theta(A)$ is reflexive is evident. ∎

Remark.- Theorem 1 is not really necessary in the argument that we have used to prove Theorem 2. In fact, the first part of Lemma 1 implies, as above, that $\sigma_B(\theta(h))$ is finite for all $h \in H$. Then, for $h, k \in H$, the map $z \rightarrow f(z) = \theta(h) + z\theta(k)$, $\mathbb{C} \longrightarrow B$, is holomorphic and so the function $z \longrightarrow \sigma_B(f(z))$ is analytic multivalued ([2]). As $\sigma_B(f(z))$ is finite for every $z \in \mathbb{R}$ we get that the same is true for every $z \in \mathbb{C}$ ([1, Theorem 3.8]). This implies – put $z = i$ – that each element of $B = \theta(A)$ has finite spectrum and therefore, from [3], we obtain that $\theta(A)$ is as in the conclusion of Theorem 2.

Corollary 1. *If A,B,and θ are as in Theorem 2, and A or B are alternative, then $\theta(A)$ is of finite dimension.*

Proof. The quadratic and alternative Jordan-Banach algebras are of finite dimension ([3, p.205]). ∎

Corollary 2 ([6, Theorem 3.5]). *Let A be a reflexive JB^*-algebra. Then A is a finite direct sum of finite-dimensional or quadratic JB^*-algebras.*

Proof. Take θ as the identity map on A and apply Theorem 2. ∎

Remark.- It is an old theorem of S. Sakai that a C^*-algebra which is reflexive as a Banach space has to be finite-dimensional. This property was used in [8] to prove the part of Theorem 3.5 which corresponds to our Corollary 2. Observe that an alternative proof of Theorem 2 can be given using Corollary 2 as starting point instead of Theorem 1 (or the analytic multivalued functions, as in the remark after Theorem 2). Details have been recently shown in [9]. This proof was pointed out in the associative case in [6],[7], and [12].

Finally, we apply to our setting a bit of geometry of Banach spaces. The definition of Banach space of cotype 2 can be seen in [4].

Corollary 3. *Let A be a JB^*-algebra and let B be a Jordan-Banach algebra of cotype 2. If $\theta : A \longrightarrow B$ is a continuous homomorphism, then $\theta(A)$ is as in Theorem 2.*

Proof. By [4, Theorem 11] θ factors through a Hilbert space, and so θ is weakly compact. ∎

Acknowledgements.

I very much thank A. Rodríguez Palacios for his interesting and useful comments, as well as his valuable references.

REFERENCES

[1] B. Aupetit, Analytic multivalued functions in Banach algebras and uniform algebras, Adv. in Math. **44** (1982), 18–60.

[2] B. Aupetit and A. Zraibi, Propriétés analytiques du spectre dans les algèbres de Jordan-Banach, Manuscripta Math. **38** (1982), 381–386.

[3] M. Benslimane and E. M. Kaidi, Structure des algèbres de Jordan-Banach non commutatives complexes régulières ou semi-simples à spectre fini, Journal of Algebra **113** (1988), 201–206.

[4] Cho-Ho Chu, B. Iochum and G. Loupias, Grothendieck's theorem and factorization of operators in Jordan triples, Math. Ann. **284** (1989), 41–53.

[5] J. E. Galé, Banach algebras generated by analytic semigroups having compactness properties on vertical lines, Proc. Centre Math. Anal. ANU **21** (1989), 126-143.

[6] J. E. Galé, T. J. Ransford and M. C. White, Weakly compact homomorphisms, to appear in Trans. Amer. Math. Soc.

[7] M. Mathieu, Weakly compact homomorphisms from C^*-algebras are of finite rank, Proc. Amer. Math. **107** (1989), 761-762.

[8] R. Payá-Albert, J. Pérez-González and A. Rodríguez-Palacios, Type I factor representations of non-commutative JB^*-algebras, Proc. London Math. Soc. **48** (1984), 428–444.

[9] J. Pérez, L. Rico and A. Rodríguez-Palacios, Full subalgebras of Jordan-Banach algebras and algebra norms on JB^*-algebras, preprint.

[10] L. Rico, *Sobre álgebras de Jordan normadas completas primas con zócalo no cero*, Doctoral dissertation, University of Granada, Spain, 1988.

[11] A. Rodríguez-Palacios, Non-associative normed algebras spanned by hermitian elements, Proc. London Math. Soc. **47** (1983), 258–274.

[12] G. A. Willis, private communication.

Nonassociative
Algebraic
Models

ISBN 1-56072-050-6

Periodic Power-Associative Rings Order Relation

S. González and C. Martínez*
Departamento de Matemáticas, Universidad de Zaragoza, 50009 Zaragoza, Spain

Abstract

In this paper we will consider periodic power-associative rings of prime characteristic $p \neq 2$, and strictly power associative if $p = 3$. If A is a such ring, it was proved by Osborn that A^+ is a periodic Jordan ring.

By using the order relation defined for a reduced Jordan ring, we prove in this paper that we can define an order relation for these rings in terms of the power-associative product in a similar way to the known cases (associative, alternative and Jordan). Finally we get some results about the given ring by using properties of its idempotent elements and some known properties about the order.

Introduction

Abian ([1] and [2]) proved that the relation $x \leq y$ if and only if $xy = x^2$ is a partial order in a reduced associative ring and this relation extends the usual order in Boolean rings.

* Supported by the DGICYT (Ps. 87-0054)

In [3] a partial order was defined for reduced Jordan rings satisfying a property (P) by: $x \leq y$ if and only if $xy = x^2$ and $x^2y = x^3 = xy^2$. The property (P) was the one given by: (P) $(x,x,y) = 0$ implies $(xy,x,y) = 0$ for every x,y.

With the definitions of hyperatom, hyperatomic and orthogonally complete modified in the natural way for Jordan rings, a structure theorem was also proved in this case: If R is a Jordan ring, then R is a direct product of Jordan division rings if and only if the above relation $\leq$ is a partial order on R and R is hyperatomic and orthogonally complete.

In [4] it was proved that the property (P) holds in every JB-algebra. Consequently the above relation $\leq$ always defines a partial order on a JB-algebra.

In [5] quadratic Jordan rings without nonzero nilpotent elements are considered. For quadratic Jordan rings the relation $\leq$ is defined by:

$x \leq y$ if and only if $V_x x = V_x y$ and $U_x x = U_x y = U_y x$.

It was proved that the above relation $\leq$ is a partial order for every reduced quadratic Jordan ring.

With the right definition, we also proved that a quadratic Jordan ring J is a direct product of quadratic Jordan division rings if and only if J id hyperatomic and orthogonally complete.

When $\frac{1}{2} \in \phi$, ϕ the ring of scalars, a quadratic Jordan ring J defines and is defined by a linear Jordan ring A. As desired, every two elements x,y satisfy $x \leq y$ in A if and only if $x \leq y$ in J; an element e is a hyperatom in A if and only if e is a hyperatom in J; A is hyperatomic (respec. orthogonally complete) if and only if J is hyperatomic (resp. orthogonally complete). So we could conclude that the relation $\leq$ is an order relation for every reduced Jordan ring and the property (P) is not needed.

In [10] Osborn studies periodic Jordan rings. If A is a power-associative ring, it is called periodic if for every a in A there is a natural number $n > 1$ such that $a^n = a$. Of course every periodic ring is a reduced ring.

If A is a periodic associative ring, A is known to be commutative and a periodic associative ring with only one idempotent element is a field (see [10]). The set of idempoten elements of A, E = E(A), is a relatively complemented lattice with $e \leq f$ if and onlyif $e = ef$, and every element a in A has associated an idempotent element e_a such that $ae_a = e_a a = a$ (if $a^n = a$, then $e_a = a^{n-1}$).

So the aim of [6] was to study a periodic associative ring A by using properties of the order relation and the lattice of idempotent elements. It was proved that:

- An element $a \in A$ is a hyperatom if and only if e_a is an atom of E(A).

- A is a direct product of periodic fields if and only if A is orthogonally complete and E(A) is atomic.

- A is hyperatomic if and onlñy if E(A) is atomic.

- If A is a direct product of periodic fields, then E(A) is atomic and complete. The converse is not true.

- A is a direct product of a finite number of periodic fields if and only if E(A) is finite.

With regard to periodic Jordan rings, it is also true that such a ring, of characteristic $p \neq 2$ and with only one idempotent element, is a periodic field (see [10]). Any periodic Jordan ring of characteristic $p \neq 2$ is a subdirect sum of simple of simple periodic Jordan rings and such a simple ring is either a field or a Jordan ring of capacity two. Then the aim of [7] was to study periodic (linear) Jordan rings of characteristic $p \neq 2$ by using properties of the order relation and the set of idempotent elements (which is not a lattice now). It was proved for a periodic Jordan ring J (char$J \neq 2$) that :

- J is associative if and only $J_{\frac{1}{2}}(e) = 0$ for every idempotent element e of J ($J_{\frac{1}{2}}(e)$ is the Peirce component).

- J is associative if and only if $R_e R_f = R_f R_e$, that is, $0 = (x, e, f)$ for every pair of idempotent elements e,f and every x in J.

- If J is orthogonally complete and E(J) is an atomic lattice with the usual relation, then J is a direct product of periodic fields and simple Jordan rings of capacity 2.

It seems natural to do a study similar to those done in [6] and [7] for quadratic Jordan rings and power-associative rings.

Quadratic Jordan rings were considered in [8]. By using a result of [10] which reduces the study of periodic rings to the study of periodic rings of prime characteristic, only the case of characteristic 2 needs be considered. Since these rings were not studied in [10], it was firstly necessary to study them. In [8] it was proved that such rings are essentially associative, that is, if J is a such ring, then $J = A^+$ for a periodic associative ring A and the product of J defines (and is defined by) the product of A. So the results of the associative case in [6] end the study of periodic quadratic Jordan rings.

The aim of this paper is to study, in a similar way, periodic power-associative rings of characteristic $p \neq 2$, and to prove that an order relation over such rings can be defined in terms of the poawer-associative product.

Previous results

A will denote here after a periodic power-associative ring of characteristic $p \neq 2$, which is strictly power-associative if $p = 3$.

The following results can be fund in [10] :

1. A^+ is a Jordan algebra.

2. If A contains only one idempotent element, then A is a periodic field.

3. If A is of capacity 2 (that is, A^+ is a periodic Jordan algebra of capacity 2) and is flexible, then A is commutative.

4. If for every idempotent $e \in A$ the subspaces $A_1(e)$ and $A_0(e)$ (Peirce components) are subrings of A, then A is a subdirect sum of periodic fields and power-associative rings of capacity 2.

5. If A is flexible, then A is a Jordan algebra.

Order relation

Let A be as above. Since A^+ is a periodic Jordan ring, we have defined an order relation on A^+ by : $x \underset{+}{\leq} y$ if and only if $x \circ y = x^2$, $x \circ y^2 = x^3 = x^2 \circ y$, where $\circ$ denotes the product in A^+ ($x \circ y = \frac{1}{2}(xy + yx)$) and we are using that the powers of an element in A and in A^+ coincide.

So we can define the following order relation in A:

Definition : Let $x, y \in A$. Then $x \leq y$ in A if and only if $x \underset{+}{\leq} y$ in A^+.

The following question is natural : Can this order relation be defined only in terms of the product in A, in a similar way to the known cases of associative and Jordan rings?

The answer is affirmative as shown by the following theorem:

Theorem 1 : The following statements are equivalent for every x, y in A :

a) $x \leq y$,

b) $\langle x, y \rangle$ is associative and $xy = x^2$,

c) $xy = yx = x^2$ and $xy^2 = y^2x = x^3 = x^2y = yx^2$,

d) $xy = yx = x^2$ and $x^2y = x^3 = y^2x$.

Proof:

b) implies c)

The element $yx - x^2$ is in $\langle x, y \rangle$, so $(yx - x^2)^2 = yxyx - yx^3 - x^2yx + x^4 = 0$, since $xyx = x^2x = x^3$ and $x^2yx = xxyx = xx^2x = x^4$. But the ring A has not nonzero nilpotent elements, so $yx - x^2 = 0$, that is, $yx = x^2$. Consequently, $x^3 = xx^2 = xxy = x^2y$, $x^3 = x^2x = yxx = yx^2$, $x^3 = xx^2 = xxy = xyy = xy^2 = y^2x$.

c) implies d) obvious.

d) implies a)

If $xy = yx = x^2$, then $x \circ y = xy = x^2$. We know, by the third-power associativity of A, that $(x, x, y) + (x, y, x) + (y, x, x) = 0$.

But $(x, x, y) = x^2y - x(xy) = x^3 - xx^2 = 0$, $(x, y, x) = (xy)x - x(yx) = x^2x - xx^2 = 0$. Consequently $(y, x, x) = 0$, that is, $(yx)x = yx^2$. So $x^3 = yx^2 = x^2y$ and $x \circ y = x^2y = x^3$.

Similarly $(y, y, x) + (y, x, y) + (x, y, y) = 0$, $(y, y, x) = y^2x - y(yx) = x^3 - yx^2 = 0$, $(y, x, y) = (yx)y - y(xy) = x^2y - yx^2 = 0$. So $(x, y, y) = 0$, that is, $xy^2 = (xy)y = x^2y = x^3$. Consequently $xy = x^3 = yx$ and $x \circ y = xy = x$. This proves that $x \underset{+}{\leq} y$ in A^+, that is, $x \leq y$.

a) implies b)

Let us suppose that $x \leq y$, that is, $x \circ y = x^2$, $x^2 \circ y = x = x \circ y^2$.

It is known that for every idempotent element e in A, A splits in : $A = A_1(e) \dot{+} A_{\frac{1}{2}}(e) \dot{+} A_0(e)$ (direct sum as vector spaces) where:

$A_1(e) = \{ a \in A : ae = ea = a \}$

$A_{\frac{1}{2}}(e) = \{ a \in A : ae + ea = a \}$

$A_0(e) = \{ a \in A : ae = ea = 0 \}$

$A_0(e)$ and $A_1(e)$ are not generally subalgebras, but $A_0 A_1 = 0 = A_1 A_0$.

If $x^{n+1} = x$, $y^{m+1} = y$, $e_x = x^n$ and $e_y = y^m$ are the associated idempotents, by using that $x \leq y$ we get : $e_x \circ y = x^n \circ y = x^{n+1} = x$, and $e_y \circ x = y^{nm} \circ x = x^{nm+1} = x^n = x$.

If $e = e_x$ and we consider the Peirce decomposition with respect to e, then $x \in A_1(e)$, $y = y_1 + y_{\frac{1}{2}} + y_0$, where $y_j \in A_j$. Then $ey = y_1 + ey_{\frac{1}{2}}$ and $ye = y_1 + y_{\frac{1}{2}}e$. Consequently $2y_1 + ey_{\frac{1}{2}} + y_{\frac{1}{2}}e = 2y_1 + y_{\frac{1}{2}} = 2e \circ y = 2x$.

Since the sum of the vector spaces $A_1(e)$ and $A_{\frac{1}{2}}(e)$ is direct, this implies that $x = y_1$ and $y_{\frac{1}{2}} = 0$. So $y = x + y_0$, $xy = x^2 + xy_0 = x^2$ and $yx = x^2 + y_0x = x^2$ by using that $0 = A_0A_1 = A_1A_0$.

Since $x \in A_1(e)$ and $y_0 \in A_0(e)$ for every n, we have also that $x^ny = x^{n+1} + x^ny_0 = x^{n+1}$, by using again that $A_1A_0 = 0$. Similarly $yx^n = x^{n+1}$. By the power-associativity, it is easy to prove that $y^n = x^n + y_0{}^n$, and so $xy^n = y^nx = x^{n+1}$. This assures that the algebra generated by x and y is associative.

Some results concerning the order relation

1. If $a \leq b$, then $ea \leq eb$ (it is an easy consequence of the corresponding result in the Jordan case).

2. If $e \leq f$, then $R_e{}^+ R_f{}^+ = R_f{}^+ R_e{}^+$, as we know for the Jordan case. Then $R_eR_f + L_eL_f + L_eR_f = R_f R_e + L_f L_e + R_f L_e + L_f R_e$.

3. $e \leq f$ if and only if $A_1(e) \subset A_1(f)$.

If $A_1(e) \subset A_1(f)$, then $e \in A_1(f)$. This means that $ef = fe = e$ and conversely $e \leq f$.

Conversely, if $e \leq f$ and $x \in A_1(e)$, then $x \in A_1{}^+(e)$, that is, $x \circ e = x = xe = ex$. By the known result in the Jordan case $x \in A_1{}^+(f)$, that is, $x \circ f = x = \frac{1}{2}(xf + fx)$.

But if $x = x_1 + x_{\frac{1}{2}} + x_0$ is the Peirce decomposition of x with respect to f ($x_1 \in A_1(f)$, $x_{\frac{1}{2}} \in A_{\frac{1}{2}}(f)$, $x_0 \in A_0(f)$), then $xf = x_1 + x_{\frac{1}{2}}f$, $fx = x_1 + fx_{\frac{1}{2}}$ and consequently $2x = xf + fx = 2x_1 + x_{\frac{1}{2}}f + fx_{\frac{1}{2}} = 2x_1 + x_{\frac{1}{2}}$. This implies $x_0 = 0 = x_{\frac{1}{2}}$ and $x = x_1 \in A_1(f)$.

Results on A

Theorem 2 : A is associative if and only if $(x, y, e) = 0 = (e, x, y)$ for every idempotent element e and for every $x, y \in A$.

Proof:

Firstly notice that the third - power associative linearized identity and the conditions in Theorem 2 imply that $(x, e, x) = 0$ for every x in A.

Since $(x, y, e) = 0 = (e, x, y)$, $A_1(e)$ and $A_0(e)$ are subalgebras for every idempotent element. In fact, if $x,y \in A_0(e)$, then $(xy)e = x(ye) = x0 = 0$ and $e(xy) = (ex)y = 0y = 0$, so $xy \in A_0$. Similarly, if $x, y \in A_1(e)$, then $(xy)e = x(ye) = xy$ and $e(xy) = (ex)y = xy$, so $xy \in A_1$.

By using the previous result 4 (see [10] , theorem 16.5) we know that A is a subdirect product of periodic fields and periodic power-associative rings of capacity 2. By using the result in the Jordan case (see [8]), it suffices to prove that $A_{\frac{1}{2}}^+(e) = 0 = A_{\frac{1}{2}}(e)$ for every idempotent element.

Let $x \in A_{\frac{1}{2}}(e)$. Then $(x,e,e) = 0$ implies $(xe)e = xe$, $(e, x, e) = 0$ implies $(ex)e = e(xe)$ and $(e, e, x) = 0$ implies $e(ex) = ex$. But $x = ex + xe = e(ex + xe) + (ex + xe)e = e(ex) + e(xe) + (ex)e + (xe)e$, that is, $(ex)e = e(xe) = 0$. Consequently $xe = (xe)e + e(xe)$ and $xe \in A_{\frac{1}{2}}(e)$. Similarly $ex = (ex)e + e(ex)$ and $ex \in A_{\frac{1}{2}}(e)$.

But we know by the Jordan case that if $z \in A_{\frac{1}{2}}(e)$, then $z^2 \in A_1(e) + A_0(e)$. So $[z^2,e] = 0$ for every z in $A_{\frac{1}{2}}(e)$. In particular $[(xe)^2, e] = 0 = [e, (ex)^2]$.

But $(xe)^2e = (xe)((xe)e) = (xe)(xe) = (xe)^2$ and $e(xe)^2 = (e(xe))(xe) = 0(xe) = 0$. So $(xe)^2 = 0$ and this implies $xe = 0$, since there are no nonzero nilpotent elements. In a similar way, we can prove that $(ex)^2 = 0$ and consequently $ex = 0$. So $x = ex + xe = 0$ and we have proved that $A_{\frac{1}{2}}(e) = 0$ as we wanted.

If e is a hyperatom of A^+, then $A_{\frac{1}{2}}^+(e) = 0$ and consequently $A_{\frac{1}{2}}(e) = 0$. So A $= A_1(e) \dot{+} A_0(e)$. We know that e is the only idempotent element in $A_1(e)$ and $A_1^+(e)$ is a periodic field. If $a, b \in A_1(e)$, they generate in $A_1^+(e)$ a finite field, which is generated by a single element c. Since c generates the same subalgebra in $A_1(e)$ and $A_1^+(e)$, b and a are both in the algebra generated by c in $A_1(e)$. This implies that $ab = ba = a \circ b$, that is $A_1(e)$ is a subring , a periodic subring with only one idempotent. So it is a periodic field. Since $A_1A_0 = 0 = A_0A_1$, we conclude that $A_1(e)$ is an ideal of A.

Corollary 1 : If A is simple, then either A is a periodic field or A^+ is without hyperatoms, that is, either A is a periodic field or A^+ is a subdirect product of periodic Jordan rings of capacity 2.

Proposition 1 : If E(A) is finite, then $A = F_1 \dot{+} \ldots \dot{+} F_r \dot{+} B$, where $F_1 , \ldots , F_r$ are periodic fields and ideals of A and B is a vector subspace such that B^+ is a direct sum of simple Jordan algebras of capacity 2.

Proof :

Let $e_1 , \ldots , e_r$ the hyperatoms of A^+ ($r = 0$ is possible). As we have before seen $A_1(e_i)$ is an ideal of A is an ideal of A and is a periodic field. If $F_i = A_1(e_i)$, we consider $F = F_1 \dot{+} \ldots \dot{+} F_r$, where the sum is clearly direct. If $B = \{ x \in A : xe_1 = xe = \ldots = xe_r = 0 = e_1x = \ldots = e_rx \}$, then we have $A = F_1 \dot{+} \ldots \dot{+} F_r \dot{+} B$ ($A^+ = F_1^+ \dot{+} \ldots \dot{+} F_r^+ \dot{+} B^+$), where B^+ is without hyperatoms and only with a finite number of idempotents. So, as we know by the Jordan case, B^+ is a direct sum of a finite number of simple Jordan algebras of capacity 2.

Corollary 2 : If E(A) is finite and A^+ is associative, then A is associative and consequently a direct sum of a finite number of periodic fields.

Proof :

It suffices to note that, using the notation of proposition 1, $B^+ = 0$, so $B = 0$ and $A = F_1 \dot{+} \ldots \dot{+} F_r$ is associative.

REFERENCES

[1] A. Abian, Direct product decomposition of commutative semisimple rings, Proc. Amer. Math. Soc. **24** (1970), 502-507.

[2] A. Abian, Order in a special class of rings and a structure theorem, Proc. Amer. Math. Soc. **52** (1975) , 45-49,

[3] S. González and C. Martínez, Order relation in Jordan rings and a structure theorem, Proc. Amer. Math. Soc. **98** (3) (1986), 379-388.

[4] S. González and C. Martínez, Order relation in JB-algebras, Comm. in Algebra **15** (9) (1986) , 1869-1876.

[5] S. González and C. Martínez, Order relation in quadratic Jordan rings and a structure theorem, Proc. Amer. Math. Soc. **104** (1) (1988), 51-54.

[6] S. González and C. Martínez, Periodic associative rings and order structure, Algebras, Groups and Geometries, to appear.

[7] S. González and C. Martínez, Periodic Jordan rings and order structure, Comm. in Algebra **18** (7) (1990) , 2021-2037.

[8] S. González and C. Martínez, Periodic Jordan rings of characteristic two, Mathematical Contemporany, to appear.

[9] H. C. Myung and L. Jiménez, Direct product decomposition of alternative algebras, Proc. Amer. Math. Soc. **47** (1975) , 53-59.

[10] M. Osborn, Varieties of Algebras, Advances in Mathematics **8** No. 2 (1972), 163-369.

Nonassociative
Algebraic
Models

ISBN 1-56072-050-6

Sur les Algèbres Réelles de Jordan non Commutatives de Division Linéaire de Dimension 8

Kaidi El Amin Mokhtar
Université Mohammed V, Faculté des Sciencies,
Département de Mathématiques, B.P. 1014 Rabat, Maroc

Rochdi Abdellatif
Université Hassan II, Faculté des Sciences II,
Département de Mathématiques, B.P. 6621 Casablanca, Maroc

Abstract: In this paper we give a new class of real noncommutative Jordan linear division algebras of dimension eight, including the algebras obtained by classical processes such as the Cayley-Dickson process and mutations.

1. Introduction

Il est bien connu que les algèbres réelles associatives normées de division sont de dimension finie et sont à isomorphisme près $\mathbb{R}$, $\mathbb{C}$ ou $\mathbb{H}$ (l'algèbre réelle des quaternions de Hamilton) [2].

Wright [10] a conjecturé en 1953 que les algèbres réelles normées de division linéaire sont de dimension finie. Milnor et Bott [3] renforcent à notre avis cette conjecture en démontrant en 1958 que l'espace vectoriel réel $\mathbb{R}^n$ posséde un produit bilinéaire intègre si et seulement si $n = 1, 2, 4$ ou 8.

Plus tard en 1977 Kaidi [6] établit la conjecture de Wright pur les algèbres faiblement alternatives (i.e. de Jordan non commutatives et $(x,x,[x,y]) = 0$) englobant les algèbres alternatives et les algèbres de Jordan en démontrant que les algèbres réelles faiblement alternatives normées unitaires de division linéaire sont à isomorphisme près $\mathbb{R}$, $\mathbb{C}$, $\mathbb{H}$ ou $\mathbb{O}$ (l'algèbre réelle des octonions de Cayley-Dicskon).

Cette conjecture est également établie pour les algèbres de Jordan non commutatives vérifiant la proprieté d'Osborn (i.e. deux éléments quelconques n'appartenant pas à la même sous-algèbre de dimension deux engendrent une sous-algèbre de dimension quatre) mais elle n'est pas encore établie pour les algèbres de Jordan non commutatives.

Dans ce texte nous donnons une caractérisation des algèbres réelles de Jordan non commutatives de division linéaire de dimension 8 qui peuvent s'obtenir par mutation et par extension cayleyenne puis nous généralisons le procédé de Cayley-Dickson afin de construire une nouvelle classe d'algèbres réelles de Jordan non commutatives de division linéaire de dimension 8 contenant strictement les algèbres précédentes.

2. Notation et prérecquis

2.1. Définitions

Soient K un corps commutatif de caractéristique nulle, A une K-algèbre et $\lambda \in K$. On apelle mutation λ de A et on note $A^{(\lambda)}$ l'algèbre ayant le même espace vectoriel sous-jacent et pour produit $x^{\lambda}.y = \lambda xy + (1-\lambda)yx$. A est dite de division linéaire si pour tout $x \in A - \{0\}$ et pour tout $z \in A$, les équations en y : $xy = z$ et $yx = z$ admettent une solution unique dans A. On dit que deux éléments x et y d'une K-algèbre A vérifient la propriété d'Osborn, s'ils appartiennent à la même sous-algèbre de dimension 2 ou s'ils engendrent une sous-algèbre de dimension 4. Soient (B,s) une K-algèbre cayleyenne et $\gamma \in K$, on note $E_\gamma(B)$ l'extension cayleyenne de B d'indice γ [4] A.III.16 . Si $K = \mathbb{R}$ ou $\mathbb{C}$ une K-algèbre A est dite normée si l'espace vectoriel A est muni d'une norme $\| \ \|$ sous-multiplicative (i.e. $\|xy\| \leq \|x\|\|y\|$).

Pour des raisons de simplicité nous supposerons dans toute la suite que A est unitaire d'unité e.

On notera [x,y] le commutateur de deux éléments x et y de A, (x,y,z) l'associateur de trois éléments x, y et z de A .

2.2 Proposition Soit A une $\mathbb{R}$-algèbre de Jordan n.c. normée de division linéaire. Alors A est quadratique.

Preuve: [6] p. 31 et p. 94.

2.3 Théorème Les algèbres réelles de Jordan n.c. de division linéaire et vérifiant la propriété d'Osborn sont de dimension finie 1, 2, 4 ou 8.

Preuve: [6] p. 122.

3. Algèbres réelles de Jordan non commutatives de division linéaire de dimension 8

A l'aide d'un résultat qui nous a été communiqué par J. Cuenca nous avons obtenu le théorème suivant:

3.1 Théorème [5] Soit A une $\mathbb{R}$-algèbre de Jordan non commutative de division linéaire de dimension 8. Alors A possède une sous-algèbre de dimension 4, de plus les deux propriétés suivantes sont equivalentes:

i) A vérifie la propriété d'Osborn.

ii) Deux sous-algèbres de A de dimension 4 sont isomorphes.

Preuve: La démonstration de ce résultat peut être consultée dans [8] p. 48.

3.2 Corollaire Les algèbres réelles de Jordan n.c. de division linéaire et vérifiant la proprieté d'Osborn sont à isomorphisme près $\mathbb{R}$, $\mathbb{C}$, $\mathbb{H}^{(\lambda)}$ ou $\mathbb{O}^{(\lambda)}$ avec $\lambda \in \mathbb{R} - \{1/2\}$.

Preuve: cf [8] p. 49.

3.3 Théorème Soit A une $\mathbb{R}$-algèbre de Jordan n.c. de division linéaire de dimension 8. Alors les deux propriétés suivantes sont équivalentes:

i) Il existe un élément $v \in A - \mathbb{R}e$ tel que pour tout $x \in A$, v et x vérifient la propriété d'Osborn.

ii) $A \cong E_{-1}(\mathbb{H}^{(\lambda)})^{(\mu)}$ où $\lambda, \mu \in \mathbb{R} - \{1/2\}$.

Preuve: [8] p. 64.

Note: Dans [8] p. 62 nous montrons également que si de plus $\lambda \neq 0,1$ alors l'algèbre $E_{-1}(\mathbb{H}^{(\lambda)})^{(\mu)}$ possède exactement deux classes d'isomorphie de sous-algèbres de dimension 4, à savoir $\mathbb{H}^{(\mu)}$ et $(\mathbb{H}^{(\lambda)})^{(\mu)} = \mathbb{H}^{(\alpha)}$ où $\alpha = 2\lambda\mu - \lambda - \mu + 1$.

Nous donons maintenant, afin de généraliser le procédé d'extension cayleyenne, la définition suivante:

3.4 Définition Soient (B,s) une K-algèbre cayleyenne et $\gamma, \alpha, \delta \in K$. On appelle extension cayleyenne "généralisée" de (B,s) d'indice (γ, α, δ) et on note $E_{\gamma,\alpha,\delta}(B)$ l'algèbre ayant pour espace vectoriel sous-jacent $B\times B$ et pour produit:

$$(x,y)(x',y') = (\alpha xx'+(1-\alpha)x'x+\gamma\bar{y}'y,\ y\bar{x}'+y'x+\tfrac{1}{2}\delta[y', y]) \text{ où } x, x', y, y', \bar{x}'= s(x') \in B.$$

3.5 Propriétés fondamentales

1) Il est clair que l'algèbre $E_{\gamma,\alpha,\delta}(B)$ est quadratique d'élement neutre $(e,0)$, associée à la même forme bilinéaire que celle de $E_\gamma(B)$. Cette forme bilinéaire étant symétrique, donc $E_{\gamma,\alpha,\delta}(B)$ est cayleyenne, munie de la même conjugaison cayleyenne et des mêmes norme et trace que celles de $E_\gamma(B)$.

2) L'application $B^{(\alpha)} \to E_{\gamma,\alpha,\delta}(B)$, $x \to (x,0)$ est un monomorphisme d'algèbres.

3.6 Remarque $E_{\gamma,\alpha,\delta}(B) = E_\gamma(B)$ si et seulement si B est commutative ou si $(\alpha,\delta) = (1,0)$. On notera par la suite $E_{\gamma,1,\delta}(B)$ par $E_{\gamma,\delta}(B)$.

3.7 Proposition Soient (B,s) une K-algèbre cayleyenne, $\gamma, \alpha, \delta \in K$ et $E_{\gamma,\alpha,\delta}(B)$ l'extension cayeleyenne "généralisée" de B d'indice (γ, α, δ), alors:

i) $E_{\gamma,\alpha,\delta}(B)$ est associative si et seulement si B est associative et commutative.

ii) si $\alpha \neq 1/2$, $E_{\gamma,\alpha,\delta}(B)$ est flexible si et seulement si B est flexible.

Preuve: i) Soient x et y deux éléments de B, on a $((x, 0); (0, e); (y, 0)) = (0, [x, y])$. Ainsi si $E_{\gamma,\alpha,\delta}(B)$ est associative, alors B est commutative et donc $B^{(\alpha)} = B$ est associative car sous-algèbre de $E_{\gamma,\alpha,\delta}(B)$ (3.5.2).

Réciproquement si B est commutative et associative alors $E_{\gamma,\alpha,\delta}(B) = E_{\gamma}(B)$ (3.6) est associative [4] A.III.17.

ii) Soient x, y, x', y' éléments de B. On a:

$$((x,y)(x',y'))(x,y) = \Big(\alpha(\alpha xx' + (1-\alpha)x'x + \gamma\bar{y}'y)x + (1-\alpha)x(\alpha xx' + (1-\alpha)x'x + \gamma\bar{y}'y) + \gamma\bar{y}(y\bar{x}' + y'x + \delta/2[y',y]) \,,\, (y\bar{x}' + y'x + \delta/2[y',y])\bar{x} + y(\alpha xx' + (1-\alpha)x'x + \delta\bar{y}'y) + \delta/2[y, y\bar{x}' + y'x + \delta/2[y',y]] \Big)$$

$$(x,y)((x',y')(x,y)) = \Big(\alpha x(\alpha x'x + (1-\alpha)xx' + \gamma\bar{y}y') + (1-\alpha)(\alpha x'x + (1-\alpha)xx' + \gamma\bar{y}y')x + \gamma(x\bar{y}' + \bar{x}'\bar{y} + \delta/2[\bar{y}',\bar{y}])y \,,\, y(\alpha\bar{x}\bar{x}' + (1-\alpha)\bar{x}'\bar{x} + \gamma\bar{y}'y) + (y'\bar{x} + yx' + \delta/2[y,y'])x + \delta/2[y'\bar{x} + yx' + \delta/2[y,y'] \,,\, y] \Big).$$

Si B est flexible, alors $B^{(\alpha)}$ est flexible ([8] p. 18) et l'on vérifie facilement pour tout x, y, z, t éléments de B les quatre identités suivantes:

1) $\bar{y}(yx)=(x\bar{y})y$

2) $(xy)\bar{y}=(x\bar{y})y$

3) $\bar{y}[x,y]=[\bar{x},\bar{y}]y$

4) $x(\bar{y}z) + \bar{z}(yx) = (\bar{z}y)x + (x\bar{y})z$

De plus on a les deux identités triviales suivantes:

1') $[\bar{x},\bar{y}] =[x,y]$

2') $[z,y]\bar{x} + [y,y\bar{t}+zx] = [y,z]x + [z\bar{x} + yt,y]$

On en déduit que $E_{\gamma,\alpha,\delta}(B)$ est flexible.

Réciproquement si $E_{\gamma,\alpha,\delta}(B)$ est flexible alors $B^{(\alpha)}$ est flexible car sous-algèbre de $E_{\gamma,\alpha,\delta}(B)$. Donc B est flexible ([8] p. 10). cqfd.

3.8 Remarques:

i) Si $\alpha \neq 1/2$, $E_{\gamma,\alpha,\delta}(B)$ est de Jordan n.c. si et seulement si B est de Jordan n.c.

ii) L'extension cayleyenne "généralisée" d'une algèbre cayleyenne associative n'est pas nécessairement alternative. En effet l'algèbre $E_{-1,1}(\mathbb{H})$ n'est pas alternative.

iii) Si $\gamma \neq 0$ et si le polynôme $X^2 + \gamma$ possède une racine ω dans K, alors l'application:

$$\begin{array}{ccc} E_{\gamma,\alpha,\delta}(B) & \rightarrow & E_{-1,\alpha,\delta\omega^{-1}}(B) \\ (x,y) & \rightarrow & (x, \omega y) \end{array}$$ est un isomorphisme d'algèbres.

Nous allons donner maintenant une classe d'algèbres réelles de Jordan n.c. de division linéaire de dimension 8 englobant les algèbres citées précédemment.

3.9 Proposition Soient γ, δ, λ, μ des réels arbitraires. Alors l'algèbre réelle A $= E_{\gamma,\delta}(\mathbb{H}^{(\lambda)})^{(\mu)}$ est de Jordan n.c. De plus A est de division linéaire si et seulement si λ, $\mu \in \mathbb{R}-\{1/2\}$, $\gamma < 0$ et $|\delta| < 2\sqrt{-\gamma}$.

Preuve: La première partie de la proposition découle de la remaque 3.8 i) précédente et du fait que la mutation d'une algèbre de Jordan n.c. est également une algèbre de Jordan n.c. [8] p. 18. Afin d'établir la seconde partie on va tout d'abord montrer qu'une condition nécessaire et suffisante pour que l'algèbre $B = E_{-1,\delta}(\mathbb{H}^{(\lambda)})$ soit de division linèaire est que $\lambda \in \mathbb{R}-\{1/2\}$ et $|\delta| < 2$.

C.S.: Supposons $\lambda \in \mathbb{R}-\{1/2\}$ et $|\delta| < 2$. On pose alors: $\lambda = \frac{\alpha+1}{2}$ où $\alpha \in \mathbb{R}^*$ et l'on a la table de multiplication de l'algèbre B relativement à la base $\{e, x_1, ..., x_7\}$ où $\{e, x_1, x_2, x_3\}$ est la base canonique de l'espace vectoriel $\mathbb{H}$, $x_4 = (0,e) = f$ et $x_{i+4} = x_i f$; $i = 1, 2, 3$:

$B=E_{-1,\delta}(\mathbb{H}(\frac{\alpha+1}{2}))$	e	x_1	x_2	x_3	x_4	x_5	x_6	x_7
e	e	x_1	x_2	x_3	x_4	x_5	x_6	x_7
x_1	x_1	$-e$	αx_3	$-\alpha x_2$	x_5	$-x_4$	$-\alpha x_7$	αx_6
x_2	x_2	$-\alpha x_3$	$-e$	αx_1	x_6	αx_7	$-x_4$	$-\alpha x_5$
x_3	x_3	αx_2	$-\alpha x_1$	$-e$	x_7	$-\alpha x_6$	αx_5	$-x_4$
x_4	x_4	$-x_5$	$-x_6$	$-x_7$	$-e$	x_1	x_2	x_3
x_5	x_5	x_4	$-\alpha x_7$	αx_6	$-x_1$	$-e$	$-\alpha x_3-\delta\alpha x_7$	$\alpha x_2+\delta\alpha x_6$
x_6	x_6	αx_7	x_4	$-\alpha x_5$	$-x_2$	$\alpha x_3+\delta\alpha x_7$	$-e$	$-\alpha x_1-\delta\alpha x_5$
x_7	x_7	$-\alpha x_6$	αx_5	x_4	$-x_3$	$-\alpha x_2-\delta\alpha x_6$	$\alpha x_1+\delta\alpha x_5$	$-e$

Afin de montrer que l'algèbre B est de division linéaire, on va démontrer que l'opérateur de multiplication à gauche par l'élément $\lambda_0 e + \sum_{i=1}^{7}\lambda_i x_i$ de B est inversible quand les λ_i sont des réels non tous nuls. La matrice de cet opérateur relativement à la base $\{e, x_1, ..., x_7\}$ est:

$$M = \begin{matrix} \lambda_0 & -\lambda_1 & -\lambda_2 & -\lambda_3 & -\lambda_4 & -\lambda_5 & -\lambda_6 & -\lambda_7 \\ \lambda_1 & \lambda_0 & -\alpha\lambda_3 & \alpha\lambda_2 & -\lambda_5 & \lambda_4 & \alpha\lambda_7 & -\alpha\lambda_6 \\ \lambda_2 & \alpha\lambda_3 & \lambda_0 & -\alpha\lambda_1 & -\lambda_6 & -\alpha\lambda_7 & \lambda_4 & \alpha\lambda_5 \\ \lambda_3 & -\alpha\lambda_2 & \alpha\lambda_1 & \lambda_0 & -\lambda_7 & \alpha\lambda_6 & -\alpha\lambda_5 & \lambda_4 \\ \lambda_4 & \lambda_5 & \lambda_6 & \lambda_7 & \lambda_0 & -\lambda_1 & -\lambda_2 & -\lambda_3 \\ \lambda_5 & -\lambda_4 & \alpha\lambda_7 & -\alpha\lambda_6 & \lambda_1 & \lambda_0 & \alpha\lambda_3+\delta\alpha\lambda_7 & -\alpha\lambda_2-\delta\alpha\lambda_6 \\ \lambda_6 & -\alpha\lambda_7 & -\lambda_4 & \alpha\lambda_5 & \lambda_2 & -\alpha\lambda_3-\delta\alpha\lambda_7 & \lambda_0 & \alpha\lambda_1+\delta\alpha\lambda_5 \\ \lambda_7 & \alpha\lambda_6 & -\alpha\lambda_5 & -\lambda_4 & \lambda_3 & \alpha\lambda_2+\delta\alpha\lambda_6 & -\alpha\lambda_1-\delta\alpha\lambda_5 & \lambda_0 \end{matrix}$$

Il s'agit de démontrer qu'avec les conditions précédentes et les λ_j non tous nuls, M est inversible. On pose:

$$N = \begin{matrix} \lambda_0 & \lambda_1 & \lambda_2 & \lambda_3 & \lambda_4 & \lambda_5 & \lambda_6 & \lambda_7 \\ -\lambda_1 & \alpha^{-2}\lambda_0 & \alpha^{-1}\lambda_3 & -\alpha^{-1}\lambda_2 & \lambda_5 & -\alpha^{-2}\lambda_4 & -\alpha^{-1}\lambda_7 & \alpha^{-1}\lambda_6 \\ -\lambda_2 & -\alpha^{-1}\lambda_3 & \alpha^{-2}\lambda_0 & \alpha^{-1}\lambda_1 & \lambda_6 & \alpha^{-1}\lambda_7 & -\alpha^{-2}\lambda_4 & -\alpha^{-1}\lambda_5 \\ -\lambda_3 & \alpha^{-1}\lambda_2 & -\alpha^{-1}\lambda_1 & \alpha^{-2}\lambda_0 & \lambda_7 & -\alpha^{-1}\lambda_6 & \alpha^{-1}\lambda_5 & -\alpha^{-2}\lambda_4 \\ -\lambda_4 & -\lambda_5 & -\lambda_6 & -\lambda_7 & \lambda_0 & \lambda_1 & \lambda_2 & \lambda_3 \\ -\lambda_5 & \alpha^{-2}\lambda_4 & -\alpha^{-1}\lambda_7 & \alpha^{-1}\lambda_6 & -\lambda_1 & \alpha^{-2}\lambda_0 & -\alpha^{-1}\lambda_3 & \alpha^{-1}\lambda_2 \\ -\lambda_6 & \alpha^{-1}\lambda_7 & \alpha^{-2}\lambda_4 & -\alpha^{-1}\lambda_5 & -\lambda_2 & \alpha^{-1}\lambda_3 & \alpha^{-2}\lambda_0 & -\alpha^{-1}\lambda_1 \\ -\lambda_7 & -\alpha^{-1}\lambda_6 & \alpha^{-1}\lambda_5 & \alpha^{-2}\lambda_4 & -\lambda_3 & -\alpha^{-1}\lambda_2 & \alpha^{-1}\lambda_1 & \alpha^{-2}\lambda_0 \end{matrix}$$

On obtient:

$$MN = \begin{matrix} \Delta_1 & 0 & 0 & 0 & 0 & 0 & 0 & 0 \\ 0 & \Delta_\alpha & 0 & 0 & 0 & 0 & 0 & 0 \\ 0 & 0 & \Delta_\alpha & 0 & 0 & 0 & 0 & 0 \\ 0 & 0 & 0 & \Delta_\alpha & 0 & 0 & 0 & 0 \\ 0 & 0 & 0 & 0 & \Delta_1 & (1-\alpha^{-2})(\lambda_0\lambda_1+\lambda_4\lambda_5) & (1-\alpha^{-2})(\lambda_0\lambda_2+\lambda_4\lambda_6) & (1-\alpha^{-2})(\lambda_0\lambda_3+\lambda_4\lambda_7) \\ 0 & 0 & 0 & 0 & \delta\alpha(\lambda_3\lambda_6-\lambda_2\lambda_7) & \Delta_\alpha+\delta(\lambda_2\lambda_6+\lambda_3\lambda_7) & \delta\alpha^{-1}\lambda_0\lambda_7-\delta\lambda_1\lambda_6 & -\delta\alpha^{-1}\lambda_0\lambda_6-\delta\lambda_1\lambda_7 \\ 0 & 0 & 0 & 0 & \delta\alpha(\lambda_1\lambda_7-\lambda_3\lambda_5) & -\delta\alpha^{-1}\lambda_0\lambda_7-\delta\lambda_2\lambda_5 & \Delta_\alpha+\delta(\lambda_1\lambda_5+\lambda_3\lambda_7) & \delta\alpha^{-1}\lambda_0\lambda_5-\delta\lambda_2\lambda_7 \\ 0 & 0 & 0 & 0 & \delta\alpha(\lambda_2\lambda_5-\lambda_1\lambda_6) & \delta\alpha^{-1}\lambda_0\lambda_6-\delta\lambda_3\lambda_5 & -\delta\alpha^{-1}\lambda_0\lambda_5-\delta\lambda_3\lambda_6 & \Delta_\alpha+\delta(\lambda_1\lambda_5+\lambda_2\lambda_6))) \end{matrix}$$

où $\Delta_\alpha = \alpha^{-2}(\lambda_0^2+\lambda_4^2)+\sum_{i=1}^{3}\lambda_i^2+\lambda_{i+4}^2$

Si les λj sont non tous nuls on pose $d = \dfrac{\det MN}{\Delta_1\Delta_\alpha^3}$ et l'on a:

$$d = \begin{vmatrix} \Delta_1 & (1-\alpha^{-2})(\lambda_0\lambda_1+\lambda_4\lambda_5) & (1-\alpha^{-2})(\lambda_0\lambda_2+\lambda_4\lambda_6) & (1-\alpha^{-2})(\lambda_0\lambda_3+\lambda_4\lambda_7) \\ \delta\alpha(\lambda_3\lambda_6-\lambda_2\lambda_7) & \Delta_\alpha+\delta(\lambda_2\lambda_6+\lambda_3\lambda_7) & \delta\alpha^{-1}\lambda_0\lambda_7-\delta\lambda_1\lambda_6 & -\delta\alpha^{-1}\lambda_0\lambda_6-\delta\lambda_1\lambda_7 \\ \delta\alpha(\lambda_1\lambda_7-\lambda_3\lambda_5) & -\delta\alpha^{-1}\lambda_0\lambda_7-\delta\lambda_2\lambda_5 & \Delta_\alpha+\delta(\lambda_1\lambda_5+\lambda_3\lambda_7) & \delta\alpha^{-1}\lambda_0\lambda_5-\delta\lambda_2\lambda_7 \\ \delta\alpha(\lambda_2\lambda_5-\lambda_1\lambda_6) & \delta\alpha^{-1}\lambda_0\lambda_6-\delta\lambda_3\lambda_5 & -\delta\alpha^{-1}\lambda_0\lambda_5-\delta\lambda_3\lambda_6 & \Delta_\alpha+\delta(\lambda_1\lambda_5+\lambda_2\lambda_6) \end{vmatrix}$$

On multiplie la deuxième ligne par λ_5. On lui ajoute le produit de la troisième ligne par λ_6 et le produit de la quatrième ligne par λ_7. On obtient après une division par Δ_α:

$$\frac{\lambda_5 d}{\Delta_\alpha} = \begin{vmatrix} \Delta_1 & (1-\alpha^{-2})(\lambda_0\lambda_1+\lambda_4\lambda_5) & (1-\alpha^{-2})(\lambda_0\lambda_2+\lambda_4\lambda_6) & (1-\alpha^{-2})(\lambda_0\lambda_3+\lambda_4\lambda_7) \\ 0 & \lambda_5 & \lambda_6 & \lambda_7 \\ \delta\alpha(\lambda_1\lambda_7-\lambda_3\lambda_5) & -\delta\alpha^{-1}\lambda_0\lambda_7-\delta\lambda_2\lambda_5 & \Delta_\alpha+\delta(\lambda_1\lambda_5+\lambda_3\lambda_7) & \delta\alpha^{-1}\lambda_0\lambda_5-\delta\lambda_2\lambda_7 \\ \delta\alpha(\lambda_2\lambda_5-\lambda_1\lambda_6) & \delta\alpha^{-1}\lambda_0\lambda_6-\delta\lambda_3\lambda_5 & -\delta\alpha^{-1}\lambda_0\lambda_5-\delta\lambda_3\lambda_6 & \Delta_\alpha+\delta(\lambda_1\lambda_5+\lambda_2\lambda_6) \end{vmatrix}$$

On multiplie la troisième colonne par λ_5 et on la retranche du produit de la deuxième colonne par λ_6 puis on multiplie la quatrième colonne par λ_5 et on la retranche du produit de la deuxième colonne par λ_7. On obtient:

$$\frac{\lambda_5^3 d}{\Delta_\alpha} = \begin{vmatrix} \Delta_1 & (1-\alpha^{-2})(\lambda_0\lambda_1+\lambda_4\lambda_5) & (1-\alpha^{-2})\lambda_0(\lambda_2\lambda_5-\lambda_1\lambda_6) & (1-\alpha^{-2})\lambda_0(\lambda_3\lambda_5-\lambda_1\lambda_7) \\ 0 & \lambda_5 & 0 & 0 \\ \delta\alpha(\lambda_1\lambda_7-\lambda_3\lambda_5) & -\delta\alpha^{-1}\lambda_0\lambda_7-\delta\lambda_2\lambda_5 & \lambda_5\Gamma_\alpha+\delta\alpha^{-1}\lambda_0\lambda_6\lambda_7 & \delta\alpha^{-1}\lambda_0(\lambda_5^2+\lambda_7^2) \\ \delta\alpha(\lambda_2\lambda_5-\lambda_1\lambda_6) & \delta\alpha^{-1}\lambda_0\lambda_6-\delta\lambda_3\lambda_5 & -\delta\alpha^{-1}\lambda_0(\lambda_5^2+\lambda_6^2) & \lambda_5\Gamma_\alpha-\delta\alpha^{-1}\lambda_0\lambda_6\lambda_7 \end{vmatrix}$$

où $\quad \Gamma_\alpha = \Delta_\alpha + \delta\sum_{i=1}^{3}\lambda_i\lambda_{i+4}.$

Si $\lambda_5 \neq 0$, on a:

$$\frac{\lambda_5^2 d}{\Delta_\alpha} = \begin{vmatrix} \Delta_1 & (1-\alpha^{-2})\lambda_0(\lambda_2\lambda_5+\lambda_1\lambda_6) & (1-\alpha^{-2})\lambda_0(\lambda_3\lambda_5+\lambda_1\lambda_7) \\ \delta\alpha(\lambda_1\lambda_7-\lambda_3\lambda_5) & \lambda_5\Gamma_\alpha+\delta\alpha^{-1}\lambda_0\lambda_6\lambda_7 & \delta\alpha^{-1}\lambda_0(\lambda_5^2+\lambda_7^2) \\ \delta\alpha(\lambda_2\lambda_5-\lambda_1\lambda_6) & -\delta\alpha^{-1}\lambda_0(\lambda_5^2+\lambda_6^2) & \lambda_5\Gamma_\alpha-\delta\alpha^{-1}\lambda_0\lambda_6\lambda_7 \end{vmatrix} =$$

$$= \Delta_1(\lambda_5^2\Gamma_\alpha^2 - \delta^2\alpha^{-2}\lambda_0^2\lambda_6^2\lambda_7^2 + \delta^2\alpha^{-2}\lambda_0^2(\lambda_5^2+\lambda_6^2)(\lambda_5^2+\lambda_7^2)) + \delta\alpha(1-\alpha^{-2})\lambda_0 [(\lambda_3\lambda_5-\lambda_1\lambda_7)^2(\lambda_5^2+\lambda_6^2)\delta\alpha^{-1}\lambda_0 + (\lambda_1\lambda_6-\lambda_2\lambda_5)^2(\lambda_5^2+\lambda_7^2)\delta\alpha^{-1}\lambda_0 + (\lambda_3\lambda_5-\lambda_1\lambda_7)(\lambda_1\lambda_6-\lambda_2\lambda_5)(\lambda_5\Gamma_\alpha+\delta\alpha^{-1}\lambda_0\lambda_6\lambda_7-\lambda_5\Gamma_\alpha+\delta\alpha^{-1}\lambda_0\lambda_6\lambda_7)] =$$

$$= \lambda_5^2\Delta_1(\Gamma_\alpha^2+\delta^2\alpha^{-2}\lambda_0^2\sum_{i=5}^{7}\lambda_i^2) + \delta^2(1-\alpha^{-2})\lambda_0^2 \left[\lambda_5^2((\lambda_1\lambda_7-\lambda_3\lambda_5)^2 + (\lambda_1\lambda_6-\lambda_2\lambda_5)^2) + (\lambda_6(\lambda_1\lambda_7-\lambda_3\lambda_5) - \lambda_7(\lambda_1\lambda_6-\lambda_2\lambda_5))^2\right].$$

Donc

$$\frac{d}{\Delta_\alpha} = \Delta_1\Gamma_\alpha^2 + \delta^2\lambda_0^2[(\lambda_1\lambda_7-\lambda_3\lambda_5)^2 + (\lambda_1\lambda_6-\lambda_2\lambda_5)^2 + (\lambda_2\lambda_7-\lambda_3\lambda_6)^2] +$$

$$+ \delta^2\alpha^{-2}\lambda_0^2\left[\Delta_1\sum_{i=5}^{7}\lambda_i^2 - (\lambda_1\lambda_7-\lambda_3\lambda_5)^2 - (\lambda_1\lambda_6-\lambda_2\lambda_5)^2 - (\lambda_2\lambda_7-\lambda_3\lambda_6)^2\right].$$

Finalement :

det M det N = det MN =

$$= \Delta_1\Delta_\alpha^4\Big(\Delta_1\Gamma_\alpha^2 + \delta^2\lambda_0^2[(\lambda_1\lambda_7-\lambda_3\lambda_5)^2 + (\lambda_1\lambda_6-\lambda_2\lambda_5)^2 + (\lambda_2\lambda_7-\lambda_3\lambda_6)^2 +$$
$$\alpha^{-2}((\sum_{i=1}^{3}\lambda_i\lambda_{i+4})^2 + (\lambda_0^2+\lambda_4^2+\sum_{i=5}^{7}\lambda_i^2)\sum_{i=5}^{7}\lambda_i^2)]\Big).$$

Ce résultat englobe évidemment le cas $\lambda_5 = 0$.

D'autre part on a:

$\Gamma_\alpha = \alpha^{-2}(\lambda_0^2+\lambda_4^2) + \sum_{i=1}^{3}(\lambda_i^2 + \delta\lambda_i + \lambda_{i+4}^2)$ et pour tout $i \in \{1, 2, 3\}$ le trinôme $\lambda_i^2 + \delta\lambda_{i+4}\lambda_i + \lambda_{i+4}^2$ est positif ou nul car $|\delta| < 2$. On en déduit que $\Gamma_\alpha \neq 0$ car les λ_i sont non tous nuls.

Ainsi det M $\neq 0$ et donc l'algèbre B est de division linéaire.

C.N. Les conditions $\lambda \in \mathbb{R} - \{1/2\}$ et $|\delta| < 2$ sont évidemment nécessaires pour que l'algèbre B soit de division linéaire.

Soit maintenant γ un réel strictement négatif. Alors l'algèbre $C = E_{\gamma,\delta}(\mathbb{H}^{(\lambda)})$ qui est isomorphe à $E_{-1,\delta/\sqrt{-\gamma}}(\mathbb{H}^{(\lambda)})$ (3.8 iii)), est de division linéaire si et seulement si $\lambda \in \mathbb{R} - \{1/2\}$ et $|\delta/\sqrt{-\gamma}| < 2$ i.e. $\lambda \in \mathbb{R} - \{1/2\}$ et $|\delta| < 2\sqrt{-\gamma}$.

Finalement l'algèbre $A = C^{(\mu)}$ est de division linéaire si et seulement si $\mu \in \mathbb{R}-\{1/2\}$ et C est de division linéaire [1] p. 589. cqfd.

Note: Dans [8] p. 68 nous avons montré que si $\delta \neq 0$ alors l'algèbre $E_{-1,\delta}(\mathbb{H})$ possède au moins 3 classes d'isomorphie de sous-algèbres de dimension 4.

Questions ouvèrtes:

i) Quelles sont les conditions nécessaires et suffisantes sur le triplet $(\alpha, \delta, \lambda)$ pour que l'algèbre $E_{-1,\alpha,\delta}(\mathbb{H}^{(\lambda)})$ soit de division linéaire?

ii) Est que les algèbres réelles de Jordan n.c. de division linéaire de dimension 8 s'obtiennent par mutation et par extension cayleyenne généralisée?

iii) Est ce que les algèbres réelles de Jordan n.c. de division linéaire, normées (complètes) sont de dimension finie?

Les auteurs remercient vivement le referee pour ses remarques et suggestions qui ont permi une rédaction plus élégante.

REFERENCES

[1] A.A. Albert, "Power-associative Rings", Trans. Amer. Math. Soc. **64** (1948), 552-593.

[2] F.F. Bonsall and J. Duncan, *Complete Normed Algebras*, Springer-Verlag 1973.

[3] R. Bott and J. Milnor, "On the parallelizability of the spheres", Bull, Amer. Math. Soc. **64** (1958), 87-89.

[4] N. Bourbaki, *Elementes de Mathématiques - Algèbre*, Chapitres I-III, Hermann (1970).

[5] J.A. Cuenca Mira, communication privée.

[6] A.M. Kaidi, "Bases para una teoría de las álgebras no asociativas normadas". Tesis Doctoral, Universidad de Granada, Spain 1977.

[7] J.M. Osborn, "Quadratic Division Algebras", Trans. Amer. Math. Soc. **105** (1962), 202-221.

[8] A. Rochdi, "Sur les algèbres non associatives normées de division", Thèse de 3º cycle, Faculté des Sciences de Rabat. Maroc, 1987.

[9] G.M. Benkart, D.J. Britten and J.M. Osborn, "Real Flexible Division Algebras", Can. J. Math. **39**, no.3, (1982), 550-588.

[10] F.B. Wright, "Absolute valued algebras", Proc. Nat. Acad. Sci. USA **39** (1953), 330-332.

Nonassociative
Algebraic
Models

ISBN 1-56072-050-6

Lattice Isomorphisms of Jordan Algebras

Jesús A. Laliena*
Departamento de Matemáticas, Colegio Universitario de La Rioja, U. Zaragoza, 26001 Logroño, Spain

Abstract

We study the algebraic structure of Jordan algebras J over a field F of characteristic not two whose lattice of subalgebras has small length, and also algebras whose lattice of subalgebras is isomorphic to a Jordan matrix algebra $H(F_n, t)$ with t the standard involution. If $n \geq 3$, it is shown that J is also a Jordan matrix algebra $n \times n$ over F.

1. Preliminary

In the following, all Jordan algebras will have finite dimensional over a field F of characteristic not 2.

Let J, J′ be two Jordan algebras. We denote by L(J), L(J′) their lattices of subalgebras. A lattice isomorphism or L-isomorphism between J and J′ is a one-to-one map Ψ from L(J) onto L(J′) such that

*Partially supported by the DGICYT (PS 87-0054)

$$\Psi(A \vee B) = \Psi(A) \vee \Psi(B), \quad \Psi(A \cap B) = \Psi(A) \cap \Psi(B) \quad \text{for all } A, B \leq J,$$

where $A \vee B$ is the subalgebra of J generated by A and B.

Lattice isomorphisms of Lie, associative and alternative algebras have already been studied in [1], [2], [3] and [5]. Here, we study Jordan algebras that are L-isomorphic to $H(F_n, t)$. This is the first step in studying Jordan algebras that are L-isomorphic to a semisimple Jordan algebra.

Let J be Jordan algebra. We denote by

(X): the vector subspace over F spanned by X, where X is some subset of J
l (J): the length of the longest chain of subalgebras of J.

We recall that if J is a simple Jordan algebra over the field F with finite dimension, then J is one of the following algebras:

(1) A division algebra.
(2) The Jordan algebra of a nondegenerate symmetric bilinear form, f, on a vector space V over an extension field K with dim V › 1. We will denote this algebra by $F1 + V_f$.
(3) A Jordan matrix algebra $H(D_n, J_A) = \{ (x_{ij}) \in D_n \ / \ A^{-1} (j(x_{ij})) A = (x_{ij}) \}$ with $n \geq 2$, $A = \begin{pmatrix} a_1 & & 0 \\ & a_2 & \\ 0 & & \ddots \ a_n \end{pmatrix}$ and $a_i, a_i^{-1} \in N(D)$, the nucleus of D, $j(a_i) = a_i$, where (D, j) is
   (i) either $\Delta \oplus \Delta^0$ where Δ is an associative division algebra, j the exchange involution, Δ^0 the algebra anti-isomorphic to Δ.
   (ii) an associative division algebra with involution
   (iii) a split quaternion algebra over an extension field of F with standard involution.
   (iv) an algebra of octonions over an extension field of F with standard involution and where n=3.

(See [4], Second Structure Theorem)

Let J be Jordan algebra. Let R be the solvable radical of J. Then there exists k › 0 such that $R^{2^k} = 0$ and $R^{2^{k-1}} \neq 0$, where $R^2 = RR$, $R^{2^k} = (R^{2^{k-1}})^2$. Let $R \geq R^2 \geq R^{2^2} \geq .....\geq R^{2^{k-1}} \geq 0$ be a chain of subalgebras of R. We observe R^{2^i} is ideal of $R^{2^{i-1}}$ for i = 1, ..., k. Moreover $R^{2^{i-1}} / R^{2^i}$ are zero algebras for i = 1,...,k. Therefore each subspace of them is a subalgebra. Thus $\dim (R^{2^{i-1}} / R^{2^i}) = l (R^{2^{i-1}} / R^{2^i})$ and then

$$\dim R = \sum_{i=2}^{k} \dim (R^{2^{i-1}}/R^{2^i}) = \sum_{i=2}^{k} l (R^{2^{i-1}}/R^{2^i}) = l (R).$$

2: Jordan Algebras with Length 1 and 2.

Lemma (2.1): Let J be Jordan algebra with l(J) = 1. Then dim J = 1 and J = (a) with $a^2 = a$ or $a^2 = 0$.

-Proof-

It follows from [2] and the fact that J is a power associative algebra .

Lemma (2.2): Let $J = F.1 + V_f$ with $\dim V \geq 2$. If $f(x,x)=0$ for some $0 \neq x \in V$ or $f(x,x) \in F^2$ for some $0 \neq x \in V$, then J has at least two minimal subalgebras. But if $f(x,x) \neq 0$ and $f(x,x) \notin F^2$ for all $0 \neq x \in V$, then J is a division algebra.

-Proof-

If $f(x,x)=0$ for some $0 \neq x \in V$, it is clear that (1) and (x) are minimal subalgebras. Also if $f(x,x)= \alpha^2$ with $\alpha \in F-\{0\}$ for some $0 \neq x \in V$, then (1) and ($1 - 1/\alpha \; x$) are minimal subalgebras. If $f(x,x) \neq 0$ and $f(x,x) \notin F^2$ for all $0 \neq x \in V$, let $a = \beta + \gamma x$ with $\beta, \gamma \in F$. If $\gamma = 0$, it is clear that $a^{-1} = \beta^{-1}$; and if $\beta = 0$, then $a^{-1} = \gamma^{-1} f(x,x)^{-1} x$. Now we suppose $\beta, \gamma \neq 0$. Then $a^{-1} = \delta + \omega x$, where $\omega = \dfrac{\gamma}{\gamma^2 f(x,x) - \beta^2}$ and $\delta = \dfrac{-\beta}{\gamma^2 f(x,x) - \beta^2}$

Lemma (2.3): Let J be Jordan algebra with only one minimal subalgebra. Then J is either a nilalgebra or a division algebra. Moreover it is clear if J is division algebra that J has only one minimal subalgebra.

-Proof-

If J is not a nilalgebra, it contains an idempotent $0 \neq e$. Let R be the solvable radical of J. There exists $0 \neq a \in R$ such that $a^2=0$. Thus J has two minimal subalgebras. Therefore R=0, and thus J is semisimple. Then J is a direct sum of simple algebras which have at least one minimal subalgebra. Therefore J is simple. If J is an algebra of type (2) in §1, from Lemma (2.2) F1+V is a division algebra. Further J is not simple of type (3) because (e_{ii}) for i = 1,2, with e_{ii} the $n \times n$ matrix over F with 1 in the i-th place of the diagonal and 0 in the other places of the matrix, are minimal subalgebras. Therefore J is a division algebra.

Lemma (2.4): Let $J = F.1 + V_f$ with $\dim V = n \geq 2$. Then $l(J) = n+1$.

-Proof-

Let $\{a_1, a_2, \ldots, a_n\}$ be an F-basis of V. It is clear that the chain of subalgebras: $V+F1 \geq \ldots \geq F1+Fa_1+Fa_2 \geq F1+Fa_1 \geq F.1 \geq 0$, is the longest chain of subalgebras of F1+V.

Lemma (2.5): Let $J = H(D_2, I)$ with I the standard involution. Then if D=F, l(J)=3, and if $D=F \oplus F$, l(J)=4.

-Proof-

If D = F, we have that $J_1 = \left\{ \begin{pmatrix} a & 0 \\ 0 & 0 \end{pmatrix} \text{ with } a \in F \right\}$ and $J_2 = \{ \begin{pmatrix} a & 0 \\ 0 & b \end{pmatrix}$ with $a,b \in F\}$ are subalgebras of J, and that $J \geq J_2 \geq J_1 \geq 0$ is the longest chain of subalgebras of J. Therefore l(J)=3.

If $D = F \oplus F$, then $J_1 = \left\{ \begin{pmatrix} (a,a) & (0,0) \\ (0,0) & (0,0) \end{pmatrix} \text{ with } a \in F \right\}$, $J_2 = \{ \begin{pmatrix} (a,a) & (c,0) \\ (0,c) & (0,0) \end{pmatrix}$ with $a,c \in F\}$ and $J_3 = \{ \begin{pmatrix} (a,a) & (c,0) \\ (0,c) & (d,d) \end{pmatrix}$ with $a,c,d \in F\}$ are subalgebras of J such that $J \geq J_3 \geq J_2 \geq J_1 \geq 0$ is the longest chain of subalgebras of J. Therefore l(J) = 4.

Lemma (2.6): Let J be Jordan algebra with l(J)=2 and at least two minimal subalgebras. Then dim J = 2.

-Proof-

If J is solvable, we know l(J) = dim J = 2. If J is not solvable, let R be the solvable radical of J. Since $l(J) = 2$, we have $l(R) = 0$ or 1. If $l(R)=1$, then from Lemma (2.1) dim R = dim J/R = 1, therefore dim J = 2. If $l(R)=0$, then J is semisimple. Since J has at least two minimal subalgebras and $l(J)=2$, it follows from Lemmas (2,2), (2,3) and (2,4) that $J=D_1 \oplus D_2$ with $l(D_1)= l(D_2)=1$. Therefore dim (D_1)= dim (D_2)=1, and thus $J=F\oplus F$.

Lemma (2.7): Let k be the cardinal of the groundfield F. Let J be Jordan algebra over F with $l(J)=2$. Then J is isomorphic to one of the following Jordan algebras:

TABLE I

Type	Defining relations	No. of minimal subalgebras
(i)	Extension field K of F with F as maximal subfield	1
(ii)	(a,a^2) with $a^3=0$	1
(iii a)	(e,r) with $e^2=e$ $r^2=0$ $er=0$	2
(iii b)	(e,r) with $e^2=e$ $r^2=0$ $er=r$	2
(iv)	$F\oplus F$	3
(v)	(e,r) with $e^2=e$ $r^2=0$ $er= 1/2\, r$	k+1
(vi)	(a_1,a_2) with $a_ia_j=0$ for all i and j	k+1

-Proof-

Here we follows the Barnes´s Proof for the associative case (see [2])

Suppose J is a division algebra with 1 as identity element. Then F.1 is the unique minimal subalgebra. Let $t\in J$ be such that $t\notin F.1$. Since $l(J)=2$ it follows that J= F[t], the algebra of all polynomials in t, and therefore J is commutative and associative. Hence, since J is finite dimensional, J is extension field of F and F is the unique maximal subfield.

Suppose now that J is semisimple, but not a division algebra. From Lemmas (2.3), (2.4), (2.5), (2.6), $J= F\oplus F$. If e_1 and e_2 are respectively the identity elements of the two summands, we have that (e_1), (e_2), (e_1+e_2) are all minimal subalgebras of J.

Suppose dim (R)=1, with R the solvable radical of J. Then R=(r) with $0\neq r\in R$ such that $r^2=0$. From [Corollary to Theorem 3, Chap V, 4] J contains some idempotent $0\neq e$, and thus J = (e, r). But (r)= R is an ideal. Therefore $er = \lambda r$ with $\lambda\in F$. In a Jordan algebra the identity $X^2(YX) = (X^2Y)X$ is satisfied. Thus taking $X= e+r$ $Y=e$ it follows that $e + 3\lambda^2 r = e + 2\lambda^3 r + \lambda r$, and this implies $3\lambda^2 - 2\lambda^3 - \lambda = 0$, that is, $\lambda=0$, 1 or 1/2. If $\lambda=0$, J is of type (iii a). If $\lambda=1$ J is of type (iii b) and if $\lambda=1/2$ J is of type (v). Let $(\mu e + \delta r)$ be a minimal subalgebra of J with $\mu,\delta \in F$. Then $(\mu e + \delta r).(\mu e + \delta r) = \mu^2 e + 2\mu\delta\lambda r = \gamma(\mu e + \delta r)$ for some $\gamma \in F$. Thus, if $\lambda=0$ or 1, J has only two minimal subalgebras, and if $\lambda=1/2$ all the subalgebras $(\mu e + \delta r)$ with $\delta \in F$ and (r) are minimal.

Suppose J is solvable. Then $J^2=0$ or $J^2 \neq 0$. If $J^2 \neq 0$, then $J^2 = (b)$ with $0 \neq b \in J$. Thus $J = (a,b)$ with $0 \neq a \in J - J^2$. Since $J^2 \geq J^3$ and $l(J) = 2$, then $J^3 = 0$ and therefore $ab = ba = b^2 = 0$. Also since $J^2 \neq 0$ it follows $J^2 = (a^2)$.

3: Jordan Matrix Algebras with Small Length.

Lemma (3.1): Let J be semisimple Jordan algebra with l(J)=3. Then J is isomorphic to one of the following algebras:

i) $F \oplus F \oplus F$
ii) $F \oplus K$ with K extension field of F and l(K)=2
iii) $H(F_2, J_A)$ with $A = \begin{pmatrix} 1 & 0 \\ 0 & \upsilon \end{pmatrix}$ for some $\upsilon \in F - \{0\}$
iv) $F.1 + V_f$ with dim V=2.
v) A division algebra with l(J)=3.

-Proof-

If we prove that $J \cong H(D_2, J_A)$ with l(J)=3 implies $J \cong H(F_2, J_A)$, then the result follows from Lemmas (2,3), (2,4) and (2,6). Let J be a Jordan algebra with $l(J) = 3$ and $J \cong H(D_2, J_A)$ with D an alternative algebra with involution j, $A = \begin{pmatrix} 1 & 0 \\ 0 & a \end{pmatrix}$ and $a \in D$ such that $a^{-1} \in D$ and $j(a) = a$. First we are going to show that if $x \in D$ such that $j(x)=x$, then $x \in F$. Consider the subalgebra $M = \left\{ \begin{pmatrix} F[x] & 0 \\ 0 & 0 \end{pmatrix} \right\}$ of $H(D_2, J_A)$. Then the following chain of subalgebras of $H(D_2, J_A)$ has length four

$$0 \leq \left\{ \begin{pmatrix} F & 0 \\ 0 & 0 \end{pmatrix} \right\} \leq \left\{ \begin{pmatrix} F[x] & 0 \\ 0 & 0 \end{pmatrix} \right\} \leq \left\{ \begin{pmatrix} F[x] & 0 \\ 0 & F \end{pmatrix} \right\} \leq J.$$

This is a contradiction, so that $x \in F$. Now we take the subalgebra $N = \left\{ \begin{pmatrix} \lambda & \upsilon\delta \\ \delta & \mu \end{pmatrix} \right.$ with $\left. \lambda,\delta,\mu \in F \right\}$ and we have the following chain of subalgebras

$$0 \leq \left\{ \begin{pmatrix} F & 0 \\ 0 & 0 \end{pmatrix} \right\} \leq \left\{ \begin{pmatrix} F & 0 \\ 0 & F \end{pmatrix} \right\} \leq N \leq J$$

Since $l(J) = 3$ it follows $N = J$; but $N = H(F_2, J_A)$ with $A = \begin{pmatrix} 1 & 0 \\ 0 & \upsilon \end{pmatrix}$. Thus $J = H(F_2, J_A)$.

Lemma (3.2): Let $J = H(F_2, t)$ be a Jordan matrix algebra with t the standard involution. Let $\Psi: L(J) \longrightarrow L(M)$ be an L-isomorphism of Jordan algebras. Then either $M \cong H(F_2, J_A)$, with $A = \begin{pmatrix} 1 & 0 \\ 0 & \upsilon \end{pmatrix}$ for some $\upsilon \in F-\{0\}$, or $M = F.1 + V_f$ with dim V=2.

-Proof-

Let $e = \begin{pmatrix} 1 & 0 \\ 0 & 0 \end{pmatrix}$, $f = \begin{pmatrix} 0 & 0 \\ 0 & 1 \end{pmatrix}$, $a = \begin{pmatrix} 0 & 1 \\ 1 & 0 \end{pmatrix}$. We have $e^2=e$, $f^2=f$, $ea = fa = 1/2\, a$, $a^2= e+f$, $ef=0$. Thus $(e,f) \cong F\oplus F$. From Lemma 2.7, $\Psi(e, f) \cong F\oplus F$ and the minimal subalgebras of $\Psi(e, f)$ are $\Psi(e) = (a_1)$, $\Psi(f) = (a_2)$, $\Psi(e+f) = (y)$ with a_1, a_2, y idempotents of M. Therefore $l(R)\neq 2,3$, where R is the solvable radical of M.

We observe that $(e) \vee J_1 = J$ for all subalgebras J_1 of J with $l(J_1) = 1$ and $J_1 \neq (e+f)$, (f) ; and the same for (f) *

Thus if $l(R) = 1$, then $R = (r)$ for some $r \in J$. But R is ideal of M, so dim $(R \vee \Psi(e)) = 2$. This contradicts with * because $R \vee \Psi(e) = \Psi((e)\vee(f)) \cong F\oplus F$. Therefore R=0 and M is semisimple.

Now we consider the subalgebra (e+f, a). From Lemma (2.7) we have $(e+f, a) \cong F\oplus F \cong \Psi(e+f, a)$. Thus M has at least five minimal subalgebras:

$\Psi(e)$, $\Psi(f)$, $\Psi(e+f)$, $\Psi(1/2\, e + 1/2\, f - 1/2\, a)$, $\Psi(1/2\, e + 1/2\, f - 1/2\, a)$

Therefore M is not isomorphic to $F\oplus K$ and M is not a division algebra. From * , M is not isomorphic to $F\oplus F\oplus F$, and thus from Lemma (3.1)we have the result of the Lemma.

Remark (3.3): Here we are going to study the idempotents and nilpotents with square zero in the algebra $J = H(F_2, J_A)$ with $A=\begin{pmatrix} 1 & 0 \\ 0 & \upsilon \end{pmatrix}$ for some $\upsilon \in F-\{0\}$.Let $e = \begin{pmatrix} 1 & 0 \\ 0 & 0 \end{pmatrix}$, $f = \begin{pmatrix} 0 & 0 \\ 0 & 1 \end{pmatrix}$, $a = \begin{pmatrix} 0 & \upsilon^{-1} \\ 1 & 0 \end{pmatrix}$ They are a basis of J over F.

If $x \in J$ and $x^2 = 0$, then $x = \lambda e + \mu f + \delta a$ with $\lambda, \mu, \delta \in F$ such that $\lambda^2+\delta^2\upsilon^{-1} = 0$, $\mu^2+\delta^2\upsilon^{-1} = 0$ and $(\lambda\delta + \mu\delta)\upsilon^{-1} = 0$. That is $\{x \in J$ such that $x^2 = 0\} = \{\lambda e - \lambda f \pm \lambda\sqrt{-\upsilon}\, a$ with $\lambda \in F$ and $\sqrt{-\upsilon} \in F\}$

If $x \in J$ and $x^2 = x$, then $x = \lambda e + \mu f + \delta a$ with $\lambda,\mu,\delta \in F$ such that $\lambda^2+\delta^2\upsilon^{-1} = \lambda$, $\mu^2+\delta^2\upsilon^{-1} = \mu$ and $(\lambda\delta + \mu\delta)\upsilon^{-1} = \delta$. That is, the set of all idempotents of J is $\{ (1-\mu)e + \mu f \pm \sqrt{(\mu-\mu^2)\upsilon}\; a$ with $\mu \in F$ and $\sqrt{(\mu-\mu^2)\upsilon} \in F\} \cup \{1\}$.

If now we want to find a subalgebra of J of type (iii) in Lemma (2.7), we will need to look for a nonzero idempotent x and a nonzero element r with $r^2=0$, such that $xr=r$ or $xr=0$. We can check that we have only two possibilities: $x=1$ and $r = \lambda e - \lambda f \pm \lambda\sqrt{-1}\, a$. That is, there are only two possible subalgebras of type (iii): $(1, e - f + \sqrt{-\upsilon}\; a)$, $(1, e-f- \sqrt{-\upsilon}\; a)$.

Finally, if we want to find the subalgebras of $H(F_2, J_A)$ isomorphic to $F\oplus F$, we need to find the orthogonal idempotents of $H(F_2, J_A)$. It is easy to check that these subalgebras only occur when $\upsilon = 1$ and that the subalgebras are $\{(1-\mu)e+\mu f+\sqrt{(\mu-\mu^2)}\; a, \mu e + (1-\mu)f - \sqrt{(\mu-\mu^2)}\; a\}$ with $\mu \in F$ and $\sqrt{(\mu-\mu^2)} \in F$. We observe that all the subalgebras which are isomorphic to $F\oplus F$ contains the unitary element.

Proposition (3.4): Let J be a Jordan semisimple algebra with $l(J)=3$ over a field F. Then $J = F.1 + V_f$ with $\dim V=2$ and with $f(x,x)=0$ for some $0 \neq x \in V$ if and only if J has a subalgebra of type (iii).

-Proof-

If $J = F.1 + V_f$ with f such that $f(x,x) = 0$ for $0 \neq x \in V$, it is easy to check that J has a basis $\{1,u,v\}$ such that $u^2 = \alpha = -v^2$, $uv = 0$. Thus $(1, u \pm v)$ are the unique subalgebras of type (iii) of Lemma (2.7).

Conversely if J has a subalgebra of type (iii) of Lemma (2.7), from Lemma (3.1) we have that either $J = F.1 + V_f$ with dim V= 2, or $J = H(F_2, J_A)$ with $A=\begin{pmatrix} 1 & 0 \\ 0 & \upsilon \end{pmatrix}$ for some $\upsilon \in F-\{0\}$. From Remark (3.3), if $X^2+\upsilon = 0$ has no solution in F, then it is clear that $J \neq H(F_2,J_A)$ and therefore J = F1+V with a subalgebra of type (iii), that is, with f(x,x) = 0 for some $0 \neq x \in V$. Thus now we can suppose $\sqrt{-\upsilon} \in F$. If $J = F.1 + V_f$, we know that f(x,x)=0 for some $0 \neq x \in V$. Thus for each $\lambda \in F$ there exists $v \in V$ such that $f(v,v)=\lambda$, because, since f is nondegenerate, we can find $y \in V - F.x$ with $f(x,y) \neq 0$ and $\beta \in F$ with $f(y+\beta x, y+\beta x) = 2\beta f(x,y) + f(y,y) = \lambda$. Therefore there exists $u \in V$ such that f(u,u)=1. Let $0 \neq v \in V - F.u$ and f(u,v)=0. Since f(x,x)=0 with $x=\alpha u+\gamma v$ for some $\alpha, \gamma \in F$, and $f(v,v) = \lambda \neq 0$, we have $\lambda = -(\alpha^2/\gamma^2)$. Now we can pick $w = (\sqrt{-\upsilon^{-1}})(\gamma/\alpha)\, v$ and thus {1/2 + 1/2 u, 1/2 - 1/2 u, v} is an F-basis of F1+V with the same multiplication table as {e, f, a}, the F-basis of $H(F_2, J_A)$ (see the proof of Lemma (3.2)). Therefore, if $\sqrt{-\upsilon^{-1}} \in F$ and J has a subalgebra of type (iii), then $J = F.1 + V_f \cong H(F_2,J_A)$; and f is such that f(x,x)=0 for some $0 \neq x \in V$.

Corollary (3.5): Let $J = H(F_2, t)$ with F a field such that the equation $X^2+1=0$ has a solution in F. Let $\Psi: L(J) \longrightarrow L(M)$ be an L-isomorphism of Jordan algebras. Then $M \cong J$.

Proposition (3.6): Let $J = H((F \oplus F)_2, I)$ with I the standard involution. Let $\Psi: L(J) \longrightarrow L(M)$ be an L-isomorphism of Jordan algebras. Then $M \cong J$.

-Proof-

Let $e = \begin{pmatrix} (1,1) & (0,0) \\ (0,0) & (0,0) \end{pmatrix}$, $f = \begin{pmatrix} (0,0) & (0,0) \\ (0,0) & (1,1) \end{pmatrix}$, $a = \begin{pmatrix} (0,0) & (1,0) \\ (0,1) & (0,0) \end{pmatrix}$, $b = \begin{pmatrix} (0,0) & (0,1) \\ (1,0) & (0,0) \end{pmatrix}$. It follows that ea = 1/2 a = fa , eb = 1/2b = fb , $e^2 = e$, $f^2 = f$, ab = 1/2 (e+f), $ef = 0 = a^2 = b^2$.

(i) We consider the subalgebra (e,f) of J. We have $(e,f) \cong F \oplus F$ and thus from Lemma (2.7) $\Psi(e,f) \cong F \oplus F$. Let $\Psi(e) = (y_1)$, $\Psi(f) = (y_2)$, $\Psi(e+f) = (y)$ be the minimal subalgebras of $\Psi(e,f)$, where y_1, y_2, y are idempotents of M.

(ii) Now consider the subalgebras of J: (e,a), (e,b), (f,a), (f,b). They are subalgebras of type (v) of Lemma (2,7) and since $\Psi(e)$, $\Psi(f)$ are idempotents, it follows that $\Psi(e,a)$, $\Psi(e,b)$, $\Psi(f,a)$, $\Psi(f,b)$ are also of type (v). Consider the subalgebra (e+f,a). It is of type (iii b), with minimal subalgebras (e+f) and (a). Thus $\Psi(e+f, a)$ is of type (iii) with minimal subalgebras $\Psi(e+f)$ and $\Psi(a)$. From (i) we know $\Psi(e+f) = (y)$ with y idempotent. Thus $\Psi(a)=(c)$ with $c^2=0$. In the same way taking (e+f,b) we can show that $\Psi(b)= (d)$ with $d^2=0$. Thus, since $\Psi(e,a)$, $\Psi(e,b)$, $\Psi(f,a)$, $\Psi(f,b)$ have only one minimal subalgebra spanned by an element with square zero, we have $\Psi(e) = (y_1)$, $\Psi(f) = (y_2)$, $\Psi(a) = (c)$, $\Psi(b) = (d)$ with $y_i c = 1/2c$, $y_i d=1/2d$ for i=1,2, $c^2=d^2=0$.

(iii) We know $\Psi(e+f, a)$ has the minimal subalgebras $\Psi(e+f)$ and $\Psi(a)$ with $\Psi(e+f) = (y)$ and $\Psi(a) = (c)$, and since $\Psi(e+f, a)$ is of type (iii), we know also that yc=0 or yc=c. In the same way working with $\Psi(e+f, b)$, we have y d=0 or y d=d.

Suppose that $y_1 = y + y_2$. Then $1/2\, c = y_1 c = (y + y_2) c = y c + 1/2\, c$. If yc= c, then $1/2\, c = 3/2\, c$, contradiction. Therefore yc=0, and similarly yd=0. We take the subalgebra $(a)\vee(b)= (e+f, a, b)$ of J. Then $\Psi(a)\vee\Psi(b) = (c) \vee (d)$ is subalgebra with length three. We are going to prove that dim $((c)\vee(d))=3$. If $(c)\vee(d)$ is semisimple, then dim $((c)\vee(d))=3$ since $c^2=d^2=0$, from Lemma (3.1). If $((c)\vee(d))$ is not semisimple with radical R such that $l(R)=1$, then dim (R)=1 and $l(((c)\vee(d)) / R) = 2$. But (c) or (d) are not contained in R because dim R=1. Thus $((c)\vee(d)) / R$ has at least two minimal subalgebras, (y+R) and (c+R). Therefore, from Lemma (2.6), dim $(((c)\vee(d)) / R) = 2$ and this implies that dim $((c)\vee(d)) = 3$. If R is with $l(R) = 2$, then $l(((c)\vee(d)) /R) = 1$ and therefore $l((c)\vee(d)) = l(R) + l(((c)\vee(d)) /R) = \dim(R) + \dim(((c)\vee(d))/ R)= 3$. It is not possible that $l(R)=3$ because $y^2 = y$. Thus dim $((c)\vee(d))=3$ in all cases. Since $\Psi(b)=(d)$ is not contained in $(y,c)= \Psi(e+f, a)$, then $\{y, c, d\}$ is a F-basis of $(c)\vee(d)$. Now since $y d = y c = c^2= d^2=0$ it follows that $cd= \lambda y+\mu c+\delta d$ with $\lambda,\mu,\delta \in F$ and $\lambda\neq 0$. If $(c)\vee(d)$ is a semisimple algebra, then there exist a unitary element $\omega y+\nu c+\sigma d$ of $(c)\vee(d)$ with $\omega,\nu,\sigma \in F$ (see [Corollary to Theorem 7, Chap 5, 1]). Thus $(\omega y+\nu c+\sigma d)\, c= c$. But this is not possible because $\lambda \neq 0$. Therefore $(c)\vee(d)$ is not semisimple and it has nonzero solvable radical. Let $0\neq r= \alpha y+\beta c+\gamma d\in R$ with $\alpha,\beta,\gamma\in F$ be such that $r^2=0$. Then $\alpha^2+2\beta\gamma\lambda=0$, $2\beta\gamma\mu=0$, $2\beta\gamma\delta=0$. Thus we have three possibilities: $\mu = \delta = 0$, $\alpha = \beta = 0$, $\gamma = \alpha = 0$.

If $\mu = \delta = 0$, then since R is an ideal, $cr = \gamma\lambda y\in R$. From $\lambda\neq 0$ and the fact that R is solvable, it follows that $\gamma = 0$. But also $dr =\beta\lambda y\in R$ and thus $\beta=0$. Therefore $0\neq r = \alpha y\in R$ with $r^2=0$, giving a contradiction.

If $\alpha=\beta=0$, then $d\in R$. Since R is an ideal $(cd - \delta d)^2 = (\lambda y + \mu c)^2 = \lambda^2 y \in R$, which is contradiction.

If $\gamma = \alpha = 0$, then $c \in R$, and as R is an ideal $(cd - \mu c)^2 = (\lambda y + \delta d)^2 = \lambda^2 y \in R$, again a contradiction.

Therefore $y_1 \neq y + y_2$ and in the same way we can show that $y_2 \neq y + y_1$. Thus $y = y_1 + y_2$ and $yc = y_1 c + y_2 c = c$, $yd = y_1 d + y_2 d = d$. That is $y = 1$.

(iv) We have that $(c) \vee (d)$ is an algebra with length 3 and with F-basis $\{y,c,d\}$ such that y is the identity and such that $cd = \lambda y + \mu c + \delta d$ with $\lambda, \mu, \delta \in F$ and $\lambda \neq 0$. In the following we show that $(c) \vee (d)$ is semisimple. Suppose R is the radical of $(c) \vee (d)$ and let $0 \neq r \in R$ such that $r^2 = 0$, and $r = \alpha y + \beta c + \gamma d$ with $\alpha, \beta, \gamma \in F$. From the fact that R is an ideal, $\beta rc - \gamma rd = \alpha\beta c - \alpha\gamma d \in R$ and thus $\alpha = 0$ or $y \in R$. This is a contradiction because y is idempotent. Thus $\alpha = 0$ and $r = \beta c + \gamma d$. But $r^2 = 0$ implies $\beta\gamma\lambda = 0$, that is, $\beta = 0$ or $\gamma = 0$, because $\lambda \neq 0$. We suppose without loss of generality that $\beta = 0$. Then $r = \gamma d$. Thus $rc = \gamma(\lambda y + \mu c + \delta d) \in R$. Therefore $\gamma(\lambda y + \mu c) \in R$ and $\gamma(\lambda y + \mu c)y = \gamma(\lambda y + \mu 1/2.c) \in R$. This implies that $\gamma = 0$ or $y \in R$, a contradiction, because y is idempotent and $r \neq 0$.

Therefore $(c) \vee (d)$ is semisimple.Since $c^2 = d^2 = 0$, then $(c) \vee (d)$ is of type (iii) or (iv) in Lemma (3.1). But $(a) \vee (b)$ has subalgebras of type (iii); therefore $(c) \vee (d)$ has also. From Proposition (3.4) and its proof, $(c) \vee (d)$ is of type (iv) in Lemma (3.1) and we have that F.c, F.d are the unique elements with square zero in $(c) \vee (d)$ and moreover $cd \in F-\{0\}$. We can choose $c' \in (c)$ and $d' \in (d)$ such that if $\Psi(e) = (y_1)$, $\Psi(f) = (y_2)$, then $\Psi(a) = (c')$, $\Psi(b) = (d')$, $y_i c' = 1/2\, c'$, $y_i d' = 1/2\, d'$ and $c'd' = y$. Therefore $M \cong J$.

Corollary (3.7): Let $J = H((F \oplus F)_2, I)$ with I the standard involution. Let $\Psi: L(J) \longrightarrow L(M)$ be an L-isomorphism of Jordan algebras. If e,f,a,b are as in the proof of Lemma (3.6), then we can choose $y_1, y_2, c, d \in M$ such that $\Psi(e) = (y_1)$, $\Psi(f) = (y_2)$, $\Psi(a) = (c)$, $\Psi(b) = (d)$ and $y_i^2 = y_i$, $y_1 y_2 = 0$, $y_i c = 1/2\, c$, $y_i d = 1/2\, d$, $cd = y_1 + y_2$, $c^2 = d^2 = 0$.

4: The Main Theorem.

Theorem (4.1): Let $J = H(F_n, t)$ with $3 \leq n$ and let $\Psi: L(J) \longrightarrow L(M)$ be an L-isomorphism of Jordan algebras. Then $M \cong H(F_n, J_a)$ with $a = \begin{pmatrix} 1 & & 0 \\ & a_2 & \\ 0 & & \ddots \; a_n \end{pmatrix}$ and $a_i \in F$.

-Proof-

We denote by e_{ij}, if $i \neq j$, and by e_i, if $j = i$, the square matrix in F with order n and which has 1 in the (i,j) position and zero in the other positions. It is clear that $\{e_{ij} + e_{ji}\}_{\substack{i;j=1 \\ i\neq j}}^{n} \cup \{e_i\}_{i=1}^{n}$ is an F-basis of J.

(i) We consider the subalgebra $A = (e_1) \vee \ldots \vee (e_n)$ of J. We are going to prove that A $\cong \Psi(A) \cong F_1 \oplus \ldots \oplus F_n$ with $F_i = F$ for i=1,..., n.

The subalgebras with length 2 in A are isomorphic to F⊕F. Thus, from Lemma (2.7), $\Psi(A)$ is semisimple. Suppose $\Psi(A)$ has a direct simple summand that is distinct from F. Then this summand has subalgebras isomorphic to one of the following:

a) A Jordan division algebra

b) $H(D_2, J_a)$, with D an alternative algebra with involution j, and with $a = \begin{pmatrix} 1 & 0 \\ 0 & b \end{pmatrix}$ such that $b, b^{-1} \in N(D)$ and j(b)=b

c) $F.1 + V_f$ with dimV = 2.

Case a) is not possible because all the subalgebras with length two in A are isomorphic to F⊕F. Now suppose $\Psi(A)$ has a subalgebra of type c). Then $\Psi^{-1}(F1+V)$ has length three, and thus $\Psi^{-1}(F1+V) \cong F \oplus F \oplus F$. The idempotents in $F1+V_f$ are $\{\lambda 1 + u \ / \ \lambda \in F$ and $u \in V$ such that $f(u,u) = 1/4\}$, and two of those idempotents are orthogonal if and only if they are $1/2 + u$ and $1/2 - u$ with $u \in V$ such that $f(u,u) = 1/4$. We choose $\Psi^{-1}((1/2 + u) \vee (1/2 - u)) \cong F \oplus F$. In F⊕F⊕F we know that if e and f are idempotents we can find an idempotent, g, such that (g) is not contained in $(e) \vee (f) \cong F \oplus F$ and $(f) \vee (g) \cong F \oplus F$. Thus we can find in F1+V an idempotent 1/2 +v such that (1/2 +v) is not contained in $(1/2-u) \vee (1/2+u) \cong F \oplus F$ and $(1/2-u) \vee (1/2+v) \cong F \oplus F$. But then v = -u, a contradiction. Finally, we suppose $\Psi(A)$ has a subalgebra of type b), that is a subalgebra $H(D_2, J_a)$, with D an alternative algebra with involution j, and $a = \begin{pmatrix} 1 & 0 \\ 0 & b \end{pmatrix}$ such that $b, b^{-1} \in N(D)$ and j(b)=b. We observe that $F[\begin{pmatrix} b & 0 \\ 0 & 0 \end{pmatrix}]$, the ring of polynomials in $\begin{pmatrix} b & 0 \\ 0 & 0 \end{pmatrix}$ with coefficients in F, is division subalgebra of $H(D_2, J_a)$. Thus it has only one minimal subalgebra. But any subalgebra not equal to F in $F_1 \oplus \ldots \oplus F_n$ has more than one minimal subalgebra. Therefore $b \in F$. Consider the subalgebra $\{\begin{pmatrix} \lambda & \mu b \\ \mu & \delta \end{pmatrix}$ with $\lambda, \mu, \delta \in F\}$ of $H(D_2, J_a)$. It has length 3, and thus is L-isomorphic to F⊕F⊕F. But it is also isomorphic to $H(F_2, J_a)$. Therefore from the study of the idempotents in $H(F_2,J_a)$ (Remark (3.3)), we obtain a contradiction.

Thus $\Psi(A) \cong F\oplus...\oplus F$ and since $l(A) = l(\Psi(A)) = n$, $\Psi(A)$ will have n direct summands.

(ii) Now we consider the subalgebra $B = (e_i) \vee (e_j) \vee (e_k)$ of A where i, j, k are distinct. It is clearly isomorphic to $F\oplus F\oplus F$. Its lattice of subalgebras has the following property: (e_i+e_j), (e_i+e_k), (e_j+e_k) are three subalgebras of length 1 such that any two of them span B. The other subalgebras of length 1 have the property that together with any other subalgebra of length 1, they span a subalgebra of length two. These subalgebras are: (e_i), (e_j), (e_k), $(e_i+e_j+e_k) = (1_B)$.

In the same way as before for $\Psi(A)$ it is easy to show that $\Psi(B) \cong F\oplus F\oplus F$. Now we know that the identity in $\Psi(B)$ is in one of the following subalgebras: $\Psi(e_i)$, $\Psi(e_j)$, $\Psi(e_k)$, $\Psi(e_i+e_j+e_k)$, and the idempotents which span the other three subalgebras are orthogonal.

We suppose $1_{\Psi(B)}$ is not in $\Psi(e_i+e_j+e_k)$ and that $1_{\Psi(B)}$ belongs to $\Psi(e_i)$. We take the subalgebra of M: $\Psi(C) = \Psi(e_i, 1/2\, e_i + 1/2\, e_j + 1/2\, e_{ij} + 1/2\, e_{ji}, 1/2\, e_i + 1/2\, e_j - 1/2\, e_{ij} - 1/2\, e_{ji})$. Since $C \cong H(F_2, t)$, from Lemma (3.2) we have that either $\Psi(C) \cong H(F_2,J_A)$ with $A=\begin{pmatrix} 1 & 0 \\ 0 & \upsilon \end{pmatrix}$ for some $\upsilon\in F$, or $\Psi(C) \cong F.1 + V_f$ with $\dim V = 2$. In the second case we know that F.1+V has an F-basis $\{1,u,v\}$ with $u^2 = \alpha$ $v^2=\beta$, $uv=0$, where $\alpha,\beta\in F$ and $\alpha,\beta\neq 0$. Thus from Proposition (3.4), the orthogonal idempotents of F.1+V are

$$\{1/2 + \left(\sqrt{(\tfrac{1}{4}-\mu^2\alpha)1/\beta}\right)v + \mu u,\ 1/2 - \left(\sqrt{(\tfrac{1}{4}-\mu^2\alpha)1/\beta}\right)v - \mu u\ \}$$

with $\mu\in F$ and such that $\left(\sqrt{(\tfrac{1}{4}-\mu^2\alpha)1/\beta}\right)\in F$. Therefore, from Remark (3.3), we have $1_{\Psi(C)}\in \Psi(e_i + e_j)$ which is the unique subalgebra contained in $(\Psi(e_i) \vee \Psi(e_j)) \cap (\Psi(1/2\, e_i + 1/2\, e_j + 1/2\, e_{ij} + 1/2\, e_{ji}) \vee \Psi(1/2\, e_i + 1/2\, e_j - 1/2\, e_{ij} - 1/2\, e_{ji}))$. But then $\Psi(e_i) \vee \Psi(e_j)$ contains the identity in the subalgebra $\Psi(e_i + e_j)$ and also in $\Psi(e_i)$. Contradiction.

Therefore $1_{\Psi(B)}\in \Psi(e_i + e_j + e_k)$, and if y_a is the nonzero idempotent in $\Psi(e_a)$ for a = i,j,k, we have $1_{\Psi(B)} = y_i + y_j + y_k$.

(iii) Let d_{ij}, c_{ij} idempotents in M such that $(d_{ij}) = \Psi(1/2\, e_i + 1/2\, e_j + 1/2\, e_{ij} + 1/2\, e_{ji})$, $(c_{ij}) = \Psi(1/2\, e_i + 1/2\, e_j - 1/2\, e_{ij} - 1/2\, e_{ji})$ with $i \neq j$. We consider the subalgebra $\Psi(D) = \Psi(e_k) \vee (d_{ij}) \vee (c_{ij})$ of M. As for the subalgebra $\Psi(A)$, we can show that $\Psi(D) \cong F \oplus F \oplus F$; and as in (ii), $1_{\Psi(D)}$ is in one of the following subalgebras: $\Psi(e_k)$, (d_{ij}), (c_{ij}), $\Psi(e_i + e_j + e_k)$, and the other three subalgebras are spanned by orthogonal idempotents. But if $1_{\Psi(D)} \in (d_{ij})$ or (c_{ij}), then the idempotents which span $\Psi(e_k)$ and $\Psi(e_i + e_j + e_k)$ will be orthogonal, which contradicts (ii). Thus, from (ii), $1_{\Psi(D)} \in \Psi(e_i + e_j + e_k)$, and moreover $d_{ij}\, y_k = 0$.

(iv) Now we denote by y_a the nonzero idempotent in $\Psi(A)$ such that $(y_a)=\Psi(e_a)$ for $a = 1,\ldots,n$, and we set $y = \sum_{a=1}^{n} y_a$. It is clear that $y\, y_i = y_i$ and $y\, d_{ij} = d_{ij}$ for i,j =

But $\{ d_{ij}\}_{i;j=1}^{n} \vee \{ y_i\}_{i=1}^{n}$ spans all M, and y is the identity element for them. Thus it is easy to prove that $y\, m=m$ for all $m \in M$.

(v) Now we are going to show that the y_i's are connected orthogonal idempotents. Let $M= \sum_{i;j=1}^{n} M_{ij}$ be the Peirce decomposition of M relative to the y_i's. We know that (y_i, y_j, d_{ij}) with $i \neq j$ is isomorphic to $H(F_2, J_A)$ or to $F1+V_{f_{ij}}$, with $\dim V_{ij} = 2$. It is clear in the first case that y_i and y_j are connected orthogonal idempotents. In the second case we know that $y_i = 1/2 + 1/2\, v_{ij}$, $y_j = 1/2 - 1/2\, v_{ij}$ with $0 \neq v_{ij} \in V_{ij}$, such that $f_{ij}(v_{ij}, v_{ij}) = 1/4$. Let $0 \neq u_{ij} \in V_{ij}$ be such that $f(u_{ij}, v_{ij})=0$. It is clear that $u_{ij} \in M_{ij}$ and u_{ij} is invertible in $M_{ii}+M_{jj}+M_{ij}$. Thus, from the Coordinatization Theorem [4], there exists a Jordan matrix algebra $H(D_n, J_A)$ with D an alternative algebra with involution and $A = \begin{pmatrix} 1 & & 0 \\ & a_2 & \\ 0 & & \ddots \; a_n \end{pmatrix}$ such that $a_i \in D$, and an isomorphism ξ from M onto $H(D_n, J_A)$ such that $\xi(y_i) = e_i$ and $\xi(u_{1j}) = e_{1j} + a_j^{-1} e_{j1}$, where u_{1j} is the element in M_{1j} which is inversible in $M_{11} + M_{1j} + M_{jj}$ for $j= 2, \ldots, n$. Now we prove that $a_j \in F$. Suppose $a_j \in D\text{-}F$; then $\xi(y_1, y_j, d_{1j})$ contains $\xi(u_{1j})$, and thus also contains the subalgebra $(F[a_j].(e_1+e_j)) \vee (e_1) \vee (e_j) \vee (e_{1j} + a_j^{-1} e_{j1})$, which has length greater than 3. But $\xi(y_1, y_j, d_{1j})$ is isomorphic to $H(F_2, t)$ or to $F1+V_{ij}$, that are algebras of length 3. Therefore $a_j \in F$ for j=2,..,n. Now, from $l(J) = l(M) = l(H(D_n, J_A))$ it follows $D = F$.

Corollary 4.2: Let $J = H((F\oplus F)_n, I)$ with I the standard involution, and let $\Psi: L(J) \longrightarrow L(M)$ be an L-isomorphism. Then $M \cong H(D_n, J_1)$, where D is an alternative division algebra with involution j.

-Proof-

We denote by e_i the square matrix of order n in J which has (1,1) in the (i,i) position and 0 in the other positions. Also we denote by a_{ij} with $i<j$ the square matrice of order n which has (0,1) in the (i,j) position and (1,0) in the (j,i) position, and by b_{ij} with $i<j$ the square matrice of order n which has (1,0) in the (i,j) position and (0,1) in the (j,i) position. It is clear that $\{e_i\}_{i=1}^{n} \vee \{a_{ij}\}_{i:j=1}^{n} \vee \{b_{ij}\}_{i;j=1}^{n}$ is an F-basis of J.

We observe that $(e_i, e_j, a_{ij}, b_{ij}) \cong H((F\oplus F)_2, I)$, and thus, from Lemma (3.6) and Corollary (3.7), $\Psi(e_i, e_j, a_{ij}, b_{ij}) \cong H((F\oplus F)_2, I)$. Moreover there exists $c_{ij} \in \Psi(a_{ij})$, $d_{ij} \in \Psi(b_{ij})$ such that $c_{ij}\, d_{ij} = y_i + y_j$ with $y_i + y_j$ the identity in $\Psi(e_i, e_j, a_{ij}, b_{ij})$ where $y_i \in \Psi(e_i)$, $y_j \in \Psi(e_j)$ are nonzero idempotents.

Now (e_i, a_{jk}), (e_i+e_k, a_{jk}), $(e_i+e_j+e_k, a_{jk})$ are subalgebras of M of type (iii) of Lemma (2.7). From Lemmas (2.7) and Corollary (3.7) we know that $\Psi(e_i, a_{jk}) = (y_i) \vee (c_{jk})$, $\Psi(e_j+e_k, a_{jk}) = (y_j+y_k) \vee (c_{jk})$, $\Psi(e_i+e_j+e_k, a_{jk}) = \Psi(e_i+e_j+e_k) \vee (c_{jk})$ and all of them are of type (iii) of Lemma (2.7). Since $\Psi(e_i+e_j+e_k)$ is subalgebra of $\Psi(e_i) \vee \Psi(e_j+e_k)$ and from Lemma (3.6) $\Psi(e_i) \vee \Psi(e_j+e_k) = (y_i) \vee (y_j+y_k) \cong F\oplus F$, we have $\Psi(e_i+e_j+e_k) = (y_i+y_j+y_k)$. But then $(y_i+y_j+y_k).\, c_{jk} = c_{jk}$ or 0. Since from Corollary (3.7) $(y_j+y_k).\, c_{jk} = c_{jk}$, then $y_i\, c_{jk} = 0$ and $(y_i+y_j+y_k).\, c_{jk} = c_{jk}$. In the same way we can prove $y_i\, d_{jk} = 0$.

Now we take $y = \sum_{i=1}^{n} y_i$. It is easy to check $y = 1_M$, because of the preceding paragraph and also that $\{y_i\}_{i=1}^{n} \cup \{c_{ij}\}_{i:j=1}^{n} \cup \{d_{ij}\}_{i;j=1}^{n}$ spans M. Moreover, since $(c_{ij} + d_{ij})^2 = y_i+y_j$, the y_i´s are strongly connected orthogonal idempotents. Now from the Strong Coordinatization Theorem [4] there exists a Jordan matrix algebra $H(D_n, J_1)$ and an isomorphism ξ from M onto $H(D_n, J_1)$.

Corollary 4.3: Let $J = H(Q_2, I)$ with Q the split quaternion algebra over F with I the standard involution.Let $\Psi: L(J) \rightarrow L(M)$ be an L-isomorphism of Jordan algebras.Then $M \cong J$.

-Proof-

We can find a F-basis of J whose multiplication table is

TABLE II

	e_1	e_2	a_1	a_2	a	b
e_1	e_1	0	(1/2) a_1	(1/2) a_2	(1/2) a	(1/2) b
e_2	0	e_2	(1/2) a_1	(1/2) a_2	(1/2) a	(1/2) b
a_1	(1/2) a_1	(1/2) a_1	0	1/2 (e_1+e_2)	0	0
a_2	(1/2) a_2	(1/2) a_2	1/2 (e_1+e_2)	0	0	0
a	(1/2) a	(1/2) a	0	0	0	1/2 (e_1+e_2)
b	(1/2) b	(1/2) b	0	0	1/2 (e_1+e_2)	0

We note that $(e_1, e_2, a, b) \cong (e_1, e_2, a_1, a_2) \cong H((F\oplus F)_2, I)$ with I the standard involution. Thus from Corollary (3.7) $\Psi(e_1, e_2, a, b) \cong \Psi(e_1, e_2, a_1, a_2) \cong H((F\oplus F)_2, I)$ and $\Psi(e_1+e_2) = (y)$ with $y = 1_M$, $\Psi(e_i) = (y_i)$ i=1,2 with $y_1+y_2 = y$ and we can choose $c_i, c, d \in M$ i=1,2, such that $\Psi(a_i) = (c_i)$, $\Psi(a) = (c)$, $\Psi(b) = (d)$ with $c_1c_2 = 1/2\, y$, $cd = 1/2\, y$, $c_i^2 = 0 = c^2 = d^2$ for i=1,2.

Now consider the subalgebras (a_i, a), (a_i, b) i=1,2. These algebras are of type (vi) of Lemma 2.6. Since $\Psi(a_i) = (c_i)$, $\Psi(a) = (c)$, $\Psi(b) = (d)$ with $c_i^2 = 0 = c^2 = d^2$ we have that $\Psi(a_i, a)$, $\Psi(a_i, b)$ are of type (vi) also. Therefore $M \cong H(Q_2, I)$ where I is the standard involution.

REFERENCES

[1] D.W. Barnes, Lattice isomorphisms of Lie algebras, J. Austr. Math. Soc. **IV**, 470 - 475 (1964).

[2] D.W. Barnes, Lattice isomorphisms of associative algebras, J. Austr. Math. Soc. **VI**, 106-121 (1966).

[3] E.I.Chupina, Lattice definability of semisimple finite-dimensional alternative

algebras, Siber. Math. J. **28** (5), 849-856 (1988).

[4] N. Jacobson, Structure and representation of Jordan algebras, A.M.S. Providence, Rhode Island (1968).

[5] J.A. Laliena, Lattice isomorphisms of alternative algebras, J. Algebra **128**, No.2, 335-355,(1990).

[6] R.D. Schafer, An Introduction to non associative algebras, Academic Press, New York and London, (1966).

[7] D.A. Towers, Lattice isomorphisms of Lie algebras, Math. Proc. Cambridge Phil. Soc. **89**, 283-292 (1981).

[8] K.A. Zhevlakov, A.M. Slin'ko, I.P. Shestakov, A.I. Shirshov, Rings that are nearly associative, Academic Press, New York (1982).

Nonassociative
Algebraic
Models

ISBN 1-56072-050-6

Prime and Maximal Ideals in Mutations of Associative Algebras*

Fernando Montaner
Departamento de Matemáticas, Universidad de Zaragoza, 50009 Zaragoza, Spain

1. Introduction.

If A is an associative algebra and p, q are elements of A , a new algebra can be derived from A by defining on the same vector space as A a new multiplication

$$x*y = xpy - yqx$$

for $x, y \in A$. The resulting algebra is called the (p,q)-*mutation* of A and is denoted by $A(p,q)$. This construction originated in Physics and has been studied by several authors (see [4] for a survey of the subject and references).

In this paper we study the relationship between prime and maximal ideals in an associative algebra A and in its mutations. We determine all prime ideals in a given mutation and we show that under some regularity conditions on the elements p, q which define the mutations there exist a one-to-one correspondence between prime (maximal) ideals in A and prime (maximal) ideals in $A(p, q)$. Finally we apply those results to the determination of the Baer ideal and semiprime ideals of $A(p, q)$. These results extend previous work done in [2] to the non-artinian case. Since we will made extensive use of the results in that reference we recall some of them here for the reader's convenience.

First we introduce some notation. All algebras are taken over the same field F of characteristic different from two. If A is an associative algebra and $S \subseteq A$, we denote by

* This paper has been written under the direction of Professor Santos González and it will be part of the author's Doctoral Thesis. The author has been supported by the Ministerio de Educación y Ciencia (F.P.I. Grant)

$l(S)$ (respectively $r(S)$) the left (right) anihilator of S. If p, q are elements of A we denote by $R(p, q)$ the subspace of elements $x \in A$ such that $pxp = pxq = qxp = qxq = 0$, we call this subspace the *(p, q)-radical* of A. The algebra $A(p, 0)$ is called the p-homotope of A and will be denoted by $A(p)$. Similarly we will write $R(p)$ instead of $R(p, q)$ when $q = 0$. If B is an algebra, we denote by B^- the algebra over the same vector space as B with the new multiplication $[x, y] = xy - yx$ where juxtaposition denotes the multiplication of B.

Next we give some of the results of [2]. The first one shows the importance of considering the (p, q)-radical when describing the ideals of a mutation algebra.

(1.1) PROPOSITION. *Let A be a prime associative algebra, let p, q be any fixed elements of A with $p \neq q$ and let I be a nonzero ideal of $A(p, q)$. Then either I contains a nonzero ideal of A or I is contained in the (p,q)-radical of A.*

(1.2) THEOREM.*Let A be an associative algebra and let p, q be any fixed elements of A with $p \neq q$. Then $A(p, q)$ is prime (respectively simple) if and only if A is prime (respectively simple) and $R(p, q) = 0$.*

(1.3) THEOREM.*Let A be a simple associative algebra and let p, q be elements of A such that $p \neq q$. Then $R(p, q)$ is the only maximal ideal of $A(p, q)$ and it is nilpotent.*

From this result we get

(1.4) COROLLARY. *Let A be a simple associative algebra and let p, q be elements of A. Then $A(p, q)/R(p, q)$ is a simple algebra.*

If A is an artinian associative algebra and $N(A)$ is its nilradical, then $A/N(A)$ is a direct sum of simple artinian associative algebras. We can find a solvable radical $S(p, q)$ in $A(p, q)$ by taking the preimage in A of the direct sum of one ideal I_i of each summand A_i of $A/N(A)$, with I_i being either the (p_i, q_i)-radical of A_i where p_i and q_i are the respective projections of $p+N(A)$ and $q+N(A)$ in A_i if $p_i \neq q_i$, or the preimage in A_i of the center of $A_i/R(p_i)$ if $p_i = q_i$.

(1.5) THEOREM. *Let A be an artinian associative algebra and let p, q be any fixed elements of A. Then $S(p, q) = N(A)$ if and only if there is no maximal ideal M of A not containing A^2 such that $p-q \in M$ and the sets $\{x \in A \mid px, qx \in N(A)\}$ and $\{x \in A \mid xp, xq \in N(A)\}$ are contained in $N(A)$.*

In particular we get a criterion for the semisimplicity of $A(p, q)$.

(1.6) COROLLARY. *Let A be an artinian associative algebra and let p, q be any fixed elements of A. Then $S(p, q) = 0$ if and only if A is semisimple, $l(p) \cap l(q) = 0 = r(p) \cap r(q)$ and there is no maximal ideal M of A with $p\text{-}q \in M$.*

Let B be an arbitrary nonassociative algebra, we denote by Spec (B) the set of prime ideals of B and by $\mathrm{Max}(B)$ the set of maximal ideals of B not containing B^2. If $b \in B$ we write $\mathrm{Spec}_b(B) = \{P \in \mathrm{Spec}(B) \mid b \notin P\}$ and $\mathrm{Max}_b(B) = \{M \in \mathrm{Max}(B) \mid b \notin M\}$.

2. Prime and maximal ideals in mutation algebras.

We study now the relation between $\mathrm{Spec}(A)$ ($\mathrm{Max}(A)$) and $\mathrm{Spec}(A(p, q))$ where A is an associative algebra and p, q are elements of A. In what follows $x{*}y$ denotes the product of x and y in the corresponding mutation algebra and juxtaposition will denote the original associative multiplication. Our first objective is to give a result for primeness analogous to Corollary (1.4). The next result treats the simplest case

(2.1) LEMMA. *Let A be a prime associative algebra and let a be an element of A. Then $A(a)/R(a)$ is prime.*

PROOF. If I and H are ideals of $A(a)$ such that $IaH \subseteq R(a)$, then $aIaHa = 0$. Thus $A(aIaAA)(aHa)A = (AaIaA)(AaHaA) = 0$. Since A is prime it implies that either $AaIaA = 0$ or $AaHaA = 0$. If $AaIaA = 0$, then $AaIa \subseteq l(A)$. But, since A is prime, $l(A) = 0$. So $AaIa = 0$. Then $aIa \subseteq r(A)$ and as before we have $aIa = 0$. Therefore $I \subseteq R(A)$.

(2.2) PROPOSITION. *Let A be a prime associative algebra and let p, q be elements of A with $p \neq q$. Then $A(p, q)/R(p, q)$ is prime.*

PROOF. Let I, H be ideals of $A(p, q)$ such that $I * H \subseteq R(p, q)$. If $I \cap H \subseteq R(p,q)$, then $H * I \subseteq R(p, q)$. Otherwise $(I \cap H){*}(I \cap H) \subseteq I * H \subseteq R(p, q)$. Thus, taking $I = H = I \cap H$ if necessary, we can assume that $H * I \subseteq R(p, q)$. Suppose that neither I nor H are contained in $R(p, q)$. Then, by (1.1) since A is prime, there exist I' and H' ideals of A such that $I' \subseteq I$ and $H' \subseteq H$. Now notice that since A is prime $R(p, q)$ can not contain any ideal of A, thus we can assume that I and H are ideals of A. Now if $l(p) \neq 0$, then $0 \neq Hl(p) \subseteq l(p) \cap H$. Hence $IqHl(p) = (Hl(p)){*}I \subseteq I * H \subseteq R(p, q)$. Thus from the definition of $R(p, q)$ follows that $qIqHl(p)q = 0$. We can rewrite this in the form $(AqIqA)H(l(p)qA) = 0$. Since A is prime and $H \neq 0$, this implies that either $AqIqA = 0$ or $l(p)qA = 0$. In the former case we arrive at $AqA = 0$ by using the primeness of A and that $I \neq 0$, and this implies $q = 0$. By Lemma (2.1) the result is true in this case. Thus we can assume $q \neq 0$. Then the latter possibility holds and we get $l(p)q = 0$, that is $l(p) \subseteq l(q)$.

Symmetrically, considering $Hl(q)$, we obtain either $p = 0$ and (2.1) applies (to the opposite algebra) or $l(q) \subseteq l(q)$. Consequently if $p \neq 0$ and $q \neq 0$, then $l(p) = l(q)$. The same argument can be used to prove that $r(p) = r(q)$.

Now $Hl(p\text{-}q) \subseteq l(p\text{-}q) \cap H$ and from the formula $x{*}y+y{*}x = x(p\text{-}q)y + y(p\text{-}q)x$ follows that $I(p\text{-}q)Hl(p\text{-}q) \subseteq I * H + H * I \subseteq R(p, q)$. Then multiplying on both sides by p we get $p\, l(p\text{-}q)Hl(p\text{-}q)p = 0$. As before, since $p \neq 0$ and $q \neq 0$, we obtain $l(p\text{-}q)p = 0$, i.e. $l(p\text{-}q) \subseteq l(p)$. Now since $l(p) = l(q)$, we have $l(p) = l(p) \cap l(q) \subseteq l(p\text{-}q)$. Therefore $l(p) = l(q) = l(p\text{-}q)$. Symmetrically it is shown that $r(p) = r(q) = r(p\text{-}q)$.

Let x be an element of A. If $px(p\text{-}q) = 0$, then $px \in l(p\text{-}q) = l(p) = l(q)$. Hence $pxp = pxq = 0$. Now $xp, xq \in r(p) = r(q)$ so $qxp = qxq = 0$. Therefore $x \in R(p, q)$. Since the other containment is obvious we have $R(p, q) = \{x \in A \mid px(p\text{-}q) = 0\}$. Similarly we have $R(p, q) = \{x \in A \mid (p\text{-}q)xp = 0\}$.

For all $x, y \in A$ we have $4x(p\text{-}q)y(p\text{-}q)x = 2x{\circ}(x{\circ}y) - (x{\circ}x){\circ}y$ where $x{\circ}y = x(p\text{-}q)y + y(p\text{-}q)x = x{*}y + y{*}x$. Thus, if $x \in H$ and $y \in I$, then $x(p\text{-}q)y\,(p\text{-}q)x \in H{\circ}(H{\circ}I) + (H{\circ}H){\circ}I$. Since $H{\circ}I \subseteq H{*}I + I{*}H \subseteq R(p, q)$ we get that for each $x \in H$, $x(p\text{-}q)I(p\text{-}q)x \subseteq R(p, q)$. Now from the definition of $R(p, q)$ follows that $px(p\text{-}q)I(p\text{-}q)xp = 0$ for every $x \in H$. Since A is prime this implies that for every $x \in H$ either $px(p\text{-}q) = 0$ or $(p\text{-}q)xp = 0$. From the characterizations of $R(p, q)$ given above it implies $H \subseteq R(p, q)$ which contradicts the primeness of A.

The previous proposition does not contemplate the case $p = q$. In this case the mutation $A(p, q)$ is the Lie algebra attached to the associative algebra $A(p)$. Using results due to Herstein [3] we can state the following

(2.3) LEMMA. *If A is a prime associative algebra and Z is the center of A, then the algebra A^-/Z is prime or zero.*

PROOF. We assume that A is not commutative and so $Z \neq A$. Let K, L be Lie-ideals of A such that $[K, L] \subseteq Z$, and assume that neither of them is contained in Z. Then for all $y \in K$ and $z \in L$ we have $[y,[y,z]] = 0$ which by [Theorem 1, 3] implies $[y, L] = 0$, and so $[K, L] = 0$. Let S be the subring of A generated by K. Since for any x, y, z in A we have $[xy, z] = x[y, z] + [x, z]y$, S is a Lie-ideal of A. If $S \subseteq Z$, then $K \subseteq Z$. Thus by [Theorem 3, 3] There exists a nonzero ideal I of A contained in S. Similarly we can find a nonzero ideal J of A contained in the subring T generated by L. From the above formula it is clear that $[I, J] \subseteq [S, T] = 0$. Now if $x \in I, y \in J$, then $axy = yax + [ax, y] = yax = xya + [ya, x] = xya$. Therefore $IJ \subseteq Z$. If $z \in IJ$ then for all $a, b \in A$, $az \in Z$ and $0 = [az,$

$b] = z[a, b]$. Since A is prime z is not a zero divisor so $[a, b] = 0$ for all $a, b \in A$, that is A is commutative.

Let us denote by $Z(p)$ the preimage in $A(p)$ of the center of $A(p)/R(p)$. Then we have

(2.4) PROPOSITION. *Let A be a prime associative algebra and let p be an element of A. Then $A(p, p)/Z(p)$ is either prime or zero.*

PROOF. It suffices to apply Lemma (2.3) to $A(p)/R(p)$, which is a prime algebra by Lemma (2.1)

Next we introduce some more notation. If S is a subspace of A, we denote by $\mathbf{I}(A)$ the core of S, that is the largest ideal of A contained in S. If $A(p,q)$ is a mutation of A, we denote by $RS(p, q)$ the set of elements x in A such that $pxp, pxq, qxp, qxq \in S$. clearly, if I is an ideal of A, then $RI(p, q)$ is an ideal of $A(p, q)$. Thus, for fixed p and q in A we can define maps $\mathbf{R}$: {ideals of A} $\rightarrow$ {ideals of $A(p, q)$} and $\mathbf{I}$:{ ideals of $A(p, q)$} $\rightarrow$ {ideals of A}, where $\mathbf{R}(I) = RI(p, q)$. It is clear that both $\mathbf{R}$ and $\mathbf{I}$ preserve inclusions and arbitrary intersections. We have the following straightforward characterization of $RI(p, q)$

(2.5) LEMMA. *Let A be an associative algebra and let p, q be elements of A. If I is an ideal of A, then $RI(p, q)/I$ is the $(p+I, q+I)$-radical of A/I.*

Proposition (2.2) can be completed to the following

(2.6) THEOREM. *Let A be an associative algebra and let p, q be elements of A with $p \neq q$. Then A is prime (respectively simple) if and only if $A(p, q)/R(p, q)$ is prime (respectively simple) and* $\mathbf{I}(R(p, q)) = 0$.

PROOF. By (2.2) (respectively (1.4)) if $A(p, q)$ is prime (respectively simple) so is A. If A is simple it is clear that $\mathbf{I}(R(p, q)) = 0$. Now if A is prime, then putting $V = Fp+Fq$, we have $V\mathbf{I}(R(p, q))V = 0$. Thus $(AVA)\mathbf{I}(R(p, q))(AVA) = 0$, which yields $\mathbf{I}(R(p, q)) = 0$ or $V = 0$, and the latter is impossible since it would imply $p = 0 = q$.

Next assume that $A(p, q)/R(p, q)$ is prime and let I, J be ideals of A such that $IJ = 0$. Then $IJ \subseteq R(p, q)$. As in the proof of (2.2) we can assume that $JI \subseteq R(p, q)$. Thus $I{*}J \subseteq R(p, q)$, so either $I \subseteq R(p, q)$ or $J \subseteq R(p, q)$. Since $\mathbf{I}(R(p, q)) = 0$ this implies that either $I = 0$ or $J = 0$. Finally if $A(p, q)/R(p, q)$ is simple, then every ideal of A is contained in $R(p, q)$ and so it is zero.

(2.7) COROLLARY. *Let A be an associative algebra , let p, q be elements of A and let I be an ideal of A such that* $p\text{-}q \notin I$. *Then I is prime (respectively maximal not containing* A^2*) if and only if* $RI(p, q)$ *is prime (respectively maximal not containing* A^2*) and* $\mathbf{I}(RI(p, q)) = I$.

PROOF. Since by (2.5) $RI(p, q)/I$ is the $(p+I, q+I)$-radical of A/I, it suffices to apply Theorem (2.6) to A/I and the elements $p+I$, $q+I$.

We can reformulate Corollary (2.7) in terms of the maps **R** and **I**. This is what we do in the following Corollary where we use the notation introduced in the firs section.

(2.8) COROLLARY. *Let A be an associative algebra and let p, q be elements of A. Then there exist maps* $\mathbf{R} : \mathrm{Spec}_{p\text{-}q}(A) \rightarrow \mathrm{Spec}(A(p, q))$ and $\mathbf{R} : \mathrm{Max}_{p\text{-}q}(A) \rightarrow \mathrm{Max}\,(A(p, q))$ *given by* $\mathbf{R}(I) = RI(p, q)$. *Moreover, these maps satisfy* $\mathbf{IR} = \mathit{Identity}$, *and thus they are one-to-one.*

Notice that in (2.7) the condition $p\text{-}q \notin I$ is superfluous when proving that I is prime. According to (2.4) if $p\text{-}q \in I$ and $RI(p, q) \in \mathrm{Spec}(A(p, q))$, then the algebra $A(p)/RI(p)$ is prime and has zero center, therefore we also have $\mathbf{I}(\mathbf{R}(I)) = I$.

(2.9) LEMMA. *Let A be an associative algebra and let p, q be elements of A. If H is a prime ideal of* $A(p, q)$, *then* $\mathbf{I}(H)$ *is a prime ideal of A. Thus the correspondence* **I** *defines a map from* $\mathrm{Spec}(A(p, q))$ *into* $\mathrm{Spec}(A)$.

PROOF. let I, J ideals of A such that $IJ \subseteq \mathbf{I}(H)$. As in the proof of (2.1) we can assume that also $JI \subseteq \mathbf{I}(H)$. Thus $I{*}J \subseteq \mathbf{I}(H) \subseteq H$, which implies that either $I \subseteq H$ or $J \subseteq H$. Therefore either $I \subseteq \mathbf{I}(H)$ or $J \subseteq \mathbf{I}(H)$.

(2.10) PROPOSITION. *Let A be an associative algebra and let p, q be elements of A. If* $H \in \mathrm{Spec}(A(p, q))$ *and* $p\text{-}q \notin I = \mathbf{I}(H)$, *then* $H = RI(p, q)$.

PROOF. By (2.9) $I \in \mathrm{Spec}(A)$. Now H/I is an ideal of $A/I(p+I, q+I)$, so by Proposition (1.1), either H/I contains a nonzero ideal of A/I or it is contained in $R(p+I, q+I)$. By the definition of $\mathbf{I}(H)$ we have the latter case which by (2.5) implies $H \subseteq RI(p, q)$. On the other hand if $x, y \in RI(p, q)$, then pyp, qxp, $qyq \in I$, so $(x{*}y){*}z = xpypz - yqxpz - zqxpy + zqyqx \in I$. Thus $(RI(p, q){*}RI(p, q)){*}A \subseteq I \subseteq H$, and similarly $A{*}(RI(p, q){*}RI(p, q)) \subseteq I \subseteq H$. Therefore $((RI(p, q)/H){*}(RI(p, q)/H)){*}(A/H) = (A/H){*}((RI(p, q)/H){*}(RI(p, q)/H)) = 0$. Then $(RI(p, q)/H){*}(RI(p, q)/H)$ is an ideal of A/H, and since H is prime $(RI(p, q)/H){*}(RI(p, q)/H) = 0$. Hence $(RI(p, q)/H) = 0$, and $RI(p, q) \subseteq H$.

The above proposition gives a description of the image of $\mathrm{Spec}_{p\text{-}q}(A)$ under the map **R**. Let us denote by $id(p\text{-}q)$ the ideal of A generated by $p\text{-}q$, then we have

(2.11) THEOREM. *Let A be an associative algebra and let p, q be elements of A. Then the image of the map* **R** *defined in (2.8) is the set* $\mathrm{Spec}_{id(p\text{-}q)}(A(p, q))$ *of prime ideals* $P \in \mathrm{Spec}(A(p, q))$ *such that* $id(p\text{-}q)$ *is not contained in P. Moreover we have maps* $\mathbf{R}:\mathrm{Spec}_{p\text{-}q}(A) \rightarrow \mathrm{Spec}_{id(p\text{-}q)}(A(p, q))$ *and* $\mathbf{I}: \mathrm{Spec}_{id(p\text{-}q)}(A(p, q)) \rightarrow \mathrm{Spec}_{p\text{-}q}(A)$ *such that* **RI** *and* **IR** *are the respective identity maps.*

PROOF. That the image of **R** is $\mathrm{Spec}_{id(p\text{-}q)}(A(p, q))$ follows from (2.10). Lemma (2.9) implies that the map **I** is well-defined and that **RI** is the identity map. Finally by Corollary (2.8) **IR** is the identity map.

(2.12) COROLLARY. *Let A be an associative algebra and let p, q be elements of A such that $id(p\text{-}q) = A$. Then we have bijective maps* $\mathbf{R}$: $\mathrm{Spec}(A) \rightarrow \mathrm{Spec}(A(p, q))$ *and* $\mathbf{I}:\mathrm{Spec}(A(p, q)) \rightarrow \mathrm{Spec}(A)$ *preserving containments and such that* **RI** *and* **IR** *are the corresponding identity maps* .

If H is a prime ideal of $A(p, q)$ such that $p\text{-}q$ is contained in $I = \mathbf{I}(H)$, let us denote by $ZI(p)$ the subspace of A such that $ZI(p)/RI(p)$ is the center of $A(p)/RI(p) = (A/I)(p+I)/(RI(p)/I)$. Then I is a prime ideal by (2.9), so by (2.4) $A(p, q)/ZI(p) = (A/I)(p+I, q+I)/(ZI(p)/I)$ is either prime or zero. Since H is prime we can not have $ZI(p) = A(p, q)$ since it would imply that $A(p, q)*A(p, q) \subseteq H$, so $H = A(p, q)$. Now as in the proof of (2.10) we have $RI(p) \subseteq H$, and since $ZI(p)*ZI(p) \subseteq RI(p)$ we get $ZI(p) \subseteq H$. For the reciprocal containment we use the following

(2.13)LEMMA. *Let A be a prime associative algebra and let I be a prime ideal of A^-.Then either I contains a nonzero ideal of A or I is contained in the center of A.*

PROOF. If I is not contained in the center of A, then by [Theorem 6, 3] there exists an ideal M of A such that $0 \neq [M, A] \subseteq I$. Since I is prime it implies that $M \subseteq I$.

Now in the above situation, if H is not contained in $ZI(p)$, then there exists an ideal M of $A(p)$ such that $ZI(p) \neq M \neq RI(p)$ and $RI(p) \subseteq M \subseteq H$. Thus $I \neq ApMpA \subseteq M \subseteq H$. But then $I \neq I + ApMpA \subseteq \mathbf{I}(H)$, which is a contradiction.

(2.14) THEOREM. *Let A be an associative algebra, let p, q be elements of A and let P be a prime ideal of $A(p, q)$. Then $I = \mathbf{I}(P)$ is a prime ideal of A. If $p\text{-}q \notin I$, then $P = RI(p,q)$, and if $p\text{-}q \in I$, then $P = ZI(p)$.*

We also have the analogous of Theorem (2.11) for maximal ideals.

(2.15) THEOREM. *Let A be an associative algebra and let p, q be elements of A. Then the image of the map* $\mathbf{R}$ *defined in* (2.8) *is the set* $\mathrm{Max}_{id(p-q)}(A(p, q))$ *of maximal ideals* $M \in \mathrm{Max}(A(p, q))$ *not containing* $A*A$ *and such that* $id(p-q)$ *is not contained in* M. *Moreover we have maps* $\mathbf{R}$:$\mathrm{Max}_{p-q}(A) \to \mathrm{Max}_{id(p-q)}(A(p, q))$ *and* $\mathbf{I}$: $\mathrm{Max}_{id(p-q)}(A(p, q)) \to \mathrm{Max}_{p-q}(A)$ *such that* **RI** *and* **IR** *are the respective identity maps.*

PROOF. If $M \in \mathrm{Max}_{id(p-q)}(A(p, q))$, then M is a prime ideal. Thus by (2.10) $N = \mathbf{I}(M)$ is a prime ideal and $M = RN(p, q)$. Now by (2.6) A/N is simple. Thus N belongs to $\mathrm{Max}_{p-q}(A)$.

(2.16) COROLLARY. *Let A be an associative algebra and let p, q be elements of A such that $id(p-q) = A$. Then we have bijective maps* $\mathbf{R}$: $\mathrm{Max}(A) \to \mathrm{Max}(A(p, q))$ *and* $\mathbf{I}$:$\mathrm{Max}(A(p, q)) \to \mathrm{Max}(A)$ *preserving containments and such that* **RI** *and* **IR** *are the corresponding identity maps* .

Recall that the Baer ideal $\mathrm{B}(A)$ of an algebra A (not necessarily associative) can be characterized as the intersection of all prime ideals of A (see [5]). If $\mathrm{B}(A) = 0$, then A is said to be semiprime. In this case A is a subdirect sum of prime algebras. The following results extend Theorem (1.5) and Corollary (1.6).

(2.17) THEOREM. *Let A be an associative algebra and let p, q be elements of A such that $id(p-q) = A$. Then* $\mathbf{I}(\mathrm{B}(A(p, q))) = \mathrm{B}(A)$ and $R\mathrm{B}(A)(p, q)) = \mathrm{B}(A(p, q))$. *In particular $A(p, q)$ is semiprime if and only if A is semiprime and* $R(p, q) = 0$.

PROOF. Since the maps **I** end **R** preserve intersections the assertion follows from Corollary (2.12).

(2.18) COROLLARY. *Let A be an associative algebra and let p, q be elements of A such that $id(p-q) = A$. Then A is semiprime if and only if $A(p, q)/R(p, q)$ is semiprime and* $\mathbf{I}(R(p, q)) = 0$. *If I is an ideal of A, then I is semiprime if and only if $RI(p, q)$ is semiprime and* $\mathbf{I}(RI(p, q)) = I$.

PROOF. The second statement follows from the first one by considering A/I. Now if A is semiprime, then $\mathrm{B}(A(p, q)) = R\mathrm{B}(A)(p, q) = R(p, q)$ and $0 = \mathrm{B}(A) = \mathbf{I}(\mathrm{B}(A(p, q)) = \mathbf{I}(R(p, q))$. Reciprocally, if I is an ideal of A such that $I^2 = 0$, then $I^2 \subseteq R(p, q)$. Thus $I*I \subseteq R(p, q)$, hence $I \subseteq R(p, q)$. Since $\mathbf{I}(R(p, q)) = 0$, we get $I = 0$.

The upper nil-radical $\mathrm{N}(A)$ of an algebra A belonging to a variety of power associative algebras can be characterized as the intersection of all prime ideals P of A such that A/P has no nil ideals (see [5]). Although mutations of associative algebras are not in general power associative, this actually happens if the mutation has a unit element. In this case we have the following result that generalizes [Theorem (2.7), 1]

(2.19) THEOREM. *Let A be an associative algebra and let p, q be elements of A. If $A(p, q)$ has a unit element, then* $N(A) = N(A(p,q))$.

PROOF. In these conditions A has a unit element 1, p-q is invertible in A and $A(p, q)$ is isomorphic to a mutation $A(1+s, s)$ with s an element of the center of A via the map $x \to x(p-q)^{-1}$. Now since $(1+s) - s = 1$, for all ideal I of A we have $RI(1+s, s) = I$, and Corollary (2.12) implies that I is a prime ideal of A if and only if it is a prime ideal of $A(1+s, s)$. Now since the powers of an element in $A(1+s, s)$ and in A coincide, if $A(1+s, s)/I$ has no nil ideals, the same is true for A/I. This proves the containment $N(A(1+s, s)) \subseteq N(A)$. On the other hand suppose that I is a prime ideal such that A/I has no nonzero nil-ideals. If H/I is a nil ideal of $A(1+s, s)/I$, then by (1.1) either H/I contains a (necessarily nil) ideal of A/I or it is contained in the $(1+s+I, s+I)$-radical, which is zero. Therefore $A(1+s, s)/I$ does not possesses nonzero nil-ideals, which proves $N(A) \subseteq N(A(1+s, s))$. Finally $N(A(p, q)) = N(A(1+s, s))(p-q) = N(A)(p-q) = N(A)$.

As we obtained Theorem (2.17) from Corollary (2.12), from (2.16) we obtain

(2.20) THEOREM. *Let A be an associative algebra and let p, q be elements of A such that $id(p-q) = A$. Then $A(p, q)$ is a subdirect sum of simple algebras if and only if A is a subdirect sum of simple algebras and* $R(p, q) = 0$.

REFERENCES

[1] ELDUQUE A., GONZALEZ S., MARTINEZ C.,Unit element in mutation algebras. *Alg. Groups Geom.* **1** (1984), 386-389.

[2] ELDUQUE A., MONTANER F., On mutations of associative algebras. *to appear in J. Korean Math. Soc.*

[3] HERSTEIN I. N., On the Lie-structure of an associative ring. *J. Algebra* **14** (1970), 561-571.

[4] MYUNG H. C., SAGLE A. A., On Lie-admissible mutations of associative algebras. *Hadronic J.* **10** (1987), 35-51.

[5] ZHEVLAKOV K. A., SLINKO A. M., SHESTAKOV I. P., SHIRSHOV A. I., *Rings that are nearly associative*. Academic Press. New York (1982).

Nonassociative
Algebraic
Models

ISBN 1-56072-050-6

Centroid and Extended Centroid of JB*-Algebras

A. Rodríguez Palacios and A.R. Villena Muñoz
Departamento de Análisis Matemático, Facultad de Ciencias, Universidad de Granada, 18071 Granada, Spain

Abstract

We prove that the centroid of a JB^*-algebra is isomorphic to the algebra of all complex valued bounded continuous functions on the set of cores of maximal-modular inner ideals with the hull-kernel topology, and that the extended centroid is isomorphic to a certain ring of complex valued partially defined continuous functions on the same space. These results extend, respectively, the classical Dauns-Hofmann Theorem and a recent result by P. Ara on the extended centroid of C^*-algebras.

I. The Dauns-Hofmann theorem for JB*-algebras.

The celebrated Dauns-Hofmann theorem asserts that the centroid of any C^*-algebra A is isomorphic in a natural way to the algebra of all complex valued bounded continuous functions on the set of primitive ideals of A endowed with the hull-kernel topology. The power of this theorem in the theory of C^*-algebras suggests it would be useful to get a similar result for the Jordan analogues of associative C^*-algebras, the so-called JB^*-algebras. In this section we will prove such a theorem.

We recall that, although in general the concept of primitive ideal of an associative algebra cannot be characterized in terms of the Jordan product, L. Hogben and K. McCrimmon [**14**] have introduced a succedaneous of this concept for Jordan algebras, through a judicious definition of "*maximal-modular inner ideal*" of a Jordan algebra, and then by considering as primitive ideal the largest ideal of the algebra contained in such a maximal-modular inner ideal I (called the "*core*" of I). The set of cores of all maximal-modular inner ideals of a Jordan algebra can be provided with a hull-kernel topology [**11**; **Section 5**] because they are prime ideals [**14**; **Proposition 5.5**]. Now the reader can suspect what a Dauns-Hofamnn theorem for JB^*-algebras should assert.

One of the main tools in our proof is the theory of M-structure in Banach spaces. The reader is refered to [**1**] and [**4**] for the concepts of "*M-ideal*" and "*primitive M-ideal*" of a Banach space X, and also for the "*centralizer*" of X (denoted by $Z(X)$). Recall that $Z(X)$ is a commutative algebra of bounded linear operators on X which, for complex X, is a C^*-algebra under the operator norm and suitable involution. The set *Prim(X)* of all primitive M-ideals of X can be provided with a hull-kernel type topology called as usual the structure topology on $Prim(X)$. The closed sets for this topology are the "*hulls*" $h(I) = \{J \in Prim(X) : I \subset J\}$, where I is any M-ideal of X. One of the main results in the theory of M-structure is a Dauns-Hofmann type theorem for Banach spaces, which appears explicitly stated in the literature for the real case (see Theorem B in the introduction of [**1**]) and a little more ambiguosly for the complex case (see [**4**; **Theorem 3.13(ii)**] together with the comments after [**4**; **Proposition 3.17**]). For later reference, we state this result.

Proposition 1. *For each primitive M-ideal J of a complex Banach space X and each operator f in the centralizer of X, there is a unique complex number $f^{\hat{}}(J)$ such that $f(x) - f^{\hat{}}(J)x \in J$ for all x in X, and the mapping $f \mapsto f^{\hat{}}$ is a $*$-isomorphism from $Z(X)$ onto the C^*-algebra $C_b(Prim(X))$ of all complex valued bounded continuous functions on $Prim(X)$ endowed with the structure topology.*

The above proposition is specially relevant in the case of a JB^*-algebra A because the primitive M-ideals of the Banach space of A are the kernels of the "*type I factor representations*" of A [**16**; **Corollary 1.13**] (an almost algebraic concept), while the centralizer of the Banach space of A agrees with the centroid of A [**9**; **Theorem 2.8 and Proposition 3.9**]. Recall that the centroid of an algebra A (denoted $\Gamma(A)$) is the algebra of those linear operators f on A satisfying

$$f(ab) = af(b) = f(a)b$$

for all a, b in A. Note that, if A is a complete normed algebra with zero annihilator, thanks to the closed graph theorem every operator in $\Gamma(A)$ is automatically bounded (a superfluous assumption on the elements of the centroid of a JB^*-algebra in the definition given in [**9**]). We will work with noncommutative (in short: n.c.) JB^*-algebras (see [**15**] for definition) because they contain all (associative) C^*-algebras and all (commutative) JB^*-algebras, and in a very precise sense they are the widest nonassociative generalizations of C^*-algebras [**17**]. We will use without previous comment the easy fact that, if A is a n.c. JB^*-algebra, the algebra A^+ obtained by symmetrization of the product of A is a JB^*-algebra under the same norm and involution as A. Also recall that the centroid of a n.c. JB^*-algebra, with the operator norm and natural involution, is a commutative C^*-algebra [**18**; **Proposition 2.1**]. We begin by extending to n.c. JB^*-algebras the above-mentioned result in [**9**] on coincidence of the centralizer and the centroid.

Proposition 2. *Let A be a n.c. JB^*-algebra. Then $Z(A) = \Gamma(A)$.*

Proof. We have $Z(A) = Z(A^+)$ because A and A^+ agree as Banach spaces, and also $Z(A^+) = \Gamma(A^+)$ by the previously cited commutative case of our assertion. Therefore it is enough to prove that $\Gamma(A^+) = \Gamma(A)$. Since clearly $\Gamma(A)$ is contained in $\Gamma(A^+)$, only we must prove the converse inclusion. To this end observe that, for f in the centroid of an algebra and D a derivation of the algebra, the bracket $[D, f]$ lies in the centroid. So, taking into account that for any a in A the mapping $D_a : b \mapsto [a, b]$ is a derivation of A^+[**19; p. 146**], we have that $f \mapsto [D_a, f]$ is a continuous derivation of the commutative C^*-algebra $\Gamma(A^+)$. Since a commutative C^*-algebra has no nonzero (continuous) derivations (use for example the Singer-Wermer theorem [**5; Theorem 18.16**]), for f in $\Gamma(A^+)$ and for all a, b in A we have $f([a, b]) = [a, f(b)]$, which clearly shows that actually f lies in $\Gamma(A)$.■

In view of the above proposition and the fact that the primitive M-ideals of the Banach space of a C^*-algebra agree with the (algebraic) primitive ideals [**15; Lemma 6.5**], the classical Dauns-Hofmann theorem for C^*-algebras follows from Proposition 1. Unfortunately we do not know if the primitive M-ideals of the Banach space of a JB^*-algebra agree with the cores of maximal-modular inner ideals, so to obtain our desired result we still need an additional effort.

From now on, the reader must retain the fact that, for n.c. JB^*-algebras, M-ideals and closed (two-sided) ideals are the same [**15; Theorem 4.3**]. Also we recall that each closed ideal J of a n.c. JB^*-algebra A is $*$-invariant and A/J is a n.c. JB^*-algebra with respect to the quotient norm and involution. [**15; Corollary 1.11**]. In keeping with the notation used in Proposition 1, we denote by $Prim(A)$ the set of primitive M-ideals of the Banach space of A. If $\beta(A)$ denotes the set of all proper closed prime ideals of A, it is not difficult to see that $Prim(A) \subset \beta(A)$. Moreover, since $\bigcap Prim(X)=0$ for any Banach space X, $Prim(A)$ is dense in $\beta(A)$ for the hull-kernel topology, and clearly the restriction of this topology to $Prim(A)$ is the structure topology.

Proposition 3. *Let A be a n.c. JB^*-algebra. Then, for each J in $\beta(A)$ and each f in $\Gamma(A)$, there exists a unique complex number $f^{\hat{}}(J)$ such that $f(a) - f^{\hat{}}(J)a \in J$ for all a in A. Moreover, if Λ is any subset of $\beta(A)$ containing $Prim(A)$, the mapping $f \mapsto f^{\hat{}}_{|\Lambda}$ is a $*$-isomorphism from $\Gamma(A)$ onto the C^*-algebra $C_b(\Lambda)$ of all complex valued bounded continuous functions on Λ.*

Proof. As any closed ideal of A, J is a n.c. JB^*-algebra, so $J = J^2$, and so $f(J) \subset J$, because f is in $\Gamma(A)$. Now $a + J \mapsto f(a) + J$ is a well defined mapping from A/J into A/J which clearly lies in $\Gamma(A/J)$. But, since J is a closed prime ideal of A, A/J is a prime n.c. JB^*-algebra, so $\Gamma(A/J)$ is a prime commutative C^*-algebra, and so $\Gamma(A/J) = \mathbf{C}\mathrm{Id}_{A/J}$. Now the existence of the number $f^{\hat{}}(J)$ in the statement is clear, and its uniqueness is a consequence of the fact that J is a proper ideal of A. Concerning the second part of the statement we first prove that, for f in $\Gamma(A)$, $f^{\hat{}}$ is continuous on $\beta(A)$ (and so on Λ). Let C be any closed subset of $\mathbf{C}$. Since by Propositions 2 and 1 $f^{\hat{}}_{|Prim(A)}$ is continuous, we have that the set $\{I \in Prim(A) : f^{\hat{}}(I) \in C\}$ is closed in $Prim(A)$, so $f^{\hat{}}(I_0)$ lies in C whenever I_0 in $Prim(A)$ satisfies

$$\bigcap\{I \in Prim(A) : f^{\hat{}}(I) \in C\} \subset I_0.$$

Since clearly $f^{\hat{}}(J_1) = f^{\hat{}}(J_2)$ whenever J_1 and J_2 are in $\beta(A)$ with $J_1 \subset J_2$, and every M-ideal of a Banach space is the intersection of the primitive M-ideals containing it [1; **Proposition 3.5(a)**], it follows that

$$\bigcap\{I \in Prim(A) : f^{\hat{}}(I) \in C\} = \bigcap\{I \in \beta(A) : f^{\hat{}}(I) \in C\}.$$

Now, for J_0 in $\beta(A)$ with $\bigcap\{I \in \beta(A) : f^{\hat{}}(I) \in C\} \subset J_0$, there exists I_0 in $Prim(A)$ such that $J_0 \subset I_0$, so $f^{\hat{}}(J_0)(= f^{\hat{}}(I_0))$ lies in C, which proves that $f^{\hat{}}$ is continuous on $\beta(A)$. Clearly $f^{\hat{}}$ is bounded on $\beta(A)$ (hence on Λ). Finally, consider the mapping $\phi : f \mapsto f^{\hat{}}_{|\Lambda}$ from $\Gamma(A)$ into $C_b(\Lambda)$. By Propositions 2 and 1 the mapping $\phi_1 : f \mapsto f^{\hat{}}_{|Prim(A)}$ is a $*$-isomorphism from $\Gamma(A)$ onto $C_b(Prim(A))$ and, if ϕ_2 denotes the one-to-one $*$-homomorphism from $C_b(\Lambda)$ into $C_b(Prim(A))$ induced by the dense inclusion of $Prim(A)$ in Λ, we have clearly $\phi_2 \circ \phi = \phi_1$. Therefore ϕ is a $*$-isomorphism onto $C_b(\Lambda)$. and the proof is complete.■

Following [12], we define the maximal-modular inner ideals of a noncommutative Jordan algebra A as the maximal-modular inner ideals (in the sense of [14]) of the Jordan algebra A^+, and we define the core of a maximal-modular inner ideal I as the largest ideal of A contained in I. As observed in [12] these cores are prime ideals and, for complete normed A, they are also closed ideals [11; **Lemma 6.5**]. Taking into account that the closed ideals of a n.c. JB^*-algebra A depend only on the Banach space of A. it follows that, denoting by $Cor(A)$ the set of cores of maximal-modular inner ideals of A, we have $Cor(A)=Cor(A^+)$.

Proposition 4. *For a n.c. JB^*-algebra A we have that*

$$Prim(A) \subset Cor(A) \subset \beta(A).$$

Proof. In view of the above commets. it only remains to prove the first inclusion and then also we can assume that A is commutative. So, let J be a primitive M-ideal of the JB^*-algebra A. We must show that J is the core of a maximal-modular inner ideal of A. First assume that A is unital. By [16; **Corollary 1.13**] J is the kernel of a type I factor representation π of A inducing, by restriction on the domain and the range, a type I representation $\pi^{\hat{}}$ of the JB-algebra Sym$(A) := \{a \in A : a = a^*\}$ (see for example [16; **Remark 1.9**]). Now it is well known (see [6] and [7]) that there exists a pure state ϕ of Sym(A) such that Ker$(\pi^{\hat{}})$ is the largest closed ideal of Sym(A) (actually, the largest ideal because Sym(A) has a unit) contained in Ker(ϕ), and that the set $I := \{b \in Sym(A) : \phi(b^2) = 0\}$ is a maximal closed inner ideal of Sym(A) (actually, a maximal inner ideal because Sym(A) has a unit). Since Ker$(\pi^{\hat{}}) \subset I \subset$Ker$(\phi)$, we have that Ker$(\pi^{\hat{}})$ is the core of I in the JB-algebra Sym(A). Since I+iI is a proper inner ideal of A, there is a maximal inner ideal M of A with I+i$I$$\subset M$. Now Sym$(M \cap M^*)$ is a proper inner ideal of Sym(A) containing I. so Sym$(M \cap M^*) = I$ by the maximality of I, and so $M \cap M^* = I$+iI. Since, as any closed ideal of A, J is $*$-invariant and Ker$(\pi^{\hat{}})$ is the core of I. we have that $J(= Ker(\pi) = Ker(\pi^{\hat{}}) + iKer(\pi^{\hat{}}))$ is the largest ideal of A contained in I+iI= $M \cap M^*$. Since M is closed in A and all closed ideals of A are

$*$-invariant, also J is the largest ideal of A contained in the maximal (automatically modular since A has unit) inner ideal M. To conclude the proof, assume now that A is not unital. Then the unitization of A (say A_1) is a JB^*-algebra [**15; Corollary 1.10**] and, by [**1; Proposition 3.5(b)**], there exists a primitive M-ideal J_1 of A_1 such that $J = J_1 \cap A$. By the first part of the proof, J_1 is the core (in A_1) of a maximal inner ideal M_1 of A_1. Since clearly $M_1 \not\subset A$, there exists x in A with $1 - x \in M_1$ and therefore (see [**14; Remark 3.5**])$M_1 \cap A$ is a maximal-modular inner ideal of A, whose core is J.■

Now the Dauns-Hofmann type theorem for JB^*-algebras follows directly from Propositions 3 and 4.

Theorem 1. *For a n.c. JB^*-algebra A, the C^*-algebras $\Gamma(A)$ and $C_b(Cor(A))$ are isomorphic.*

II. The extended centroid of JB*-algebras.

As its name suggests, the extended centroid is a suitable enlargement of the centroid of a semiprime nonassociative algebra introduced in [**3**], whose main interest is found in the theory of structure and classification of prime algebras. Our aim in this section is to extend to JB^*-algebras the recent result by P. Ara [**2**] exhibiting the extended centroid of a C^*-algebra as a certain ring of complex valued partially defined continuous functions on the primitive spectrum of the algebra. We will use a reformulated definition of the extended centroid of a semiprime algebra A obtained in [**8**]. The elements in the extended centroid of A (noted $C(A)$) will be the so called *"maximal essentially defined centralizers"* on A. *"Essentially defined centralizer"* means linear mapping f from a suitable essential ideal of A (say *dom(f)*) into A satisfying

$$f(ab) = af(b) \quad \text{and} \quad f(ba) = f(b)a$$

for all a in A and b in dom(f), and *"maximal"* means that there are no nontrivial extensions of f with the same properties as f. For each essentially defined centralizer on A there is a unique maximal essentially defined centralizer on A which extends it, and $C(A)$ is a von Neumann regular commutative ring if we take as sum (resp.: product) of two elements in $C(A)$ the unique maximal essentially defined centralizer on A which extends the usual sum (resp.: composition) of the given elements as partially defined operators on A. If, as in the case of a n.c. JB^*-algebra, A has an involution, we can transfer in a natural way the involution of A to the extended centroid of A by considering, for f in $C(A)$, the element f^* in $C(A)$ given by

$$dom(f*) = (dom(f))^* \quad \text{and} \quad f^*(b) = (f(b^*))^*$$

for all b in dom(f^*). In the case of a n.c. JB^*-algebra A, the good behaviour of the induced involution on $C(A)$ will follow from the following

Proposition 5. *For every element a in a n.c. JB^*-algebra A, we have*

$$(1/2) \| a \|^2 \leq \| a^* a \| .$$

Proof. Since tha family of all type I factor representations of A is faithful (a consequence of [**16; Corollary 1.13**]), a standard procedure allows us to assume that A is actually a type I n.c. JB^*-factor. If A is either commutative or of the form $B^{(\lambda)}$ for suitable C^*-algebra B and $0 \leq \lambda \leq 1$, our assertion is true (see [**15; Proposition 2.2**] for the first case). Otherwise, by [**16; Theorem 2.7**], A must be a quadratic JB^*-algebra and therefore, by [**16; Theorems 3.2 and 3.4**], there are a real Hilbert space X and an anticommutative bilinear product $(x, y) \mapsto x \wedge y$ on X satisfying

$$(x \wedge y \mid z) = (x \mid y \wedge z) \quad \text{and} \quad \| x \wedge y \| \leq \| x \| \| y \|$$

such that, by defining in the Hilbert sum $\mathbf{R} \times X$ the product

$$(\alpha, x)(\beta, y) := (\alpha\beta - (x \mid y), \alpha y + \beta x + x \wedge y),$$

we have that A equals $(\mathbf{R} \times X)_{\mathbf{C}}$ with the complexification product, the involution of A is given by

$$[(\alpha, x) + i(\beta, y)]^* = (\alpha, -x) + i(-\beta, y),$$

and the norm of A is given by

$$\| (\alpha, x) + i(\beta, y) \|^2 =$$

$$\| (\alpha, x) \|^2 + \| (\beta, y) \|^2 + 2[(\| (\alpha, x) \|^2 \| (\beta, y) \|^2 - ((\alpha, x) \mid (\beta, y))^2]^{1/2}.$$

Now, for $a = (\alpha, x) + i(\beta, y)$ in A, we have

$$(1/2) \| a \|^2 =$$

$$(1/2)(\| (\alpha, x) \|^2 + \| (\beta, y) \|^2) + [\| (\alpha, x) \|^2 \| (\beta, y) \|^2 - ((\alpha, x) \mid (\beta, y))^2]^{1/2} \leq$$

$$(1/2)(\| (\alpha, x) \|^2 + \| (\beta, y) \|^2) + ((1/4)[\| (\alpha, x) \|^2 + \| (\beta, y) \|^2]^2)^{1/2} =$$

$$\| (\alpha, x) \|^2 + \| (\beta, y) \|^2 \leq \| (\alpha, x) \|^2 + \| (\beta, y) \|^2 + 2 \| \alpha y - \beta x - x \wedge y \| = \| a^* a \|,$$

as required.∎

Recall that an involution $*$ on a ring R is said to be positive definite if, for any finite family $\{r_i\}$ of elements in R, the relation $\sum r_i^* r_i = 0$ implies $r_i = 0$ for all i. Also recall that a von Neumann regular ring with involution such that $r^* r = 0$ implies $r = 0$ is called a $*$-regular ring.

Lemma 1. *Let A be a n.c. JB^*-algebra. Then $C(A)$ is a $*$-regular ring with positive definite involution.*

Proof. Suppose that $\sum_{i=1}^n f_i^* f_i = 0$ with $f_1, \ldots, f_n$ in $C(A)$. For a in $\bigcap_{i=1}^n dom(f_i)$, $a^* a$ lies in $\bigcap_{i=1}^n \text{dom}(f_i^* f_i)$ and

$$0 = \sum_{i=1}^n f_i^* f_i(a^* a) = \sum_{i=1}^n (f_i(a))^* f_i(a).$$

By [**16; Theorem 3.7**], for all $i = 1, \ldots, n$ we have $(f_i(a))^* f_i(a) = 0$, and by Proposition 5 also $f_i(a) = 0$. Finally $f_i = 0$, because $\bigcap_{i=1}^{n} \mathrm{dom}(f_i)$ is an essential ideal of A.∎

If R is a unital rational algebra with positive definite involution, we define the positive cone of R as the set of elements of the form $\sum r_i^* r_i$, where $\{r_i\}$ is any finite subset of R. Thus R becomes a partially ordered set in the usual way. An element r in R is said to be bounded if $r^* r \leq n$ for suitable positive integer n. The set of all bounded elements of R will be denoted by R_b. In order to compute the ring of bounded elements in the extended centroid of a n.c. JB^*-algebra A, observe that, if $\mathcal{I}$ denotes the family of all closed essential ideals of A, then $\{\Gamma(I) : I \in \mathcal{I}\}$ is in a natural way a directed system of commutative C^*-algebras. Indeed, if I and J are in I with $I \subset J$ and f is in $\Gamma(J)$, from the fact that $I^2 = I$ we obtain that $f(I) \subset I$, and since I is an essential ideal of A we have that $f(I) = 0$ implies $f = 0$, thus showing that the restriction mapping $\Gamma(J) \to \Gamma(I)$ is well defined and is a one-to-one $*$-homomorphism

Proposition 6. *Let A be a n.c. JB^*-algebra. Then $C(A)_b$ is $*$-isomorphic to the algebraic direct limit $\lim\limits_{I \in \mathcal{I}} \Gamma(I)$, where $\mathcal{I}$ denotes the family of all closed essential ideals of A.*

Proof. First we claim that, for any closed ideal I of A and any f in $\Gamma(I)$, we have $f(ab) = af(b)$ and $f(ab) = f(b)a$ for all a in A and b in I. Since this assertion is clearly true if I is a direct summand of A, we will reduce to this case along a standard procedure of bidualization. By double transposition we can extend f to get a w^*-continuous linear mapping $f^{tt} : I^{oo} \to I^{oo}$, where I^{oo} denotes the bipolar of I in A'' (the bidual of A). From the facts that A'' is a n.c. JB^*-algebra [**15; Theorem 1.7**], I is w^*-dense in I^{oo}, and the product of A'' is separately w^*-continuous [**15; Theorem 3.5**], it follows that f^{tt} lies in $\Gamma(I^{oo})$. But I^{oo}, as any w^*-closed ideal of A'', is a direct summand of A'' [**15; Theorem 3.9**], which concludes the proof of the claim. Now it is clear that, if I is in $\mathcal{I}$, each element in $\Gamma(I)$ can be regarded as an essentially defined centralizer on A, and, by passing to the unique maximal essentially defined centralizer on A which extends the considered one, we obtain a natural one-to-one $*$-homomorphism $\Gamma(I) \to C(A)$. Since $\Gamma(I)$ (as any C^*-algebra) agrees with the set of its bounded elements, this homomorphism actually takes values in $C(A)_b$. On the other hand, if I and J are in $\mathcal{I}$ with $I \subset J$, the diagram

$$\begin{array}{ccc} \Gamma(J) & \longrightarrow & \Gamma(I) \\ & \searrow & \downarrow \\ & & C(A) \end{array}$$

commutes and therefore we have a natural one-to-one $*$-homomorphism

$$\lim_{I \in \mathcal{I}} \Gamma(I) \to C(A)_b.$$

To conclude the proof of the Proposition, we will show that this last homomorphism is onto. To this end, let g be in $C(A)_b$, so that there are k and n in $\mathbf{N}$

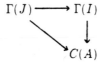

and $g_1, \ldots, g_k$ in $C(A)$ such that $g^*g + \sum_{i=1}^k g_i^* g_i = n$. For a in the essential ideal $J := \mathrm{dom}(g) \bigcap (\bigcap_{i=1}^k \mathrm{dom}(g_i))$, we have

$$na^*a = (g(a))^* g(a) + \sum_{i=1}^{k} (g_i(a))^* g_i(a)$$

so by **[16; Theorem 3.7]**

$$0 \leq (g(a))^* g(a) \leq na^*a, \quad \| (g(a))^* g(a) \| \leq n \| a^*a \|,$$

and so using Proposition 5 $\| g(a) \| \leq \sqrt{2n} \| a \|$. Now g is continuous on J and, as any element in the centroid of a semiprime normed algebra, g is a (partially defined) closed operator on A [8; **Corollary 1**]. Therefore $\mathrm{dom}(g)$ must contain the closure of J in A (say I). Since $I^2 = I$, we have $g(I) \subset I$, thus giving, by restriction on the domain and the range, an element in $\Gamma(I)$. Now I is a closed essential ideal of A and g lies in the range of the natural homomorphism $\Gamma(I) \to C(A)_b$.■

Once the above proposition has been proved, our desired description of the extended centroid of a n.c. JB^*-algebra will follow from a result by D. Handelman [**13**] and the machinery developed in Section 1, through the next lemma. For a closed ideal I of a n.c. JB^*-algebra A, denote by $U_A(I)$ the open subset of $\beta(A)$ given by

$$U_A(I) := \{J \in \beta(A) : I \not\subset J\}.$$

Lemma 2. *Let I be a closed ideal of a n.c. JB^*-algebra A. Then, for each J in $U_A(I)$, $I \cap J$ lies in $\beta(I)$, and the mapping $J \mapsto I \cap J$ from $U_A(I)$ into $\beta(I)$ is continuous with dense range.*

Proof. Let J be in $U_A(I)$. Then $I \cap J$ is a proper closed ideal of I and we must prove that it is also a prime ideal of I. If P_1 and P_2 are ideals of I with $P_1 P_2 \subset I \cap J$, then $\overline{P_1}$ and $\overline{P_2}$ are closed ideals of I with $\overline{P_1 P_2} \subset I \cap J$. But, by the transitivity of the property of being an M-ideal, $\overline{P_1}$ and $\overline{P_2}$ are actually ideals of A and, since J is a prime ideal of A, we have that either $\overline{P_1}$ (hence P_1) or $\overline{P_2}$ (hence P_2) is contained in $I \cap J$, which proves that $I \cap J$ is a prime ideal of I. The continuity of the mapping $J \mapsto I \cap J$ follows from straightforwards computations with the hull-kernel topologies. To see the density of the range of the above mapping it is enough to show that the kernel of this range in $\beta(I)$ is zero. But, since $Prim(A) \subset \beta(A)$, this kernel is contained in $\bigcap\{J \cap I : J \in Prim(A), I \not\subset J\} = \bigcap Prim(I) = 0$ (see [1; **Proposition 3.5(b)**] for the first equality).■

If, for a closed ideal I of a n.c. JB^*-algebra A, we consider the mapping $\alpha : J \mapsto I \cap J$ from $U_A(I)$ into $\beta(I)$, by the density of the range of α we obtain a natural one-to-one $*$-homomorphism $C_b(\beta(I)) \to C_b(U_A(I))$ given by $g \mapsto g \circ \alpha$. On the other hand, Proposition 3 gives natural $*$-isomorphisms

$$\Gamma(I) \cong C_b(\beta(I)) \cong C_b(Prim(I)),$$

and writing

$$W_A(I) := \{J \in Prim(A) : I \not\subset J\}$$

we have also a antural $*$-isomorphism $C_b(Prim(I)) \cong C_b(W_A(I))$ **[1; Proposition 3.5(b)]**. If we write

$$V_A(I) := \{J \in Cor(A) : I \not\subset J\},$$

we have dense inclusions

$$W_A(I) \subset V_A(I) \subset U_A(I)$$

(see Proposition 4) inducing one-to-one $*$-homomorphisms

$$C_b(U_A(I)) \to C_b(V_A(I)) \to C_b(W_A(I)).$$

Since the following diagram commutes

$$\begin{array}{ccc} \Gamma(I) \cong C_b(\beta(I)) & \longrightarrow & C_b(Prim(I)) \\ \downarrow & & \downarrow \\ C_b(U_A(I)) \to & C_b(V_A(I)) & \to C_b(W_A(I)) \end{array}$$

it follows that all the mappings in the diagram are $*$-isomorphisms, so in particular $\Gamma(I) \cong C_b(V_A(I))$ (canonically). Now observe that $\{V_A(I) : I \in \mathcal{I}\}$ (where, as above, $\mathcal{I}$ denotes the family of all closed essential ideals of A) is exactly the family of all open dense subsets of $Cor(A)$ (say $\mathcal{F}$) and that the canonical isomorphisms $\Gamma(I) \cong C_b(V_A(I))$ are compatible with the order of the direct systems of C^*-algebras $\{\Gamma(I) : I \in \mathcal{I}\}$ and $\{C_b(\Omega) : \Omega \in \mathcal{F}\}$. It follows from Proposition 6 that $C(A)_b \cong \lim_{\Omega \in \mathcal{F}} C_b(\Omega)$. But $C(A)$, being a $*$-regular ring with positive definite involution (Lemma 1), is the classical ring of quotients of its bounded ring **[13; Corollary 1.2]**. Therefore, denoting by $C(E)$ the $*$-algebra of all complex-valued continuous functions on any topological space E, we have proved

Theorem 2. *The extended centroid $C(A)$ of a n.c. JB^*-algebra A is $*$-isomorphic to the algebraic direct limit $\lim_{\Omega \in \mathcal{F}} C(\Omega)$ where $\mathcal{F}$ denotes the family of all open dense subsets of $Cor(A)$.*

Since, for a prime n.c. JB^*-algebra A. all non empty subsets of $Cor(A)$ are dense, we have

Corollary. *The extended centroid of a prime n.c. JB^*-algebra is the base field* **C**.

Concluding Remark. Using the results in **[10]**, one can prove that the primitive ideals of a C^*-algebra agree with the cores of its maximal-modular inner ideals. Therefore our Theorems 1 and 2 contain, respectively, the classical Dauns-Hofmann theorem and Ara's theorem **[2; Theorem 2.3]**.

References.

[1] E. M. Alfsen and E. G. Effros, Structure in real Banach spaces II. *Ann. of Math. 96* (1972), 129-173.

[2] P. Ara, Extended centroid of C^*-algebras. *Archiv. der Math. 54* (1990), 358-364.

[3] W. E. Baxter and W. S. Martindale 3rd., Central closure of semiprime nonassociative rings. *Commun. Algebra 7* (1979), 1103-1132.

[4] E. Behrends, *M-structure and the Banach-Stone theorem.* Lecture Notes in Math. 736, Springer-Verlag, 1979.

[5] F. F. Bonsall and J. Duncan, *Complete Normed Algebras.* Springer-Verlag, 1973.

[6] L. J. Bunce, The theory and Structure of Dual JB-algebras. *Math. Zeitschrift 180* (1982), 525-534.

[7] L. J. Bunce, Type I JB-algebras. *Quart. J. Math. Oxford* (2) *34* (1983), 7-19.

[8] M. Cabrera and A. Rodríguez, Extended centroid and central closure of semiprime normed algebras: a first approach. *Commun. Algebra 18* (1990), 2293-2326.

[9] S. Dineen and R. M. Timoney, The centroid of a JB^*-triple system. *Math. Scand. 62* (1988), 327-342.

[10] C. M. Edwards and G. M. Rüttimann, Inner ideals in W^*-algebras. *Michigan Math. J. 36* (1989), 147-159.

[11] A. Fernández, Modular annihilator Jordan algebras. *Commun. Algebra 13* (1985), 2597-2613.

[12] A. Fernández and A. Rodríguez, Primitive noncommutative Jordan algebras with nonzero socle. *Proc. Amer. Math. Soc. 96* (1986), 199-206.

[13] D. Handelman, Rings with involution as partially ordered abelian groups. *Rocky Mountain J. Math. 11* (1981), 337-381.

[14] L. Hogben and K. McCrimmon, Maximal modular inner ideals and the Jacobson radical of a Jordan algebra. *J. Algebra 68* (1981), 155-169.

[15] R. Payá, J. Pérez and A. Rodríguez, Noncommutative Jordan C^*-algebras. *Manuscripta Math. 37* (1982), 87-120.

[16] R. Payá, J. Pérez and A. Rodríguez, Type I factor representations of noncommutative JB^*-algebras. *Proc. London Math. Soc. 48* (1984), 428-444.

[17] A. Rodríguez, Nonassociative normed algebras spanned by hermitian elements. *Proc. London Math. Soc. 47* (1983), 258-274.

[18] A. Rodríguez, Jordan axioms for C^*-algebras. *Manuscripta Math. 61* (1988), 297-314.

[19] R. D. Schafer, *An introduction to nonassociative algebras.* Academic Press, 1966.

Nonassociative
Algebraic
Models

ISBN 1-56072-050-6

Primitive Nonassociative Normed Algebras and Extended Centroid

A. Rodríguez Palacios
Departamento de Análisis Matemático, Facultad de Ciencias, Universidad de Granada, 18071 Granada, Spain

Abstract.

We prove that complete normed primitive complex Jordan algebras have extended centroid equal to the base field **C**. This result is derived from a similar theorem, also proved in this paper, on general complete normed nonassociative algebras.

I. What is and how to use the extended centroid.

The concept of extended centroid is based on the one of "*partially defined centralizer*", namely: linear mapping f from a (two-sided) ideal dom(f) of an algebra A into A such that

$$f(ab) = af(b) \quad and \quad f(ba) = f(b)a$$

for all a in A and b in dom(f). If dom(f) is an essential ideal of A, we say that f is an "*essentially defined centralizer*" on A. A partially defined centralizer f on A is called "*maximal*" if, whenever g is any partially defined centralizer on A which extends f, we have $g = f$. The crucial fact for the definition of the extended centroid is the following.

Proposition 1. *For any essentially defined centralizer f on a semiprime algebra A, there is a unique maximal essentially defined centralizer on A which extends f.*

If A is a semiprime algebra, the usual sum and composition of essentially defined centralizers on A (as partially defined operators on A) are new essentially defined centralizers on A. In this way, we have:

Theorem 1 *(definition and basic properties of the extended centroid). Let A be a semiprime algebra. Then the set $C(A)$ of all maximal essentially defined centralizers on A, with sum the unique maximal essentially defined centralizer on A which extends the usual sum and product the unique maximal essentially defined centralizer on A which extends the usual composition, is a von Neumann regular commutative ring called the extended centroid of A. Moreover, the extended centroid of A is a field if and only if A is a prime algebra.*

Remark 1. i) The extended centroid was introduced by W. S. Martindale [**14**] in the case of prime associative algebras, and later it has been generalized to the case of prime and even semiprime nonassociative algebras (see [**6**] and [**2**], respectively). The above Proposition and Theorem are reformulated versions, inspired in [**12**] and explicitly given in [**3**], of basic results proved in [**2**].

ii) Note that the extended centroid of a semiprime algebra A contains as a subring the usual "*centroid*" of A, namely: the ring of all everywhere defined centralizers on A.

The concept of extended centroid is specially relevant in the case of prime algebras. In such case the extended centroid is a field extension of the base field, and the eventual fact that the extended centroid is reduced to the base field (the prime algebra is then called "*centrally closed*") has many interesting consequences. This fact is more remarkable if we recall that, by enlarging the base field (to the extended centroid) and the algebra (to get the so called "*central closure*"), every prime algebra becomes a centrally closed prime algebra. Since, sometimes, examples are more illuminating than general principles, we give here three nice results explaining the advantages of central closeability.

Theorem 2 [14]. *A centrally closed prime associative algebra A satisfies a generalized polynomial identity if and only if A contains an idempotent e such that eAe is a finite dimensional division algebra. As a consequence, centrally closed prime associative algebras satisfying generalized polinomial identities have nonzero socle.*

Theorem 3 [10]. *Every centrally closed prime Malcev algebra is either a Lie algebra or a seven-dimensional non-Lie Malcev simple algebra, namely: the quotient of a Cayley-Dikson algebra (regarded as a Malcev algebra) with respect to its center.*

Theorem 4 [4]. *Every centrally closed prime nondegenerate noncommutative Jordan algebra with nonzero socle is either quadratic, commutative, quasiassociative, or a division algebra.*

II. What is known about the extended centroid of normed semiprime algebras?

From now on, all algebras will be *complex* algebras. The fact that the centroid of any complete normed simple algebra is **C** could suggest the conjecture that every complete normed prime algebra is centrally closed. But this conjecture is quite incorrect, even if the complete normed prime algebra is additionally assumed to be a semisimple associative and commutative algebra, as shown by the example of the disc algebra (complex valued continuous functions on the closed unit disc of the complex plane which are holomorphic in the open unit disc). This fact becomes an unfortunate intrinsic limitation of the theory of complete normed algebras because, as observed in **[3]**, if a complete normed prime algebra is not centrally closed, then its central closure cannot be complete normable. Therefore it is interesting to find reasonable additional conditions on a complete normed prime algebra implying central closeability. In the associative case the following result is perhaps the most relevant which we know in this direction.

Theorem 5 *(***[13,14]*** + folklore in Banach algebra theory). Every complete normed primitive associative algebra is centrally closed.*

Still in the associative case of our question, M. Mathieu introduces the concept of "*ultraprime*" normed associative algebra, by requiring for a given normed associative algebra A the existence of a positive number K such that

$$\| M_{a,b} \| \geq K \| a \| \| b \|$$

for all a, b in A, where $M_{a,b}$ denotes the "*multiplication operator*" on A given by $x \mapsto axb$. He proves the following

Theorem 6 [15]. *Every ultraprime normed associative algebra is centrally closed.*

Moreover, the nice description of the extended centroid of an arbitrary (associative) C^*-algebra given by P. Ara must be cited. For the statement of this result we need some notation. For a topological space X, let $\mathcal{E}(X)$ denote the set of all complex valued continuous functions defined on an open dense subset of X (depending on the given function) under the equivalence $F \equiv G$ if F and G agree in the intersection of their domains. With pointwise defined operations, $\mathcal{E}(X)$ becomes a $*$-regular ring. Ara's result reads as follows.

Theorem 7 [1]. *The extended centroid of a C^*-algebra A is $*$-isomorphic in a natural way to $\mathcal{E}(Prim(A))$, where $Prim(A)$ denotes the set of all primitive ideals of A with the Jacobson topology.*

Either from Theorem 6 (see **[16; Proposition 2.3]**) or from Theorem 7, we obtain:

Corollary 1. *Every prime C^*-algebra is centrally closed.*

Concerning the nonassociative case of our question, perhaps the first known result is the following.

Theorem 8 [4]. *Every normed prime nondegenerate noncommutative Jordan algebra with nonzero socle is centrally closed.*

The natural extension of Ara's theorem to the context of noncommutative JB^*-algebras (see **[18]** for definition) can also be proved if we take as primitive ideals those in the sense of **[11]**, namely: the cores of maximal-modular inner ideals.

Theorem 9 [21]. *The extended centroid of a noncommutative JB^*-algebra A is *-isomorphic to $\mathcal{E}(Cor(A))$, where $Cor(A)$ denotes the set of cores of maximal-modular inner ideals of A endowed with the hull-kernel topology. As a consequence, prime noncommutative JB^*-algebras are centrally closed.*

Note that the above theorem can be regarded as a general nonassociative result (free of identities), because in a very precise sense noncommutative JB^*-algebras are the widest nonassociative generalizations of associative C^*-algebras [**19**]. But more explicitly, there are some known results on the extended centroid of general nonassociative normed algebras, most of them deriving from the next theorem. For an algebra A, let $M(A)$ denote the usual multiplication algebra of A, and, for normed A, let B be any algebra of operators on A with

$$M(A) \subseteq B \subseteq s - clos(M(A)),$$

where $s - clos$ denotes closure in the algebra $BL(A)$ of all bounded linear operators on A for the strong operator topology. A minimal B-invariant subspace of A, for some B as above, will be called an "*atom*" of A.

Theorem 10 [3]. *Let A be a complete normed (nonassociative) algebra with a family $\mathcal{I}$ of mutually orthogonal atoms such that the annihilator of the sum $\sum_{P \in \mathcal{I}} P$ is zero. Then A is semiprime and the extended centroid of A equals $\mathbf{C}^{\mathcal{I}}$.*

Corollary 2.*i) If the annihilator of the sum of the minimal ideals of a complete normed algebra A is zero, then A is semiprime and $C(A)$ equals a product of copies of* $\mathbf{C}$. *As a consequence, complete normed prime algebras with a minimal ideal are centrally closed.*

ii) The extended centroid of a semiprime (nonassociative) H^*-algebra is a product of copies of $\mathbf{C}$, *so prime H^*-algebras are centrally closed.*

Remark 2. While the first part of the corollary is a direct consequence of the theorem, this is not the case for the second one, which is harder (see [**3**] for details) and needs one of the main results in [**5**].

Before concluding this section, let us mention some easy, but crucial, observations in [**3**], used in the proofs of Theorems 9 and 10, and which will be useful later.

Proposition 2.*i) The domain of a closed partially defined centralizer f on a normed algebra A is invariant under $s - clos(M(A))$ and, for F in $s - clos(M(A))$ and b in dom(f), we have $f(F(b)) = F(f(b))$.*

ii) Any maximal essentially defined centralizer on a semiprime normed algebra is automatically closed.

III. Main result.

The aim of this section is to look for a satisfactory general nonassociative extension of Theorem 5. Actually, the proof of Theorem 5 can be transfered with minor changes to obtain that any complete normed nonassociative algebra, for which zero is a "*primitive*" ideal in the sense of [**20; Definition 2.1**], is centrally closed. But this result is not very interesting, because a commutative algebra satisfies the above algebraic assumption if and only if it is simple with unit, while any nontrivial anticommutative algebra fails such a property. Therefore we must find a weaker concept of primitivity for nonassociative algebras. This concept will appear later in a natural way. The first step in our extension of Theorem 5 is to refine Martindale's algebraic ingredient in the proof of this theorem as follows.

Proposition 3. *Let M be a maximal modular right ideal of an associative algebra A, and let f be a partially defined centralizer on A with $dom(f) \not\subset M$. Then, if π denotes the quotient mapping $A \to A/M$, there is an element $f^{\sim}$ in the centralizer set for the irreducible right A-module A/M such that $\pi(f(x)) = f^{\sim}(\pi(x))$ for all x in dom(f).*

Proof. First we show that $f(M \bigcap dom(f)) \subseteq M$. Otherwise, we have $A = M + f(M \bigcap dom(f))$, so there exists a left modular unit u for M which lies in $f(M \bigcap dom(f))$, and writing $u = f(m)$ with m in $M \bigcap dom(f)$, we have, for arbitrary x in dom(*f*), that $x - mf(x) = x - f(m)x = x - ux \in M$, so x lies in M, contradicting the assumption dom(*f*)$\not\subset M$. On the other hand, π (dom(*f*)) is a nonzero submodule of the irreducible right A-module A/M, so $A/M = \pi$ (dom(*f*)). Now $\pi(x) \mapsto \pi(f(x))(x \in$dom(*f*)) is a well defined mapping (say $f^{\sim}$) from A/M into A/M, and clearly $f^{\sim}$ belongs to the centralizer set for the right A-module A/M.

The second step in our argument is to apply the above proposition to some non-complete normed associative algebras (the so called normed Q-algebras) for which the analytic ingredients in the proof of Theorem 5 remain true (see [**17, 22**]). We recall that a normed Q-algebra is a normed associative algebra for which the set of quasi-invertible elements is open, and we note that this assumption is equivalent to the fact that for any a in the algebra the equality

$$\lim_{n \mapsto \infty} \{\| a^n \|^{1/n}\} = sup\{| \lambda |: \lambda \in sp(a)\}$$

holds.

Proposition 4. *Let P be a (left or right) primitive ideal of a normed Q-algebra A, and let f be a partially defined centralizer on A with $dom(f) \not\subset P$. Then there exists λ in $\mathbf{C}$ such that $f(x) - \lambda x \in P$ for all x in dom(f).*

Proof. Assume that P is a right primitive ideal of A (the case of left primitive ideals will follow by considering the opposite algebra), so that there exists a maximal modular right ideal M of A such that P is the largest (two-sided) ideal of A contained in M. Since dom(*f*) is an ideal of A with dom(*f*)$\not\subset P$, we have dom(*f*)$\not\subset M$, so it follows from Proposition 3 and the observation in the proof of [**3; Proposition 6**] that the centralizer set for any irreducible module over a normed Q-algebra equals $\mathbf{C}$(see also [**17; Theorem 6.7**]) that there exists λ in $\mathbf{C}$ such that $f(x) - \lambda x$ lies in M for all x in dom(*f*). To conclude the proof, note that the set $\{f(x) - \lambda x : x \in dom(f)\}$ is an ideal of A.

The following step in our argument is to "*unassociativize*" the above proposition. To this end, consider a nonassociative algebra A, let B any algebra of linear operators on A with $M(A) \subseteq B$, and fix a (left or right) primitive ideal of B (say $\mathcal{P}$). Then the largest B-invariant subspace of A contained in $\{a \in A : L_a, R_a \in \mathcal{P}\}$ (where L_a and R_a denote respectively the left and right multiplication operators by a on A) is a (two-sided) ideal of A. Ideals appearing in this way are called (left or right) B-primitive ideals of A, and A is said to be a B-primitive algebra if zero is a B-primitive ideal of A. Another concept we need is the one of quasi-full subalgebra of an associative algebra. A subalgebra B of an associative algebra C is said to be a quasi-full subalgebra of C if for any b in B we have $\sup\{|\lambda| : \lambda \in sp(B,b)\} = \sup\{|\lambda| : \lambda \in sp(C,b)\}$. Note that every closed subalgebra of a complete normed associative algebra is a quasi-full subalgebra and that, for any Banach space X, the Banach algebra $BL(X)$ (bounded linear operators on X) is a quasi-full subalgebra of $L(X)$ (possibly unbounded linear operators on X) **[20; Remark 1.8]**, so every closed subalgebra of $BL(X)$ is a quasi-full subalgebra of $L(X)$.

Proposition 5. *Let A be a complete normed (nonassociative) algebra, B a quasi-full subalgebra of $L(A)$ with*

$$M(A) \subseteq B \subseteq s - clos(M(A)),$$

P a B-primitive ideal of A, and f be a closed partially defined centralizer on A with $dom(f) \not\subseteq P$. Then there exists $\lambda \in \mathbf{C}$ such that $f(x) - \lambda x \in P$ for all x in $dom(f)$.

Proof. Consider the set

$$H \equiv \{F \in B : F(A) \subseteq dom(f) \text{ and } fF \in B\}.$$

Clearly H is a right ideal of B. But since we assume $B \subseteq s{-}clos(M(A))$, by Proposition 2 i), H is also a left ideal of B and the mapping $\phi : F \mapsto fF$ is a partially defined centralizer on B whose domain is H. Let $\mathcal{P}$ be the primitive ideal of B defining P as the largest B-invariant subspace of A whose elements y satisfy that L_y and R_y lie in $\mathcal{P}$. We claim that $H \not\subseteq \mathcal{P}$. Otherwise, since L_x and R_x lie in H (so in $\mathcal{P}$) for all x in dom(f) (this can be easily verified) and dom(f) is a B-invariant subspace of A (another application of Proposition 2(i)), we have that dom(f)$\subseteq P$, contradicting our assumptions. Now ϕ is a partially defined centralizer on B whose domain is not contained in the primitive ideal $\mathcal{P}$ of B, and B, as any quasi-full subalgebra of $L(A)$ contained in $BL(A)$, is a normed Q-algebra with respect to the operator norm. Therefore Proposition 4 gives the existence of a complex number λ such that $fF - \lambda F$ lies in $\mathcal{P}$ for all F in H. As a consequence, for x in dom(f),

$$L_{f(x)-\lambda x}(= fL_x - \lambda L_x) \text{ and } R_{f(x)-\lambda x}(= fR_x - \lambda R_x)$$

lie in $\mathcal{P}$. This, together with the fact that

$$\{f(x) - \lambda x : x \in dom(f)\}$$

is a B-invariant subspace of A (Proposition 2 i)), shows that for x in dom(f) $f(x) - \lambda x$ lies in P, as required.

Remark 3. Among the possible choices of the algebra B in the above proposition, there is a largest one, namely $B = s - clos(M(A))$, and a smallest one, namely the "*quasi-full multiplication algebra*" of A (the intersection of all the quasi-full subalgebras of $L(A)$ containing $M(A)$). Observe that this last algebra, noted $QFM(A)$, is defined in a purely algebraic way.

Given a nonassociative algebra A and an algebra B of linear operators on A with $M(A) \subseteq B$, define the B-radical of A (noted B-*Rad*(A)) as the largest B-invariant subspace of A whose elements y satisfie that L_y and R_y belong to the Jacobson radical of B. Clearly B-Rad(A) is the intersection of all the (left or right) B-primitive ideals of A.

Lemma. *Let A be a normed algebra and B be a subalgebra of $L(A)$ with*

$$M(A) \subseteq B \subseteq s - clos(M(A)).$$

If B-Rad(A)=0, then A is semiprime.

Proof. If P is any ideal of A such that $P^2 = 0$, $\overline{P}$ is a closed ideal of A with $\overline{P}^2 = 0$ and, as any closed ideal of A, $\overline{P}$ is invariant under $s - clos(M(A))$ so also under B. These facts make clear that, for x in $\overline{P}$ and F in B, we have $L_x^2 = L_x F L_x = 0$, from which we deduce that the ideal of B generated by L_x is a nilideal. so a quasiinvertible ideal, and so $L_x \in Rad(B)$. Analogously $R_x \in Rad(B)$. Therefore $\overline{P} \subseteq B - Rad(A) = 0$, $P = 0$, and A is semiprime, as required.

Now we state and prove our main result.

Theorem 11. *Let A be a complete normed algebra, B a quasi-full subalgebra of $L(A)$ with*

$$M(A) \subseteq B \subseteq s - clos(M(A)),$$

and assume that A is B-primitive. Then A is a centrally closed prime algebra.

Proof. Clearly B-Rad(A)=*0* so, by the above lemma, A is semiprime and we can consider the extended centroid $C(A)$ of A. Thank to Proposition 2(ii), each f in $C(A)$ is a closed partially defined centralizer on A so, applying Proposition 5 (with P equal to the zero ideal, which by assumption is a B-primitive ideal of A), we obtain $f = \lambda$ for suitable λ in $\mathbf{C}$. Therefore $C(A) = \mathbf{C}$, and A is prime according to the conclusion of Theorem 1.

Remark 4. i) The methods and concepts used throughout this section are very close to the ones developed in **[20]** and **[8]** to obtain general nonassociative extensions of Johnson's uniqueness of norm topology and Civin-Yood decomposition theorems for Banach algebras. There, the assumption $B - Rad(A) = 0$ for $B = FM(A)$ (the "*full multiplication algebra*" of A) is an algebraic nonassociative translation of the familiar associative semisimplicity, suitable to obtain the desired results. Here, the assumption that A is B-primitive for $B{=}QFM(A)$ is the analogous algebraic translation of primitivity for the nonassociative extension of Theorem 5.

ii) If A is a complete normed prime algebra with an atom, then, using the crucial Lemma 7 in [3], it is not difficult to see that A is (left) B-primitive for $B = s - clos(M(A))$. Therefore our main result contains the particularization of Theorem 10 to prime algebras.

iii) Let A and B be as in Proposition 5. From the facts that B is a normed Q-algebra and that primitive ideals of a normed Q-algebra are closed, it follows easily that B-primitive ideals of A are closed.

iv) Assume additionally that A is topologically simple and that B-Rad$(A){=}0$. Then, by iii), A is B-primitive, so centrally closed in view of our main result.

v) The ideas in the proof of Theorem 11 actually show that, if A and B are as in Proposition 5 and there exists a family $\mathcal{I}$ of non essential B-primitive ideals of A with $\bigcap \mathcal{I} = 0$, then A is semiprime and $C(A)$ has a faithful family of "*characters*" (nonzero homomorphisms from $C(A)$ into the base field $\mathbf{C}$).

vi) Generalized annihilator normed algebras are, by definition, those semiprime normed algebras for which all essential ideals are dense (see [8]). From iii) and v) above, together with [3; **Proposition 5**] and the easy fact that for any algebra A the inclusion

$$FM(A) - Rad(A) \subseteq QFM(A) - Rad(A)$$

is true, it follows that, if A is a generalized annihilator complete normed algebra and $QFM(A)$-Rad$(A){=}0$, then $C(A)$ equals a product of copies of $\mathbf{C}$. This result contains corollary 2 ii) because semiprime H^*-algebras are clearly generalized annihilator and, for such an algebra A, the equality $QFM(A)$-Rad$(A){=}0$ is true (argue as in [8; **Lemma 14**]).

IV. Application to Jordan algebras.

Following an earlier idea by E. I. Zelmanov for unital Jordan algebras, a reasonable concept of "*primitive*" Jordan algebra was introduced by L. Hogben and K. McCrimmon in [11]. A Jordan algebra is said to be primitive if it contains a maximal-modular inner ideal with zero core. We recall that the core of a given subspace of an algebra is the largest ideal of the algebra contained in the subspace, and that an inner ideal M of a Jordan algebra A is said to be maximal-modular if there exists an x in A such that M is maximal among all proper x-modular inner ideals of A. We refer to [11; **p. 158**] for the concept of x-modularity, only remarking here that an x-modular inner ideal M of A satisfies $U_{1-x}(A) \subseteq M$. As in the associative case, primitive Jordan algebras are prime [11; **Proposition 5.5**], and perhaps the most interesting consequence of our main result is the following.

Theorem 12. *Every complete normed primitive Jordan algebra is centrally closed.*

The above theorem is a direct consequence of Theorem 11 and the following proposition, which asserts that primitive Jordan algebras are B-primitive for all choices of the algebra of operators B.

Proposition 6. *Let A be a Jordan algebra, P the core of a maximal-modular inner ideal of A, and B be any algebra of linear operators on A containing $M(A)$.*

Then the largest B-invariant subspace of A contained in P is a (right) B-primitive ideal of A.

Proof. Let M denote the maximal-modular inner ideal of A whose core is P, and write

$$\mathcal{M} := \{F \in B : F(A) \subseteq M\},$$

which is clearly a right ideal of B. First we show that $\mathcal{M}$ is proper. To this end, let x be in A such that M is actually a maximal x-modular inner ideal of A, and suppose on the contrary that $\mathcal{M} = B$. Then $2L_x - U_x \in \mathcal{M}$, so for any a in A $(2L_x - U_x)(a) \in M$, and so

$$a = U_{1-x}(a) + (2L_x - U_x)(a) \in M,$$

contradicting that M is a proper inner ideal of A. On the other hand, for any F in B and a in A, we have

$$[F - (2L_x - U_x)F](a) = U_{1-x}F(a) \in M,$$

so $F-(2L_x-U_x)F \in \mathcal{M}$, which shows that $2L_x - U_x$ is a left modular unit for $\mathcal{M}$. Now $\mathcal{M}$ is a proper modular right ideal of B, and therefore there exists a maximal modular right ideal $\mathcal{N}$ of B such that $\mathcal{M} \subseteq \mathcal{N}$. Let $\mathcal{P}$ be the largest ideal of B contained in $\mathcal{N}$ (so that $\mathcal{P}$ is a right primitive ideal of B) and let Q denote the largest B-invariant subspace of A contained in $\{a \in A : L_a \in \mathcal{P}\}$ (so that Q is a right B-primitive ideal of A). The proof will be concluded by showing that also Q is the largest B-invariant subspace of A contained in P. If $Q \not\subseteq P$, then also $Q \not\subseteq M$ and, since $M + Q$ is an x-modular inner ideal of A, we have $M + Q = A$ in view of the maximality of M. As a consequence, $x = m + q$ for suitable m in M and q in Q. Now, on the one hand, M is a q-modular inner ideal of A **[11; Proposition 2.10]** and therefore, as above, $2L_q - U_q$ is a left modular unit for $\mathcal{M}$ (so also for $\mathcal{N}$). On the other hand, we have the following chain of implications

$$\left.\begin{array}{c} q \in Q \Rightarrow L_q \in \mathcal{P} \Rightarrow L_q \in \mathcal{N} \\ \Downarrow \\ q^2 \in Q \Rightarrow L_{q^2} \in \mathcal{P} \Rightarrow L_{q^2} \in \mathcal{N} \end{array}\right\} \Rightarrow 2L_q - U_q \in \mathcal{N}.$$

The fact that a left modular unit for a proper right ideal lies in the ideal is a contradiction which shows that $Q \subseteq P$. If R is any B-invariant subspace of A contained in P, then

$$\mathcal{R} := \{F \in B : F(A) \subseteq R\}$$

is a (two-sided) ideal of B contained in $\mathcal{M}$, so in $\mathcal{N}$, and so also in $\mathcal{P}$. For y in R we have clearly $L_y \in \mathcal{R}$ and therefore $L_y \in \mathcal{P}$, which shows in view of the definition of Q that $R \subseteq Q$. Now Q is the largest B-invariant subspace of A contained in P, as required.

Remark 5. i) Following [9], define the maximal-modular inner ideals of a non-commutative Jordan algebra A as the maximal-modular inner ideals of the Jordan algebra A^+ obtained by symmetrization of the product of A, and say that A is primitive if zero is the core (relative to A) of a maximal-modular inner ideal of A. Only minor changes in the above proof are needed to see that Proposition 6 is also true in the more general case of A being a noncommutative Jordan algebra. Therefore, by Theorem 11, complete normed primitive noncommutative Jordan algebras are centrally closed. Taking into account that associative algebras which are primitive in the associative sense also are primitive in the noncommutative Jordan sense (a consequence of **[11; Example 3.3]**), this last result contains Theorem 5.

ii) Let A be a complete normed (possibly noncommutative) Jordan algebra, and B be an algebra of linear operators on A such that

$$M(A) \subseteq B \subseteq s - clos(M(A)).$$

Then the cores of maximal-modular inner ideals of A are B-primitive ideals of A. This follows from Proposition 6 once we know that the cores of maximal-modular inner ideals of A are invariant under $s - clos(M(A))$ (so also under B) because they are closed ideals of A **[7; Lemma 6.5]**.

References.

[1] P. Ara, Extended centroid of C^*-algebras. *Archiv. der Math. 54* (1990), 358-364.

[2] W. E. Baxter and W. S. Martindale 3*rd*, Central closure of semiprime nonassociative rings. *Commun. Algebra 7* (1979), 1103-1132.

[3] M. Cabrera and A. Rodríguez, Extended centroid and central closure of semiprime normed algebras: a first approach. *Commun. Algebra 18* (1990), 2293-2326.

[4] M. A. Cobalea and A. Fernández, Prime noncommutative Jordan algebras and central closure. *Algebras, groups and geometries 5* (1988). 129-136.

[5] J. A. Cuenca and A. Rodríguez, Structure theory for noncommutative Jordan H^*-algebras. *J. Algebra 106* (1987), 1-14.

[6] T. S. Erickson, W. S. Martindale 3*rd* and J. M. Osborn, Prime nonassociative algebras. *Pacific J. Math. 60* (1975), 49-63.

[7] A. Fernández, Modular annihilator Jordan algebras. *Commun. Algebra 13* (1985), 2597-2613.

[8] A. Fernández and A. Rodríguez, A Wedderburn theorem for nonassociative complete normed algebras. *J. London Math. Soc. 33* (1986), 328-338.

[9] A. Fernández and A. Rodríguez, Primitive noncommutative Jordan algebras with nonzero socle. *Proc. Amer. Math. Soc. 96* (1986), 199-206.

[10] V. T. Filippov, Theory of Mal'tsev algebras. *Algebra i Logika 16* (1977), 101-108.

[11] L. Hogben and K. McCrimmon, Maximal modular inner ideals and the Jacobson radical of a Jordan algebra. *J. Algebra 68* (1981), 155-169.

[**12**] R. E. Johnson, The extended centralizer of a ring over a module. *Proc. Amer. Math. Soc. 2* (1951), 891-895.

[**13**] W. S. Martindale 3*rd*, Lie isomorphisms of prime rings. *Trans. Amer. Math. Soc. 142* (1969), 437-455.

[**14**] W. S. Martindale 3*rd*, Prime rings satisfying a generalized polinomial identity. *J. Algebra 12* (1969), 576-584.

[**15**] M. Mathieu, Ring of quotients of ultraprime Banach algebras with applications to elementary operators. *Proc. Centre Math. Anal. Austral. Nat. Univ. 21* (1989), 297-317.

[**16**] M. Mathieu, Elementary operators on prime C^*-algebras, I. *Math. Ann. 284* (1989), 223-244.

[**17**] T. W. Palmer, Spectral algebras. Preprint.

[**18**] R. Payá, J. Pérez and A. Rodríguez, Noncommutative Jordan C^*-algebras. *Manuscripta Math. 37* (1982), 87-120.

[**19**] A. Rodríguez, Nonassociative normed algebras spanned by hermitian elements. *Proc. London Math. Soc. 47* (1983), 258-274.

[**20**] A. Rodríguez, The uniqueness of the complete norm topology in complete normed nonassociative algebras. *J. Funct. Anal. 60* (1985), 1-15.

[**21**] A. Rodríguez and A. R. Villena, Centroid and extended centroid of JB^*-algebras. In *Workshop on nonassociative algebraic models*, Nova Science Publishers, New York (to appear).

[**22**] B. Yood, Homomorphisms of normed algebras, *Pacific J. Math. 8* (1958), 373-381.

SUBJECT INDEX